ROCKS and DIRT

An introduction to mineralogy, geology and soil science

by Ellen Johnston McHenry

ISBN 978-0-9887808-8-0

Published by Ellen McHenry's Basement Workshop
State College, Pennsylvania
www.ellenjmchenry.com
ejm.basementworkshop@gmail.com

Printed on demand by Lightning Source, Inc., Lavergne, Tennessee.

Other books by this author:

The Elements; Ingredients of the Universe
Carbon Chemistry
The Brain
Cells
Botany in 8 Lessons
Protozoa; A Poseidon Adventure
Mapping the World with Art

These books can be found on Amazon and at www.ellenjmchenry.com.

CONTENTS

CHAPTER 1: ELEMENTS

People have always been curious about what things are made of. It's obvious that plants are different from animals, and that rocks are nothing like water. Light and air seem to be made of nothing at all. Until just a few hundred years ago, people did not have scientific equipment advanced enough to be able to study the nature of **matter**. ("Matter" is the scientific name for "stuff." Anything and everything in the universe that is made of atoms and occupies space is matter.) Ancient people were as smart as we are today, but they had no computers, no microscopes, no spectrometers, no thermometers, no telescopes, and often not even any books. The only tools they had were their five senses. They had to come up with theories about how things worked based only on what they saw, heard, smelled, tasted, and felt, combined with some common sense and some religious and cultural beliefs.

The ancients were very clever. They had a good "hands-on" knowledge of the world and discovered and invented a lot of things. They domesticated animals, began farming, and invented ways to make houses out of whatever materials were around them. They wrote stories, poems and songs, made fabrics, pottery and jewelry, studied

the stars, made calendars, and kept track of eclipses and comets. Some of these ancient peoples discovered that when you heated certain rocks to a very high temperature, liquid stuff came out that turned into a hard metal when it cooled. The metal could then be shaped into knives, tools, or jewelry. The metals seemed to have been hiding in the rocks until the heat drove them out. Rocks that contained metals were called **ores** and became very valuable. Observing the changes that took place when you heated and melted things led people to start wondering what rocks were made of. Why did certain metals always come out of certain rocks and never out of other kinds of rocks? What made some rocks heavier than others? Why did some rocks burn like wood? Why did some rocks smell bad?

About 2,000 years ago, in the city of Alexandria, Egypt, a small group of very curious people started doing what we would call chemistry experiments. One of these curious people was a woman named Mary (or Miriam). Mary was an expert at making a certain purple dye for coloring cloth, and she invented a chemical apparatus for boiling liquids. This apparatus was upgraded and improved over the centuries, but it still bears her name, being called the "bain-marie," which is French for "Mary's Bath." Mary also wrote down some chemical recipes, including one for making gold from certain plants. Gold from plants?! This fired the imaginations of these curious people, and a whole new field of study began. This type of study would eventually be called **alchemy**.

a "bain-marie"

The word "alchemy" came into English from French, which borrowed it from Latin, which borrowed it from Arabic, which seems to have borrowed it from Greek. In Arabic, "al" means "the," and "kimiya" (from Greek "khemia") is such an old word that we don't know for sure what it meant originally. It could be related to the Greek word for Egypt, "keme," or the Greek word for mixture, "chymeia." Whatever its origins, the word "alchemy" was used right up until the early 1700s.

One of the next great alchemists was **Jabir ibn Hayyan**, who lived in Persia (now Iran) in the 700s. Jabir, or "Geber," as the Greeks called him, was an amazing person. He studied mathematics, astronomy, geography, medicine, philosophy, physics and chemistry. As far as we know, he was the first person in history to begin doing scientific experiments. He designed pieces of equipment that could boil and separate mixtures, catch steam, and collect particles. (Oddly enough, one of the main goals of his experimenting was to find out how to create life. He wrote down recipes that he believed would make snakes and scorpions!) He began working with substances that

Jabir ibn Hayyan

we now call "acids," and found one that seemed to be able to dissolve gold. He was also the first person to begin to classify rocks and minerals according to their chemistry. He discovered that some substances would turn to steam when they were heated, such as mercury and sulfur. Other substances would melt into a liquid, such as gold, silver and copper. There were also substances, such as ordinary stones, that would neither melt nor turn to steam, and could only be crushed into a powder. Jabir was more interested in the chemical properties of substances, particularly what they did when they were heated, than he was with their color or texture or weight. He had discovered a key to figuring out what matter was made of.

I love to burn things!

Jabir also studied the writings of the Greeks, and from them he accepted the idea that water, fire, air and earth were essential elements, the "stuff" from which other stuff is made. It might seem silly to think that a tree could be made of a combination of water, air and earth, but that's what they believed. (However, we need to bear in mind that sometimes true chemistry is odd, too. Doesn't it sound equally strange to say that water, a liquid, is made of two gases, hydrogen and oxygen?) Jabir's experiments led him to believe that there were more than four elements. He believed that mercury and sulfur were also elements. No matter what you did to these substances, they always remained the same. They could be combined with other things, but they could not be broken apart. Jabir had made an incredibly important discovery. He had correctly identified two of the basic chemical elements. All of the chemical elements have now been discovered, and the complete list is a chart called the Periodic Table.

Periodic Table of Elements

Hey- we did a whole book on that chart!

Yep, we sure did! The readers can check it out on this website:

www.ellenjmchenry.com

THE ELEMENTS

Can you find mercury (Hg) and sulfur (S)? Use the large version on page 141.

People had been using copper, iron, silver, gold, lead, mercury and sulfur for a long time. Coppersmiths probably had not given much thought to what their copper was made of. They knew which ore rocks contained copper, they knew how to heat the ore to get the copper out, and they knew how to fashion the copper into useful things. The alchemists were the first to ask the question: "What is copper?"

"The Philosopher's Stone" by Joseph Wright, 1771

The Arabic alchemists that came after Jabir added "salt" to their list of elements. This was a step backward, but they had no way of knowing this. Salt was not an element, but it would not be separated into its elements, sodium and chlorine, until the 1800s. The early Arabic alchemists made another mistake—they thought that it might be possible to turn one element into another. They still recognized that some substances were basic elements, but thought that perhaps this did not rule out turning one element into another.

In the early days of the Renaissance in Europe (the 1200s), Europeans began reading all these ancient Arabic alchemy texts. They were especially fascinated by the idea that one metal might be turned into another. Perhaps there was a way to turn copper or lead into gold! They began fashioning their own chemistry equipment and thus began the type of alchemy see in paintings: old men surrounded by pots full of strange brews. European alchemists were interested in chemistry, in general, but they were particularly intent upon discovering something they called the "philosopher's stone."

To them, a philosopher was exactly what the Greek word meant—someone who loved knowledge. Philosophy to them had nothing to do with morals or ethics; it was simply the pursuit of knowledge of any kind. The stone didn't have to be a literal stone, but any substance that would change one metal into another, and specifically a "base" (not valuable) metal into gold. The reputation of this mythical stone grew over time and it was eventually believed that not only could it change metals into gold, but it could also cure all diseases. In an age without modern medicine, it was no wonder that everyone was so desperate to find this miraculous stone!

While European alchemists were hard at work trying to turn things into gold, they made some interesting discoveries. Like Jabir, they began noticing that some substances seemed very resistant to change, as if they couldn't be changed. The list of basic elements began to grow. The list included the substances that the ancients had already discovered—gold, silver, copper, mercury, lead, sulfur, iron, and tin—plus some new ones that you might not be familiar with: antimony, arsenic and bismuth. Bismuth was usually discovered by miners who were looking for lead or silver. Bismuth would be found sandwiched between lead, on top, and silver, below. Since bismuth was more shiny than lead, but not as shiny as silver, this led the miners to believe that lead would eventually become silver, and bismuth was an intermediate stage, halfway between lead and silver. When the miners found a layer of bismuth they would say, "Oh no, we came too soon!"

Bismuth sandwich:

The elemental nature of bismuth was first noticed by a German scientist, **Georgius Agricola**, in the early 1500s. Agricola spent a large part of his adult life studying mines and the stuff that came out of them. He wrote one of the first books on mining techniques. Agricola believed that bismuth was a completely different substance from either lead or silver. It was, and always would remain, bismuth.

It was Agricola, and scientists immediately after him, who began to realize that there were actually three levels of classification needed. First was *rocks*, the most obvious category because everyone could seem them lying on the ground or being carried out of mines. The second was **elements**—the substances that could not be broken down into anything else. But there was also a third category: **minerals**. Minerals (the stuff that comes out of mines) were a combination of two or more elements. Sometimes minerals appeared in their pure form, but other times they were hidden in ore rocks, and had to be melted out. A rock might contain several minerals. This concept would become one of the established facts about rocks and minerals: **Elements make minerals, and minerals make rocks.**

ELEMENTS   MINERALS   ROCKS

The next great leap in understanding the elements came in Germany in 1669. An alchemist by the name of Hennig Brandt was trying to make gold. Yes, after all those centuries of failure, alchemists were still hoping to change one element into another. By this time, alchemists had tried just about every rock and mineral on the planet. So Hennig decided to boil something that was yellowish-gold but wasn't a mineral. He collected hundreds of gallons of urine and boiled it down until it became a thick paste. (Yes, it smelled really bad.) When he began heating this paste as hot as he could get it, something amazing happened—it began to glow! It glowed with a bright, white light. Hennig had discovered a new element, the element we call **phosphorus**.

This caused a major shift in the thinking of scientists all over the world. Organic things were made of elements, too, not just rocks and minerals. Phosphorus would be found in both living and non-living things. How many other elements were there? The race to discover new elements was on!

In the early 1700s, the word "alchemy" was replaced by the word "chemistry." The new "modern" scientist of the 1700s was no longer interested in trying to turn things into gold. Discovering new elements or inventing a new process for manufacturing chemical products had replaced the search for the philosopher's stone.

The science of **mineralogy** was born in the 1700s, as there were more men like Agricola who devoted their careers to studying minerals and the elements they were made of. Mineralogists of the 1700s determined that zinc, cobalt and nickel were also elements.

Miners named several elements. "Nickel" means "demon" in German. Ore that contained nickel seemed to be cursed because it would not produce the copper that the miners wanted. "Cobalt" is German for "goblin." The presence of cobalt in silver ore made it harder to extract the silver. The names "copper" and "zinc" came from ancient miners so long ago that no one knows their original meaning.

During the 1800s, chemists tested every mineral sample they could get. They now understood that minerals were made of elements and were eager to discover unknown elements. They improved their equipment and invented new tools and techniques. One of the most important new technologies of the 1800s was electricity. Electricity is the only way that some compounds can be separated into their elements, so these elements could not possibly have been discovered until the age of electricity. The list of elements grew rapidly during this century. By the late 1800s, there were 63 known elements.

In the late 1800s, chemists were starting to make charts of the chemical elements. These elements were the basic building blocks of every type of matter in the universe, so they needed to be printed into a chart for students to study. Several men began working on ways to classify the elements, but only one is well remembered today: Russian chemist **Dmitri Mendeleev**. *(men-dell-LAY-ev)* Mendeleev's stroke of genius was to realize that there were quite a few elements that had not

Dmitri Mendeleev   *Mendeleev's table of elements*

yet been discovered. He made a rectangular chart and left blank spaces in places where he believed there was an element missing. He turned out to be absolutely right in every case. Soon, those missing elements were discovered just as Mendeleev predicted. Today we call this chart the Periodic Table of the Elements, and we've already seen it on pages 2 and 3. There are many more elements today than in Mendeleev's day, but the basic structure of the table has remained the same.

After most of the elements had been discovered, a new question arose: **"What are elements made of?"** This question could not be answered by regular chemists. No amount of boiling or evaporating could tell you what an element was actually made of. The basic building blocks of elements are particles too small to see. This question had to be answered by a new type of scientist—someone who could combine experiments with logic and math. Our knowledge about atoms comes as much from logical reasoning as it does from experiments.

John Dalton

The first scientist to officially propose a theory about elements was **John Dalton**. In 1803, Dalton wrote that be believed that elements were made of individual particles called **atoms**. Every atom of an element was identical. An atom was the smallest piece of an element that could not be divided. In other words, if you start with a lump of gold the size of a marble, you can easily cut it in half. You could easily cut those halves in half again. By this time you might have lumps the size of peas, but you could cut those in half again. If you kept cutting those pieces in half, you would eventually get pieces so small you'd need a magnifier to see them. Let's say you do have a magnifier and can go on cutting them in half. The pieces will keep getting smaller and smaller and smaller. How small can they get? Eventually you will get down to an individual piece of gold that can't be divided any further. That's what Dalton called an atom. Dalton did not have an idea how small an atom might be, but he was right in thinking that you'd reach a point where if you divided it gain, you'd no longer have gold. Dalton didn't have any ideas about what particles might make up an atom, but he was right about the existence of atoms.

Around the turn of the 20th century (early 1900s), the work of three scientists came together to give the world the first theory about what an atom is made of. **Ernest Rutherford, J. J. Thomson** and **Neils Bohr** are all considered to be physicists, not chemists. They each made a different contribution to figuring out the mystery of the atom. If you want to know the particulars, you can easily find more information on the Internet. The end result of their research was a proposal that atoms consisted of a central nucleus made of positively charge **protons** and neutral **neutrons**, surrounded by negatively charged **electrons**. Bohr was the one who came up with a model called the solar system model because it resembled the planets orbiting the sun.

Since then, it has been discovered that electrons don't exactly orbit the nucleus. They move so fast that they occupy a cloud-like area. However, the solar system model continues to be used because it is very helpful in explaining how atoms stick together.

the "solar system" model

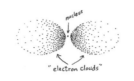 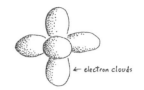

electron cloud models

Electrons are arranged in shells around the nucleus. The electrons in the out-ermost shell are the ones that can interact with the environment, so they are the ones that give the atom most of its chemical properties. In this copper (Cu) atom, the electron highlighted in red is the one that will interact with other atoms.

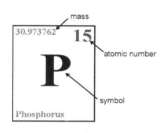

Each atom has a unique number of protons, neutrons and elecrons. It is the number of protons that determines which element it is. If an atom has one proton, it is hydrogen. If it has two protons, it's helium. If it has three, it's lithium, and so on down the Periodic Table. The number of protons is called the **atomic number**. It is the most important number in those squares on the Periodic Table. You will also see a number called the **atomic mass**. This is a lot like weight. Each element also has a unique letter abbreviation called its **symbol**.

Elements in their pure form have the same number of electrons as protons. This is to balance out the positive and negative charges. However, the arrangement of the electrons in those rings is critically important, and some arrangements are better than others. For example, having just one lonely electron in the outer ring is really awful. Even two in the outer ring is less than ideal, and atoms like these will do just about anything to fix their elec-tron problems. The best case scenario is when an unhappy atom finds another atom that is also unhappy, and their problems are exactly opposite. When one atom wants to get rid of an electron and another wants to get one, this works out perfectly and the two atoms stick together and become the solution to each other's problems. A good example of this is table salt, NaCl, where sodium (Na) gives an electron to chlorine (Cl). Sometimes three or more atoms will join together to solve each other's electron problems. We call these clumps of atoms **molecules**.

Sometimes atoms of the same element will join together to make a molecule. In the air around us we have two great examples: N_2, a pair of nitrogen atoms, and O_2, a pair of oxygen atoms. When atoms of different elements join to form a molecule, we call this a **compound**. Compounds are always made of at least two different types of elements. NaCl, salt, is a compound. We'll meet lots of compounds in the next chapter.

O_2 is a molecule.

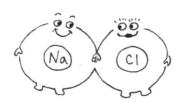

NaCl is both a molecule and a compound.

Activity 1.1 Comprehension self-check

Can you remember the answers to these questions? If not, go back to the text and see if you can find the answers. When you are done, you can check your answers using the answer key at the back of the book.

1) True or false? People started doing chemistry experiments for the very first time during the 1500s. _____

2) If Jabir Ibn Hayyan had found a rock to study, what would he have been most interested in?

  a) the color of the rock   b) the temperature at which the rock melted   c) the location of the rock

3) What four things did ancient peoples consider as elements? _____

4) Is salt an element (listed on the Periodic Table)? _____

5) What two amazing things did people believe the philosopher's stone would do? _____

_____

6) What element is often found sandwiched between lead and silver? b_____

7) Georgius Agricola was the first scientist to realize that there are three levels of classification:

    _____ make _____, and _____ make _____.

8) What element did Henning Brandt discover in urine? _____

9) Nickel is German for _____. Why? _____

10) What did Dmitri Mendeleev invent?   _____

11) Who was the first person to use the word "atom"? _____

12) What is an atom's atomic number? (What does the number mean?) _____

13) How many protons does an atom of phosphorus have? _____

14) True or false? Elements in their pure form have an equal number of protons and electrons. _____

15) True or false? All compounds are molecules, but not all molecules are compounds. _____

Activity 1.2 Videos

This book has a playlist on YouTube! Go to **www.YouTube.com/TheBasementWorkshop**. This channel belongs to the author of this book, so the videos on all the playlists have been previewed and chosen very carefully. Find the link that says SHOW ALL PLAYLISTS, if the "Rocks and Dirt" playlist is not visible. Click around until you find "Rocks and Dirt." The videos will be posted in order, with chapter 1 videos first, then chapter 2, etc. Unfortunately, YouTube has not provided any way of adding notes that the viewers can see, so there's no way to let you know exactly what videos go with each chapter. But you can figure it out. The videos about elements go with chapter 1. Then chapter 2 will feature videos on minerals.

Activity 1.3 Elements that make minerals

The crust of the earth is basically made of only a few dozen elements. In fact, if you weighed the planet and then found out how much of that weight came from each element, you'd discover that 75% of the weight comes from just two elements: silicon and oxygen. Of that remaining 25%, 24% is made of aluminum, iron, calcium, sodium, potassium, and magnesium. That means that only 1% is left for all the rest of the elements. (When we look at living things, hydrogen, oxygen and carbon are extremely abundant.) Let's make sure we're familiar with the symbols for these common elements in the earth's crust. Use a Periodic Table (or Google) to find the names that go with these atomic symbols.

1) H _____   7) P _____   13) Cu _____

2) C _____   8) S _____   14) Cl _____

3) O _____   9) Fl _____   15) Si _____

4) Na _____   10) Ca _____   16) K _____

5) Mg _____   11) Fe _____   17) Au _____

6) Al _____   12) Ag _____   18) Zn _____

CHAPTER 2: MINERALS

Now that we understand what an element is, we can explore the world of minerals. Geologists have come up with a very tight definition of what a mineral is. There are five "rules" that must be satisfied in order for something to be considered a mineral:

1) It must be inorganic (non-living).
2) It must be naturally occurring (not man-made).
3) It must be a solid (no liquids or gases).
4) It must have a chemical recipe (formula).
5) It must form some kind of crystalline (geometric) structure.

Hey guys— better listen up!. These rules are for you!.

The order of these rules is not important. When you see this list in other books or on websites, the numbers might be different. They can be listed in any order. What's important is to understand what they mean, so let's take a short paragraph to explain each one.

1) It must be inorganic (non-living). This seems pretty obvious. No one has to tell you that a mouse is not a mineral. However, this rule isn't quite as stupid as it sounds. Sugar crystals look very much like the mineral crystals found in sand, but sugar can't be classified as a mineral because sugar is made by plants, which are living things. Pearls also look like they could be minerals. They are hard and shiny and made of the same elements that many minerals are. However, since pearls must be made by a living thing (an oyster), they can't be minerals. Seashells are made of the same elements that are found in calcite crystals, but seashells are not minerals because a little animal (a mollusk) made them. (Sometimes the word "abiogenic" is used instead of the word "inorganic" because "a" means "not," "bio" means "life," and "genic" means "created by.")

Sugar is NOT a mineral.

Amber is fossilized tree sap.

There are two notable exceptions to this rule: coal and amber. Coal probably came from ancient plants that were buried under great pressure, and amber is fossilized tree sap. These two substances are often listed in books as "organic minerals." Technically, there's no such thing as an organic mineral, but apparently coal and amber have special privileges. Mineral collectors often have specimens of amber in their collections.

2) It must be naturally occurring (not man-made). Metals such as steel and bronze are man-made mixtures of metals, so they can't be minerals even though they are hard, shiny and metallic like many minerals. Plastics can't be minerals, either, since they are made in factories.

3) It must be a solid. Nothing tricky here. Substances like air and water can't be minerals. (However, oddly enough, if you freeze water and make it into a solid (ice) it then qualifies as a mineral under all five rules!)

4) It must have a chemical recipe (formula). Sometimes this rule is phrased like this: "It must have a definite chemical composition." We've substituted the word "recipe" for "definite chemical composition" because it is much easier to understand. A mineral has a unique recipe made from one or more of the elements on the Periodic Table. The best way to understand this is to give some examples. The recipe for the mineral **quartz** is SiO_2. If the atoms were all the same size and could be measured using cooking scoops, you could cook up a batch of quartz in your imaginary chemical kitchen using two scoops of oxygen atoms for every one scoop of silicon atoms. If you wanted to make a batch of the mineral **calcite**, you'd use this recipe: $CaCO_3$. That would be like one scoop of calcium atoms, one scoop of carbon atoms, and three scoops of oxygen atoms. (If you don't see a number listed below a letter, that means the number is 1.) A mineral called **talc** is made with magnesium, silicon and oxygen, using this recipe: $Mg_3Si_4O_{10}$. Recipes can be as simple as one ingredient, such as Cu (copper) or they can be quite complicated, like the recipe for turquoise: $CuAl_6(PO_4)_4(OH)_8 4H_2O$.

5) It must have a crystalline (geometric) structure. You have probably seen pictures of gemstones such as diamonds, that have beautiful geometric shapes. Perhaps you or someone you know has a mineral collection that includes crystals such as quartz, calcite, amethyst or fluorite. The atoms in a mineral line up in very regular ways, forming a pattern called a **lattice**. Lattice patterns are based on simple shapes such as squares and hexagons, but they can often look quite complicated. Sometimes the overall shape of a mineral specimen will look very much like its lattice structure, but other times it won't resemble the lattice at all. The lattice shown here is made of nothing but carbon atoms locked tightly together to make a **diamond**.

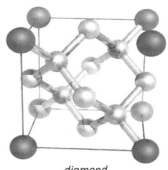
diamond

There are approximately 4,000 kinds of minerals. We've already mentioned some of them: quartz, diamond, calcite, amethyst, talc, turquoise, salt, ice, and elements such as copper, sulfur, silver and gold. Some minerals have names that are thousands of years old, like jasper, onyx, beryl, chalcedony, opal, and agate. Many minerals are named after the person who first identified them, such as Thomsonite, Smithsonite, Allanite and Johannsenite. The rest have names that sound like science words, such as hematite, dioptase, pyrite, and zircon. (Some names even sound a bit funny, like crocoite, wulfenite, apatite, and bauxite.) Mineral books are full of odd names!

Now, 4,000 items is a really large number. Mineralogists had to come up with a way to classify and organize them. Should they sort them by color? By shape? By location? What would be the most useful categories? The best solution turned out to be sorting them by their chemistry. This would not have been possible many centuries ago, but starting in the 1800s, chemists began inventing ways of finding out what elements were present in a given mineral sample. By the 1900s, mineral analysis became very accurate and the "recipe" for every mineral was discovered. Mineral recipes tended to be based on just a handful of common elements, such as carbon, silicon, sulfur, oxygen, phosphorus, chlorine and fluorine. From these elements they formed the names of the mineral groups: silicates, carbonates, sulfides, sulfates, oxides, halides, and phosphates.

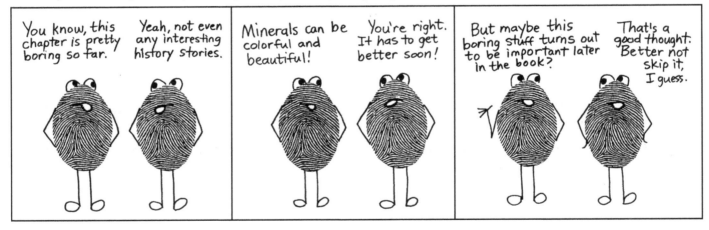

Yes, this stuff turns out to be important when we start talking about rocks. Rocks are made of minerals. If we're already familiar with lots of minerals, studying rocks will be much easier.

Minerals are put into groups according to what element their chemical recipes are based on. Quartz, SiO_2, is based on silicon, so it is a silicate. A mineral called pyrite (also known as "fool's gold") has the recipe FeS_2, where S is sulfur, so it is classified as a sulfide. Anything with a recipe that includes (CO_3) is a carbonate. Calcite is made of $CaCO_3$, so it is a carbonate. You may not have heard of the mineral corundum, Al_2O_3, which is an oxide, but you've certainly heard of the gemstones that are made of corundum: rubies and sapphires. This photo shows a famous sapphire, called the Logan Sapphire, which is in the Museum of Natural History in Washington, DC. (Sapphires are blue because there are small amounts of the elements titanium and iron mixed into the Al_2O_3.)

Minerals are classified according to the elements in their chemical recipes. These are the basic categories you will find in any book about minerals. Some books add a few more categories, these eight are enough for us. Also, we've trimmed the examples down to just minerals that are very well-known and are easy to pronounce. Mineral guide books will give you quite a few more examples.

You don't have to sit and memorize this list! The list is mostly as a source of information you can come back to later. Just skim through it now, and then remember it is here for future reference.

NATIVE ELEMENTS are pure elements.
　　Examples: sulfur (S), copper (Cu), silver (Ag), gold (Au), bismuth (Bi), diamond (C), graphite (C)

SILICATES are based on silicon, Si.
　　Examples: quartz (SiO_2), zircon ($ZrSiO_4$), topaz ($Al_2SiO_4(OH,F)_2$), beryl ($Be_3Al_2Si_6O_{18}$), talc ($Mg_3Si_4O_{10}(OH)_2$),
　　　　and all the different types of feldspar with their various recipes (ex: $KAlSi_3O_8$ and $CaAl_2Si_2O_8$)

CARBONATES are based on the element carbon, but the carbon is attached to three oxygens: CO_3.
　　Examples: calcite ($CaCO_3$), dolomite ($CaMg(CO_3)_2$), smithsonite ($ZnCO_3$), malachite ($Cu_2CO_3(OH)_2$),

SULFIDES are based on sulfur, S.
　　Examples: pyrite (FeS_2), chalcopyrite ($CuFeS_2$), galena (PbS), cinnabar (HgS)

SULFATES are also based on sulfur, but associated with oxygen as SO_4.
　　Examples: epsom salt ($MgSO_4$), barite ($BaSO_4$), gypsum ($CaSO_4$),

OXIDES are based on oxygen, O.
　　Examples: hematite (Fe_2O_3), corundum (Al_2O_3), rutile (TiO_3), pitchblende, (UO_2, where U is uranium)

HALIDES are based on elements called "halogens," which include chlorine, Cl, and fluorine, F.
　　Examples: salt (NaCl), fluorite (CaF_2), sylvite (KCl)

PHOSPHATES are based on the element phosphorus, but attached to four oxygens: PO_4.
　　Examples: apatite ($Ca_5(PO_4)_3$), turquoise ($CuAl_6(PO_4)_4(OH)_8 4H_2O$)

The largest of these groups is the silicate group. We'll take an entire chapter to discuss the silicates because they are the most abundant minerals in the earth's crust, and are an important ingredient in the magma that comes out of volcanoes.

When we turn the page, we won't be talking about chemistry anymore. We'll be discussing things like color, texture, hardness, heaviness, and shininess. But don't forget about this chemistry! We'll go back to chemistry again in chapter 3.

Mineral guide books use those chemical categories as their basic organizational structure, sort of like chapters. The order of the chemical categories will be different in each book. Once you are into a category, such as carbonates, the minerals will be listed in seemingly random order, with a short paragraph about each. The paragraphs on each mineral will have exactly the same format: a list of characteristics using some familiar words such as color and hardness, and some less familiar words such as streak, cleavage, and habit. There is also usually a small picture beside each paragraph, showing a geometric shape. The paragraph will also probably tell you where the mineral is found in the world and what it is used for. Though the order of the characteristics is different in each book, the mineral's crystal shape is usually at the top of the list, so we'll tackle that explanation first.

CRYSTAL STRUCTURES (also called "SYSTEMS" or "HABITS")

Atoms stick together in different ways. Remember from chapter 1 that the electrons are arranged in shells around the nucleus. The electrons in the outermost shell are the ones that interact with other atoms. The number of electrons in that outer shell will be the most important factor in determining what the interactions will be. Sometimes atoms give away their electrons, and sometimes they share. Certain atoms prefer pairing up with certain other atoms. Given a chance, sodium atoms will always rush to pair up with chlorine atoms. Sulfur and carbon atoms will gladly attach to some oxygens to make little groups like SO_4 and CO_3.

As trillions upon trillions of atoms pair up, the whole mass of atoms starts to be visible as some kind of substance. If trillions of sodiums and chlorines get together, they will interlock into an alternating pattern that will result in a cubic shape. If trillions of hydrogens and oxygens get together they will make lots of little H_2O molecules that can interlock into 6-sided geometric shapes (snowflakes) if the temperature gets cold enough. Trillions and trillions of carbon atoms will interlock to form either flat, gray sheets of graphite, or beautiful, clear diamonds, depending on the pressure and temperature surrounding them.

Scientists in the 20th century were able to analyze minerals using techniques (often with x-rays) that allowed them to determine exactly how the atoms were connected. They found that minerals' crystal structures fell into basically six categories. Inside each category, there are some variations (more variations than shown here).

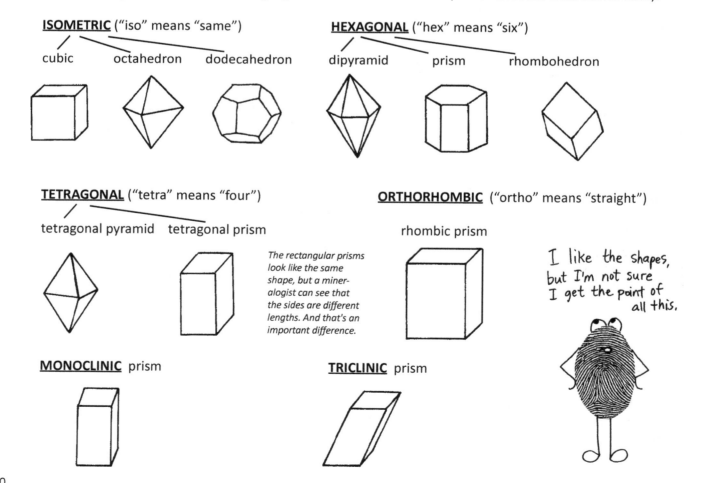

ISOMETRIC ("iso" means "same")

cubic   octahedron   dodecahedron

HEXAGONAL ("hex" means "six")

dipyramid   prism   rhombohedron

TETRAGONAL ("tetra" means "four")

tetragonal pyramid   tetragonal prism

The rectangular prisms look like the same shape, but a mineralogist can see that the sides are different lengths. And that's an important difference.

ORTHORHOMBIC ("ortho" means "straight")

rhombic prism

I like the shapes, but I'm not sure I get the point of all this.

MONOCLINIC prism

TRICLINIC prism

In reality, many mineral specimens don't actually look like these shapes. A few do, however, including pyrite, halite, quartz, fluorite, calcite and galena.

pyrite

fluorite

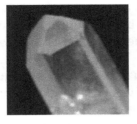

quartz

calcite

galena

The reason that minerals make these shapes is due to that interlocking pattern of atoms we keep talking about. Here on the right an artist has drawn the cubic pattern found in NaCl crystals. The green dots are chlorine (Cl) atoms and the gray dots are sodium (Na) atoms. If you look at just the pattern of dots, it is not hard to see how this structure could form a cube. What is harder is to see how this pattern could form other shapes. The artist has helped us out a bit with imagining another shape: an octahedron. Ordinary table salt tends to be cubic, while rock salt can take this octahedron shape. Both shapes result from this cubic lattice pattern. Since an octahedron really doesn't look cubic, we use the more general word, isometric, as our category title. Isometric means that the shape is symmetric (mirror image) from front to back, top to bottom, and side to side.

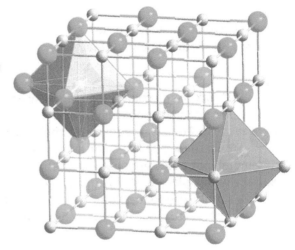

Most minerals don't have such simple shapes. Or if it does show a simple shape, there are lots of tiny ones, not one large crystal. Some minerals even have shapes with rounded edges, like the form of barite that is often called the "desert rose." Sometimes the same mineral can take different shapes, depending on the temperature, pressure and amount of moisture present when it was formed, or if other elements are mixed in.

barite ("desert rose")

mineral specimens without nice crystals

quartz crystals without distinct sides

Sometimes a mineral's crystal shape is not apparent at all, as is the case with chalk. Chalk is indeed considered to be a mineral, but it is soft and powdery. The mineral called talc is also very soft, so soft, in fact, that you can scratch it with your fingernail! As with everything in life, most things fall into categories pretty well, but you always have your duckbill platypuses that seem to defy classification.

Another mineral that is hard to classify is afghanite. Its chemical recipe is $(Na,K)_{22}Ca_{10}[Si_{24}Al_{24}O_{96}](SO_4)_6Cl_6$. It has so many elements that it's hard to know which chemical group it belongs in. Fortunately, the mineral doesn't care.

afghanite

COLOR

Finally, something easy! Mineral books will always list the color(s) of each mineral. Some minerals are always the same color. Pyrite (fool's gold) is a shiny gold color, galena is dark gray, and pure sulfur is yellow. Other minerals have many color variations. Calcite and quartz have variations that are clear, white, yellow, brown, purple, green or pink. Corundum can be yellow, brown, green, blue or red. (When corundum is blue we call it a sapphire and when it is red we call it a ruby.)

Ruby is red corundum.

Along with color, the guide will also tell you if the mineral is **opaque**, **transparent** or **translucent**. Opaque means that light cannot pass through it. A piece of wood is opaque. Transparent means that light can go through it and you can also see through it. Glass is transparent. Translucent means that light can go through it, but you can't see through it. A thin piece of paper is translucent. Quartz and calcite can be transparent or translucent. Pyrite and galena are definitely opaque.

LUSTER

Luster is the way the surface of the mineral looks, without respect to color. The luster can be shiny, dull, waxy, glassy, metallic, pearly, greasy, etc.

STREAK

"Streak" is the color the mineral makes when it is rubbed onto a surface. Often, a piece of white porcelain (similar to the back of a floor tile) is used to do the streak test. Porcelain seems to be the ideal surface for streaking minerals.

A piece of pyrite leaves a black streak.

You'd think that the color that rubs off a mineral would be the same as the mineral, but oddly enough, it is often different. For example, if you scratch a surface with a piece of shiny gold pyrite, the mark it makes is black, not gold. (That's a simple way to test your "gold" to see if it is just fool's gold.) Some minerals are so hard that they don't leave a mark.

HARDNESS

Minerals can be very soft, like chalk, or extremely hard, like diamond. In 1812, a scientist named Friedrich Mohs invented a way to compare the relative hardness of minerals. He chose 10 minerals that would form a hardness scale, which today is called the Mohs hardness scale. The softest mineral on the scale, number 1, is talc. Talc is so soft that, as we previously mentioned, you can scratch it with your fingernail. Just up from that, at number 2, is gypsum. A piece of gypsum can put a scratch into a piece of talc, but a piece of talc can't scratch a piece of gypsum. By trying to scratch a mineral with a mineral, you can evaluate the hardness of each one. Number 10 on the scale is diamond, of course. Right under that, at number 9 is corundum (rubies and sapphires). Here are the others:

1	2	3	4	5	6	7	8	9	10
talc	gypsum	calcite	fluorite	apatite	feldspar	quartz	topaz	corundum	diamond

This scale is not a precise measurement of hardness. If you want an exact measure, you have to use a machine called a sclerometer which measures microscopic scratches cut into the mineral by a diamond blade. A sclerometer still puts talc at number 1, but diamond would be a 1600. Corundum scores only a 400, so you can see how much harder diamond is than any other mineral. Diamond is 4 times as hard as corundum.

Hardness is one clue that can help to identify unknown minerals. Sometimes geologists will also use a penny, a steel knife blade and a steel nail to help them identify hardness, because these items are 3.5, 5.5 and 6.5 on the Mohs scale. Better yet, you can go to websites like geology.com. and get yourself a set of hardness picks. They have conveniently installed the correct hardnesses on the numbered tips and you can just scratch away on your minerals.

CLEAVAGE

Cleavage means how something cracks or breaks. (A large meat knife is sometimes called a meat cleaver.) Some minerals split into angular little pieces, some break into flat sheets and some just crumble. How a mineral breaks is determined by its atomic lattice pattern. Often the crack will occur right along a line in the pattern.

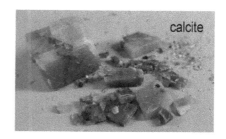
calcite

Calcite is a very interesting mineral to break. Calcite crystals often have a rhombohedral shape. If you smash a calcite rhombohedron, you get a bunch of little rhombohedrons. If you smash those little rhombohedrons you get even smaller rhombohedrons. This would continue on down to the microscopic level because the atomic lattice itself is rhombohedral. You could have a rhombohedral piece so tiny that you can't see it.

Some minerals exist as flat sheets, so they almost peel apart. Biotite and muscovite (forms of mica) look like stacks of paper-thin, almost transparent sheets, pressed tightly together. The sheets come apart easily.

biotite

People who cut gemstones take advantage of the natural tendency of crystals to cleave along the weakest planes of the lattice. With a few firm taps in the right places, the crystal's overall shape can be improved, not ruined.

a craftsman working on a diamond

"Sea of Light Diamond" (Iran)

"Hope Diamond" (Washington DC)

DENSITY

Density is another characteristic that gets a number assigned to it. The number 1 on the density scale belongs to liquid water. The density of everything in the world is compared to water. If something is less dense than water, it floats. If something is more dense than water, it sinks. But what, exactly, is density?

The word "dense" basically means a whole lot of something in a given area. ("Sparse" is the opposite of dense and means very few things in an area.) How can we apply this definition to minerals? A dense mineral has many atoms packed into a given area. Sometimes, the size of the atoms factors in, as well, since large atoms take up more space than small ones.

All minerals (except for ice) will sink in water, so they all have a density greater than 1. One of the most dense minerals is gold, which can have a density as high as 19.3. Another heavy mineral (because it contains lead) is galena, but it is only a 7.5. Most minerals have densities between 3 to 6. The least dense minerals are the salts, which can sometimes be as light as 1.5 (which still sinks in water). Density isn't a very exciting property of minerals, but it can be very helpful in figuring out the identity of an unlabeled mineral specimen.

So if someone calls you "dense" just tell them that means you have more brain cells packed into your head!

NOTE: Your mineral guide book might use another word for density: specific gravity. Specific gravity is defined as the density of the mineral divided by the density of water. The density of water is 1, so that means you are always dividing by 1, which doesn't change your number. Aren't scientists silly sometimes?

OTHER PROPERTIES

A few minerals have some very unusual characteristics.

1) MAGNETISM: Magnetite is highly magnetic. Other minerals containing iron can be slightly magnetic, but magnetite takes the prize for magnetism.

2) BIREFRINGENCE: "Bi" means "two" and "re-fringence" has to do with light being bent by the crystal. Birefringent calcite is clear so you can see through it, but what you see is two images, not one. It's like crystal double vision.

3) FIBER OPTICS: A mineral called ulexite demonstrates **fiber optic** qualities. Ulexite is fairly transparent, so if you place it on a piece of newspaper, you can see the words and pictures on the paper. What is odd is that those words and picture look like they come up to the surface of the rock.

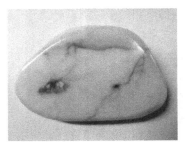 

4) RADIOACTIVITY: Pitchblende (uranium oxide) contains radioactive elements such as uranium, radium and polonium. The famous scientist Marie Curie discovered the elements radium and po-lonium by boiling down big pots of pitchblende for months on end. Radioactive means that the nucleus of the atom is falling apart, losing pieces that go flying off. These pieces can be dangerous to the cells of living things.

USES FOR MINERALS

Detailed guide books will often give a sentence or two about what the mineral is used for. Here are a few samples that you might find interesting.

BARITE: a source of the element barium which is used for medical imaging and in the oil drilling industry

BAUXITE: a main source of the element aluminum

BERYL: as a gemstone, also as a source of the element beryllium for aircraft metals, fluorescent lamps, x-ray tubes

CALCITE: lenses, fertilizer, cement, polishing powders, filters,

CINNABAR: a main source of the element mercury (which is a liquid at room temperature)

CORUNDUM: unattractive crystals are crushed and used as abrasives, nice crystals get make into gemstones

DIAMOND: jewelry, tips on cutting and drilling equipment

FLUORITE: used as "flux" in metal manufacturing, also used by industries that make plastics and optical devices

GALENA: an excellent source of lead (Galena was a main source of lead for musket shot in N. America in the 1700s.)

GRAPHITE: pencils, and lubricants (to make things slippery)

GYPSUM: plaster and plasterboard, Portland cement, fertilizer, and in pottery

HALITE (rock salt): used in several ways in the food industry, used to melt ice on roads, used in other industries

HEMATITE: an important source of iron, also used as a pigment for paints

MAGNETITE: an important source of iron (which is used to make steel)

MALACHITE: green pigment for paint, jewelry, decorations

PYRITE: a source of sulfur for sulfuric acid which is used in many industrial processes

SPHALERITE: a main source of the elements zinc, gallium, cadmium and indium, which are used in industry

Activity 2.1 Comprehension self-check

Can you remember the answers to these questions? If not, go back to the text and see if you can find the answers. When you are done, you can check your answers using the answer key at the back of the book.

1) True or false? Since sugar has a crystalline form, it can be classified as a mineral. _____

2) True or false? Salt is a mineral. ___

3) There are two exceptions to the rule about minerals not being organic: _____ and _____

4) Why is bronze NOT a mineral? _____

5) True or false? Most minerals have a chemical recipe, but some don't.

6) Diamonds are made of nothing but this element: _____

7) When a mineralogist sends a Valentine card, it might begin like this: Rubies are ____, sapphires are _____.

8) What mineral is made of a square lattice of sodium and chlorine atoms? _____

9) What is the correct name for "fool's gold"? _____

10) What is the correct name for the mineral called the "desert rose"? _____

11) A quartz crystal has six sides. What category does it belong in: hexagonal or tetragonal? _____

12) Which one of these is NOT isometric? cube, octahedron, rhombohedraon, dodecahedron? _____

13) Name a mineral that doesn't appear to have a crystal structure at all: _____

14) What does "translucent"mean? _____

15) Name a mineral that can be transparent: _____ Name a mineral that is opaque: _____

16) What is an easy way to prove that fool's gold is not gold? _____

17) What is the softest mineral on the Mohs hardness scale? _____ The hardest? _____

18) If you try to scratch a piece of quartz using a piece of calcite, will it work? _____

19) Other than diamond, what is the hardest mineral? _____

20) The words shiny, dull, waxy and metallic describe what property of minerals? _____

21) True or false? A mineral will usually break (cleave) right along a line (plane) in its atomic lattice pattern. _____

22) Name a mineral whose cleavage pattern is flat sheets. _____

23) What has a density of exactly 1? _____

24) Name a mineral that is extremely dense: _____

25) The more correct term for density is _____ _____.

Activity 2.2 Videos

Go to the "Rocks and Dirt" playlist at: **www.YouTube.com/TheBasementWorkshop**, and watch the videos on minerals. Stop when you get to the quartz videos and save them for the next chapter.

Activity 2.3 "Odd one out"

In each list of four minerals, three belong to the same chemical group and one is the "odd one out." Which one does not belong? Write the correct category for the three others on the line.

1) _____ copper, silver, turquoise, gold
2) _____ quartz, apatite, zircon, beryl
3) _____ gypsum, calcite, dolomite, malachite
4) _____ pyrite, galena, cinnabar, bismuth
5) _____ salt, gypsum, calcite, barite
6) _____ corundum, feldspar, pitchblende, hematite
7) _____ barite, sylvite, fluorite, salt

Gypsum can sometimes look like this.

Activity 2.4 A word puzzle

Here is a fun way to do more practice with mineral names. The answers to these clues are all in the chapter.

ACROSS:
2) Used to make plaster, cement and fertilizer
3) The most magnetic mineral
8) A green mineral used for jewelry and as a pigment in paint.
9) Its name means "field stone," it comes in bluish-green and pink and it is number 6 on the Mohs hardness scale.
10) Number 8 on the Mohs hardness scale
11) One of the softest minerals, number 1 on the Mohs
15) A radioactive mineral that contains uranium.
17) Contains the element fluorine.
18) Most aluminum is extracted from this.
19) Number 5 on the Mohs hardness scale.
20) Has the chemical formula FeS_2.
22) A mineral that contains the element boron and shows the unusual property of fiber optics
23) Is used as a source of zinc, gallium, cadmium and indium.

DOWN:
1) A form of pure carbon, but in hard, clear crystals.
2) A mineral that contains a lot of lead, and was used to make lead ammunition for musket rifles in the 1700s.
4) A form of pure carbon that forms flat sheets. You use this mineral every day when you write with pencils.
5) A source of the element barium.
6) Contains the element beryllium.
7) This red mineral is an excellent source of iron. The first part of its name is Greek for blood (because it is red).
12) A mineral ore for mercury, having the formula HgS
13) Formula $CaCO_3$, pure crystals can look like rhombohedrons.
14) Both sapphires and rubies are made of this mineral.
16) The mineral name for salt.
18) A mineral that comes in flat, shiny, black sheets
21) SiO_2

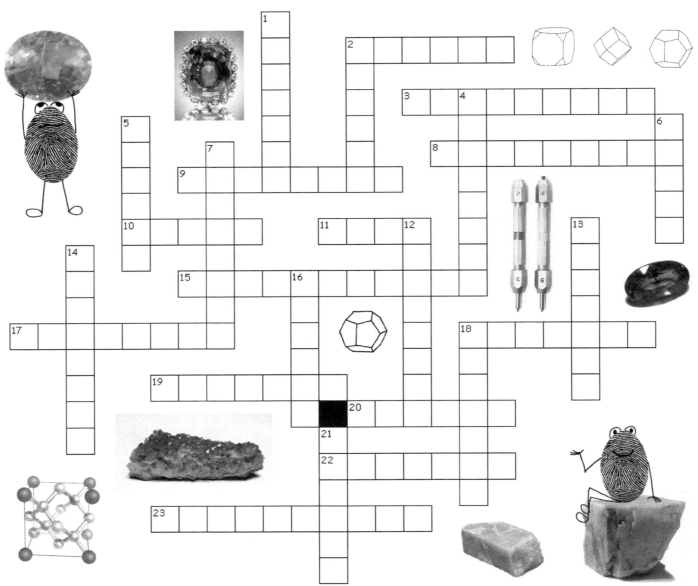

16

Feldspar is often pink or greenish-blue.

CHAPTER 3: THE SILICATES

Most of our planet is made of the elements **silicon** and **oxygen**, with help from the elements iron, magnesium, calcium, and carbon. The silicate family of minerals is by far the most abundant, so it deserves an entire chapter. To understand this family of minerals, we'll start by reviewing some facts about atoms.

We'll need to look at a Periodic Table. (You can use the table on page 141, or you can use one from the Internet or in another book; it doesn't matter very much what table you look at because they are all the same.) Look at the element silicon, atomic number 14. The most important thing to notice is that it is directly below the element carbon, number 8. All the elements in this column (up and down) have 4 electrons in their outermost shell. Remember, it is this outermost shell that interacts with the environment, so the elements in this column will behave in a similar fashion. Not identical, but very similar. This outer shell would like to have 8 electrons, but it only has 4. So silicon and carbon will be out looking to share or borrow 4 electrons from other atoms. When they find an atom, or several atoms, that are willing to bond with it, they form very strong bonds called **covalent bonds**. Covalent bonds are hard to break and provide a "backbone" for the mineral structure.

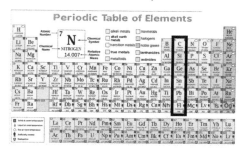

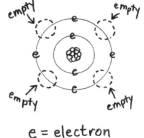

A carbon atom has 4 electrons in its outer shell and wants 4 more.

e = electron

Same with silicon

Silicon has exactly the same problem, even though it is a slightly bigger atom.

Another type of bonding that atoms can do is called **ionic bonding**. Ionic bonds occur when an atom is willing to share its electrons, but not in such a strong, permanent way. Ionic bonds are more easily broken than covalent bonds are. (In fact, you can break the ionic bonds in salt crystals (NaCl) just by putting the crystals into water. The water molecules themselves are strong enough to break apart these ionic bonds. We call this "dissolving.") A combination of covalent and ionic bonds are found in silicate minerals.

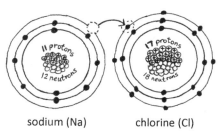

sodium (Na)   chlorine (Cl)

CH_4

SiO_4

The simplest way that carbon and silicon atoms can solve their electron problem (needing to make 4 bonds) is to find 4 small atoms that would like to share 1 electron. In the case of carbon atoms, the element most likely to pair up with them is hydrogen. Hydrogen atoms want to share just 1 electron, so carbon pairs with 4 hydrogens to make CH_4, a molecule that we call methane ("natural gas"). In the case of silicon, the element most likely to pair up with it is oxygen. A silicon atom will attract 4 oxygens to make SiO_4. The SiO_4 molecule doesn't have a nice name, like methane does. We simply call it the **silicon tetrahedron**. The word tetrahedron means "4-sided." (A tetrahedron might remind you of the pyramids of Egypt, but if you count the sides (including the bottom) of the pyramids, they have 5 sides, not 4.)

Both these molecules have 1 central atom connected to 4 other atoms. They both form a tetrahedron shape. They both have strong covalent bonds. They both can form long chains, too. Methane molecules can drop some of their hydrogens in order to bond to other carbons, forming chains we call hydrocarbons. Hydrocarbons can be just a few carbons long, or up to thousands of carbons long. Small hydrocarbon chains include propane, gasoline (petrol), diesel fuel, and wax. Longer hydrocarbon chains are found in plastics, starches and fats.

In a similar fashion, a silicon tetrahedron can drop one of its oxygens in order to bond with another tetrahedron. They can keep doing this and form a long line of silicon tetrahedrons.

Let's take a closer look at what is going on here. We need to look at the oxygen atoms. Unlike hydrogen, oxygen atoms want to make 2 bonds. Oxygen is number 8 on the Periodic Table. (Find oxygen on the table.) All the elements in this column (under oxygen) have 6 electrons in their outer shell. They'd like to have 8, so they are out looking to borrow 2 electrons from one or more atoms in their neighborhood. When an oxygen atom solves its problem by bonding with two hydrogen atoms (who each have 1 electron to share) we call that molecule H$_2$O, water. If there are no other atoms around, oxygens will pair up with themselves and each share a pair of electrons with the other. This isn't ideal, but it is better than nothing. The oxygen in the air around is doing this, making molecules of O$_2$. Bearing in mind that oxygen atoms want to borrow 2 electrons, not 1, how does this affect the chemistry of the silicon tetrahedron?

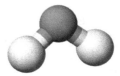

Oxygen is always red.

Those oxygens at each corner of the tetrahedron are sharing 1 electron with silicon. That means that the oxygen atoms would like to find 1 more electron. Very often there are atoms of iron, Fe, and magnesium, Mg, floating in the environment. (Magma, hot liquid rock, often contains the elements iron and magnesium.)   Fe and Mg atoms have an electron arrangement that allows them to share 2 electrons with other atoms. (Iron can even share 3 sometimes.) An Fe atom can share its electrons with one or more of the oxygens in a silicon tetrahedron. This makes the oxygens very happy and they create ionic bonds with the Fe and Mg atoms. Remember, ionic bonds are not as strong as covalent bonds, but they are strong enough to hold all these atoms together and make the resulting substance as hard as rock.

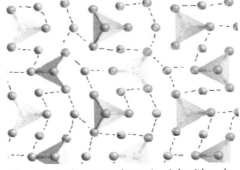

The tetrahedrons are shown in pink with red oxygens at the points. The blue balls represent metal atoms, perhaps Fe or Mg. The dashed lines represent electrical attraction keeping the atoms in place.

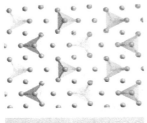

If the silicon tetrahedrons stay separate, we get a mineral called **olivine**. It looks a little bit like a soup, if you imagine the tetrahedrons and the atoms to be bits of chopped vegetables. However, it can't be a soup because it is a solid, not a liquid. Chemists call this type of chemical structure a **solid solution**. (Stone soup!) Olivine is a very easy mineral to remember because it is olive green. Sometimes it is clear, sometimes it is translucent or even opaque. Why is it this color? The arrangement of the atoms just happens to absorb all the colors of the rainbow except for olive green. This color of green light gets reflected back, so we can see it. (The exact details of how the light interacts with the atoms is very complicated. It is enough to know that green light bounces back at you, and all the other colors are absorbed and thus disappear.)

Have you ever read the book "Stone Soup"?   Yeah, it's a classic!

Now let's put some tetrahedrons into long lines—many long lines, like the strong metal rods they put into concrete to reinforce it. All those iron and magnesium atoms will be like the concrete that keeps the metal bars in place. Minerals that have this structure are called **pyroxenes**. *(pie-rocks-eens)* There are different types of pyroxenes, each one defined by tiny "impurities" of other elements mixed in, such as aluminum, titanium, chromium, sodium, calcium, zinc and lithium. The only type of pyroxene you may have heard of is jadeite, from which the gemstone jade is made. (Augite and diopside are shown below but you don't need to remember their names.)

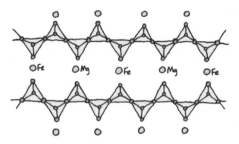

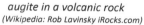
augite in a volcanic rock
(Wikipedia: Rob Lavinsky iRocks.com)

diopside
(Rob Lavinsky)

jadeite

jade carving

The next thing we can do with the silicon tetrahedrons is to put two chains together to make a double chain. In our picture, below, you can see that when the two chains link together, they look like a series of hexagons. We still have atoms of iron and magnesium in the spaces between the chains. Minerals with this structure are **amphiboles**. *(am-fih-boles)* The Greek word root "amphi" means "ambiguous, " (hard to tell what it is). Amphiboles are hard to identify, even for geologists. The only amphiboles you may have heard of are **hornblende** and **asbestos**.

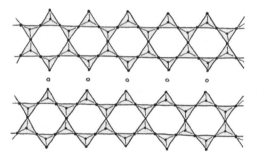

Long lines create the needly texture of asbestos.

hornblende

asbestos
(Wikipedia: Aram Dulyan, London)

microscopic view of asbestos fibers

Mineral identification kits often contain a sample of hornblende. It's one of those samples you hardly notice in the collection—just a dark rock that does not look like anything special. What is significant about hornblende is that is often a main ingredient in a very important rock: granite.

Asbestos is one mineral you definitely won't find in a mineral kit. It was discovered in the 1980s that asbestos fibers can get into your lungs and cause lung diseases including cancer. The cells that are your clean-up crew in your body can't get rid of asbestos—it is indigestible. Worse yet, its shape is like sharp little needles so it is constantly poking and irritating your cells. The nice thing about asbestos is that it is fireproof, which is why it had become so popular as insulation in public buildings and wrapped around electrical wires. Billions of dollars have been spent trying to safely remove asbestos from buildings and appliances.

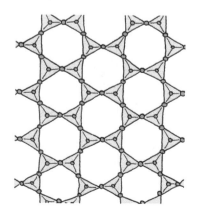

Now we can take a bunch of double chains and connect them together to make a flat sheet, then stack the sheets. We will still have iron and magnesium and other kinds of elements in the middle between these sheets, helping them to stick together, but they won't stick very well, and will peel. You are probably already guessing that biotite might belong to this group, and you are right. Minerals made of sheets of silicon are called **micas**. Biotite and muscovite are types of mica. Biotite is always dark because it contains iron and magnesium, and muscovite is always clear or light-colored because it has lots of aluminum and potassium. (The potassium atoms in muscovite form an entire layer by themselves, helping the silicon layers to peel apart. If you've ever seen cheese slices packaged with pieces of paper between the slices, think of the paper as the layers of potassium atoms.)

Unlike asbestos, the mica minerals are harmless. They have physical characteristics that make them useful in many branches of technology. They are used to make parts for navigation compasses, optical filters, helium-neon lasers, missile systems components, medical electronics, optical instrumentation, radar systems, radiation detectors, and electronic insulators. Very high quality sheet mica can sell for as much as $2000 per kilogram (two pounds).

Biotite is dark.   *Muscovite is light.*

The last variation of silicate structure is called a **framework**. As the name suggests, a framework structure looks like it has supporting beams going in every direction. We can't build a framework, however, by just putting a bunch of sheets together. To make a framework out of the sheets, we would have to take apart all the silicon tetrahedrons and completely redesign them. Ideally, every tetrahedron is joined to four other tetrahedrons. In reality, there are often some other elements, such as potassium and sodium, lurking in and around the tetrahedrons.

Minerals that fall into the framework category include **feldspar**, **quartz** and a number of **gemstones**. The word feldspar is German for "field stone." Apparently this mineral is very common in the fields where the Germanic peoples settled in Europe. Centuries ago, those Germanic peoples discovered that these field stones did not contain any useful ores. You could heat those field stones all day long and not a drop of any metal came out of them—very disappointing. Feldspar could be used as building stone, but that was about it. Nowadays, feldspar is a bit more useful. Modern industries can grind up some types of feldspar and use the powder for making ceramics, paints, rubber, glass and plastics. A few types of feldspar can also be a source of aluminum.

Feldspars are also rich in the elements potassium, K, and sodium, Na. Geologists have a theory that while the minerals were forming in a pool of cooling magma, all the Fe and Mg atoms were used up by the olivine. As long as there was Fe and Mg available, olivine and pyroxenes could form. Once the Fe and Mg were gone, the silicon tetrahedrons had to start forming sheets and frameworks. There were still K and Na atoms floating around, and some of them got incorporated into the frameworks. All this information shows up in the chemical recipe for feldspar: $(K,Na)AlSi_3O_8$. The K and Na inside the parentheses means that it could be either one. There is some aluminum, Al, and, of course, the silicon framework, Si_3O_8. (Si_3O_8 means that we have fewer oxygens per silicon, since the tetrahedrons are connected. SiO_4 can be expressed as Si_2O_8, so we can compare fractions.)

That rock makes a nice chair.

See these parallel lines?

To identify feldspar, there are three things to look for: 1) The color is usually white, pink or greenish-blue. 2) It has a cleavage pattern that makes the chunks similar in shape, looking like a box that has one or two sides folded in slightly. 3) There are usually parallel lines running down at least one side.

Quartz is also a framework silicate. The word quartz comes from an old German word for "hard." In its most pure form, its formula is SiO_2. Why not SiO_4, like olivine? In quartz, the tetrahedrons are all surrounded by other tetrahedrons, so oxygen atoms double up, attaching to more than one silicon atom. The framework geometry requires fewer oxygen atoms per silicon atom. When you see very large quartz crystals it can be hard to believe that they just came out of the ground that way. They are so perfect, with their 6-sided hexagonal shape. Yet, they do naturally form that way. (The YouTube playlist has a video of people finding huge crystals.)

If there are no impurities (other elements) in quartz, the crystals are very clear. If you add a tiny amount of certain other elements, quartz can take on many different colors. The most well-known example of colored quartz is purple amethyst. It is thought that small amounts of iron cause the purple color. Iron is also the coloring agent in citrine quartz. Rose quartz gets its pink color from the elements titanium and manganese. Brown quartz, often called smoky quartz, gets its color not from extra elements, but from radiation. Being near a source of radioactivity causes little holes to open in the framework structure and this changes the way that light is absorbed and reflected.

clear quartz

amethyst

citrine

rose quartz

smoky quartz

Quartz is not only beautiful as a gemstone, it is also very useful in modern technology. Quartz has a characteristic that the other silicates do not, called the **piezoelectric effect**. (*pie-EE-zo*, or *pea-AY-zo*) "Piezo" means "pressure." When the crystal framework is squeezed, is produces a small amount of electrical charge. Also, the converse is true—when a small amount of electrical voltage is applied to it, the crystal changes shape. This phenomenon was first discovered several hundred years ago, but not officially documented until the year 1880, by Pierre Curie, husband of Marie Curie, who discovered several radioactive elements.

This is just one small piece of a quartz framework crystal. We see just 3 Si atoms (pink with a + charge) and 3 O atoms (red with a - charge). This hexagon would be connected to many others. The hexagons are electrically balanced, even though some of the atoms themselves are more positive or negative.

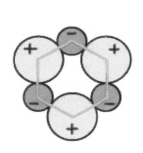

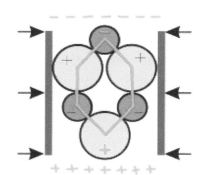

In this picture, the crystal is being squeezed very hard. The hexagon is bent out of shape slightly. One end of the hexagon becomes more negative, and the other becomes more positive. This allows electrical current (moving electrons) to flow from one side to the other. This happens to all the hexagons in the crystal.

Quartz crystals are used in devices such as sonars, clocks, and ultrasound machines. Small electrical voltages are sent through a tiny quartz crystal. The crystal's lattice changes shape, back and forth, about 30,000 times per second, generating a high-pitched sound wave that can be used for sonar (like a dolphin does) or for medical imaging (like those baby pictures everyone gets during pregnancy). Also, since quartz crystals vibrate at exactly the same rate every second, these vibrations can be used in small time-keeping devices such as watches.

The red arrow shows where the quartz crystal is located inside a watch. It looks like a tuning fork.

Quartz is definitely the most useful framework silicate. Some other types aren't really much good for anything except sitting there and looking pretty. However, people will pay a lot of money for these do-nothing gemstones, so we'd better mention them! When you start adding aluminum to the silicate framework, it seems to make

the crystals harder. For example, the mineral **topaz**, number 8 on the Mohs hardness scale, has this chemical recipe: $Al_2SiO_4(F,OH)_2$. You know what that SiO_4 is in the middle—the silicon tetrahedron. The aluminum atoms tuck in and around those tetrahedrons, and there are also either fluorine (F) atoms or OH's. (An OH is like a "broken" water molecule, with one H missing.) When you see letters in parentheses, that means it can be either one. Sometimes you might find fluorine and other times you might find OH's. Topaz naturally comes in clear, yellow, blue, pink, purple and orange. Clear topaz can be treated with heat or radiation to produce strong colors. If you buy a topaz gemstone in a jewelry store nowadays, it has probably been heated or exposed to radiation. Naturally brilliant colors are rare.

If you mix both aluminum and beryllium (atomic number 4) into a silicon framework, you get a mineral called **beryl**, $Be_3Al_2Si_6O_{18}$. If the piece of beryl is green and transparent, it is worth a lot of money because what you have is an **emerald**. Beryl comes in other colors, too, but they are not as well-known. Beryllium is a rare element, so there are not too many places in the world where you can find emeralds.

Garnet is the name for a whole group of gemstones. You can play "mix and match" with quite a few elements and still have a garnet. The recipe for garnet allows for this mixing and matching. They use the letters X and Y to stand for the elements you can choose. The recipe looks like this: $X_3Y_2(SiO_4)_3$. For X, you can choose one of these elements: iron, magnesium manganese, or calcium. (Notice that there are two elements with similar names: magnesium and manganese.) For Y, you can choose one of these: iron, aluminum, manganese, vanadium, or chromium. There are many possible combinations of these elements, so there are many types of garnets with a wide range of colors. Very clear, brightly colored garnets are used for gemstones in jewelry. Garnets that are not so pretty get ground up and made into industrial sandpaper.

A red garnet attached to the piece of quartz where it grew. This type of garnet is $Ca_3Al_2Si_3O_{12}$.
(credit: Rob Lavinsky, iRocks.com, Wikipedia)

Tourmaline is another group of crystals that are based on silicon but have added elements. Tourmalines always have the element boron, number 5 on the Periodic Table. In addition to boron, they might also have iron, magnesium, manganese, calcium, sodium, aluminum, chromium, vanadium and lithium. Lots of options to choose from! The recipes lets you choose: $XY_3Z_6Al_6(Si_6O_{18})$ $(BO_3)_3(F, OH)_4$. You can see your silicon framework in there, as (Si_6O_{18}). You've got some aluminum to make it hard, and some boron (B) to make it officially tourmaline. You've got the same ending that topaz does, with your choice between fluorine and OH (a "broken" water molecule). At the beginning, you get to choose not two, but three different elements. The list of elements is similar to the choices for garnet, but you use 1 "scoop" of whatever X is, 3 "scoops" of Y and 6 "scoops" of whatever Z is. That's how you cook up a batch of tourmaline.

When corundum has a bit of chromium (Cr) mixed in, it turns red. Blue is caused by small quantities of iron (Fe) and titanium (Ti).

We've already met **corundum**. You may remember that when corundum is red we called it a ruby and when it is blue we call it a sapphire. Sometimes corundum is brown or tan or a yucky in-between color that no one wants as a gemstone. These ugly bits of corundum end up as industrial sandpaper. It makes a great abrasive because it is very hard, number 9 on the Mohs scale.

Corundum doesn't actually belong in this chapter because its chemical formula is Al_2O_3. The silicon has been completely replaced by aluminum! Remember, when aluminum is added to a silicate it makes it harder. Here we have nothing but aluminum. Is it harder? Yes. Corundum is the hardest substance on earth, except for diamond. (And diamond is made from nothing but carbon—no silicon at all.)

We end this chapter with the most colorful members of the silicate family: the **cryptocrystalline** silicates. Many of the shiny, colored rocks you find in gift shops fall into this category. "Crypto" means "hidden." The crystals in these silicates are so small that they are hidden from our eyes. You won't see any 6-sided shapes or flat sheets. The crystals might be made of only a few dozen or a few hundred atoms. They still have the general formula, SiO_2, like quartz, but they also have some added elements ("impurities") that give them various colors. They are classified mainly by their colors or by what their patterns look like. Cryptocrystalline silicates are sometimes called by their ancient name: **chalcedony** (*KAL-seh-done-ee*, or *kal-SED-on-ee*).

Agates *(AG-its)* have rings of various colors, going from the outside into the center. (Although some agates do have a hole in the center.) Agates probably formed in the same was as geodes, starting as an empty air bubble in hot lava, into which seeped hot water saturated with silcon, oxygen and other elements. Each time a little more seeped in, another ring was added.

Geodes probably began as an air bubble inside hot rock. The bubble was surrounded by hot liquid that was rich in silicon. The silicon liquid leaked in gradually and cooled inside the bubble, making crystals.

Chrysoprase gets its color from small amounts of the element nickel. (The formula is still SiO_2, because the amount of nickel is so small.) "Chyrso" is Greek for "gold," and "prase" comes from the Greek word "prasinon" meaning "green."
(Photo: "Xth-floor" Wikipedia)

Carnelian was used in ancient times for signet rings that were pressed into hot wax. Hot wax does not stick to carnelian, and fine details can be carved into it, making it ideal for this purpose.

Landscape jasper can have all kinds of fantastic patterns. If you see a rock with swirls and rings and interesting patterns, it is probably jasper.

Banded jasper comes in many colors. This one is called tiger jasper for obvious reasons. This particular sample is not polished, just dipped in water to make the colors show up well in the photo.

Onyx *(awn-ix)* often comes in black, though this sample is red. Oddly enough, the name comes from the Greek word for fingernail.

Heliotrope (also known as "bloodstone") is dark green with specks of red hematite. In ancient times it was said to have magical powers, even invisibility.
(Photo: Benutzer Ra'ike,Wiki)

Opals have water as part of their chemical recipe. They came from magma that had a lot of water mixed into it. Opals are the national gemstone of Australia.

Chalcedony is the stone of which **petrified wood** is often made. Water that was saturated (completely filled) with silicon seeped into the dead tree and filled in the spaces where the living cells had once been. This process occurred not too long after the tree died, but before it had time to rot. The trees had to have been immersed in mineral-rich liquid. It's not a process that can occur in a dry environment as far as we know.

Activity 3.1 Comprehension self-check

Can you remember the answers to these questions? If not, go back to the text and see if you can find the answers. When you are done, you can check your answers using the answer key at the back of the book.

1) The crust of the earth is made of mostly these two elements: _____ and _____.

2) Both carbon and silicon have how many electrons in their outer shell? _____

3) How many electrons would carbon and silicon like to have in their outer shell? _____

4) When silicon bonds to oxygen, it makes this type of bond: _____

5) When sodium bonds to chlorine they make this kind of bond: _____

6) What element does carbon bond with to make a methane (natural gas) molecule? _____

7) How many electrons are in oxygen's outer shell? _____ How many electrons does it want to gain? _____

8) Name two metal atoms that might bond with the oxygens on a silicon tetrahedron: _____ and _____

9) Name a type of mineral that is a solid solution, having tetrahedrons surrounded by metal atoms: _____

10) What group of minerals has as its basic structure, long lines of tetrahedrons? p_____

11) Name a mineral from this double-chain group that you would never find in a mineral collection: _____

12) Name a mineral from this double-chain group that would likely be found in a mineral collection: _____

13) Name the group of minerals that has as its structure sheets of tetrahedrons: _____

14) A mineral from this sheet group that is dark in color is: _____ And light in color? _____

15) Do these sheet silicates have any uses in technology? _____

16) True or false? Ancient peoples used feldspar as a source (an ore) of metals such as iron and aluminum. ___

17) Besides silicon and oxygen, feldspars have these elements: _____ , _____ and _____

18) Do feldspars contain any iron or magnesium? _____ What is a possible reason for this? _____

_____

19) Feldspar and quartz have this basic structure: f_____

20) True or false? The word "quartz" means "hard." ___

21) Which one of these is NOT a type of quartz? a) amethyst   b) gypsum   c) citrine

22) What happens if you squeeze a quartz crystal? _____

23) When you add this mineral to a silicon crystal, it seems to make it harder: _____

24) This gemstone is number 8 on the Mohs hardness scale: _____

25) This gemstone has both aluminum and beryllium added to the silicon structure. Its mineral name is _____ and when it is a green gemstone it is called _____.

26) What happens to garnets that are not good enough to be gemstones? _____

27) What type of gem always has the element boron in it? t_____

28) Rubies and sapphires are actually this mineral: _____

29) What is the very old name for what we call cryptocrystalline silicates? c_____

30) What does "crypto" mean? _____

31) Which one of these probably started out as an air bubble? a) jasper   b) onyx   c) carnelian   d) geode

32) Which cryptocrystalline sometimes has patterns that look like imaginary landscapes? _____

33) Which cryptocrystalline has water as part of its chemical recipe? _____

34) Which cryptocrystalline was used for signet rings because wax would not stick to it? _____

35) In ancient times this was called the "invisibility stone." _____

Activity 3.2 Videos

Go to the "Rocks and Dirt" playlist at: **www.YouTube.com/TheBasementWorkshop**, and watch the videos on quartz and other silicates.

Activity 3.3 Recipe review

Can you match each picture to its chemical recipe? (You can use information in the text to help you.)

A   B   C   D   E   F

G   H   I (cinnabar)   J   K (graphite)   L (diamond)

M   N   O (halite)   P   Q   R

1) Au _____   6) FeS$_2$ _____   11) (Fe,Mg)SiO$_4$ _____   16) SiO$_2$ _____
2) Ag _____   7) BaSO$_4$ _____   12) KAlSi$_3$O$_8$ _____   17) CaCO$_3$ _____
3) S _____   8) HgS _____   13) Ca$_3$Al$_2$Si$_3$O$_{12}$ _____   18) organic _____
4) C _____   9) NaCl _____   14) SiO$_2$ _____
5) C _____   10) CaF$_2$ _____   15) SiO$_2$ _____

Activity 3.4 A picture puzzle

Look carefully at the ends of these petrified logs then look at the picture at a living branch that was cracked. Then think about how stone splits. When did these logs break—before or after they were petrified?

Activity 3.5   Read about talc

Talc is a member of the silicate family, but isn't like any of the others. You'll remember that talc is number 1 on the Mohs hardness scale. You can scratch it with your fingernail. It also feels "soapy." If you pick up a mineral specimen and it feels soapy and you can scratch it, it must be talc. Gypsum is also very soft, but doesn't feel soapy. The reason talc feels soapy is because it is related to clay, and you know how slimy wet clay feels. Talc is classified as a "clay mineral," which is a subset of the silicate group. Like the cryptocrystalline silicates, talc has extremely small crystals. Strangely, these crystals can be tiny pieces of sheets, like we find in mica. The microscopic sheets can slip around above and below each other.

Talc's recipe is $Mg_3Si_4O_{10}(OH)_2$. Geologists think that the (OH) on the end came from water molecules (H_2O) as the talc was forming. It may have come from olivine or pyroxene or amphibole that was under pressure, and had lots of water and carbon dioxide in the environment.

The word "talc" is very old. It can be traced back centuries, through Latin, Arabic and Persian. The name doesn't mean anything; it is simply the name for this mineral. Ancient peoples sometimes carved it (since it was so soft) or used it as a lubricant, since it is slippery. In modern times, many more uses have been discovered for talc. The most well-known use is as bath powder or "baby powder." (This use for talc is being discouraged now, as there are concerns about long-term effects on the lungs.) It is used to make paper, plastics, paint, medicines, cosmetics, ceramics and even as an ingredient in some foods.

Recently, there has been some concern over whether powdered talc can cause lung disease such as lung cancer. Lab rats were forced to breathe talc dust for six hours a day, five days a week, for over a year. Only then did some of them start to get lung cancer. So yes, talc can cause health problems if you breathe enough of it, like the rats did. However, talc is still considered to be "generally safe" because the amount of talc a human would be exposed to is extremely small.

Activity 3.6   What is a carat?

If you decide to learn more about gemstones, you will certainly run into this word: **carat**. Gems are measured in carats, not in ounces or grams. The word carat is very old. English borrowed it from Italian, which borrowed it from Arabic, which borrowed it from Greek. The original meaning seems to have been "carob seed." Carob seeds were used as a standard of weight in the ancient world. In modern times, a carat has been defined as 200 milligrams, which is the same as 0.2 g (two tenths, or one fifth of a gram). That means 1 gram equals 5 carats.

So, could you measure other things in carats? Sure, why not? How many carats would these objects be?

1) An ice cube that weighs 25 grams: _____ carats
2) A carrot that weighs 100 grams: _____ carats
3) A marble that weighs 10 grams: _____ carats
4) A banana that weighs 235 grams: _____ carats
5) A piece of talc that weighs 6 ounces: _____ carats   (1 ounce is about 28 grams)

Carat is one of the four ways that gems are evaluated. All four ways start with the letter C: carat, color, clarity, cut. Carat is how much it weighs, color is important because some are more rare than others, clarity means how transparent it is (clear is good), and cut means the shape of the gem and how perfect its facets (sides) are.

CHAPTER 4: ROCKS

Now that we know a lot about minerals we are ready to start building rocks. Rocks are made out of minerals, so as we study rocks we will meet many of the minerals we learned about in the last two chapters. Understanding minerals is the key to understanding rocks.

Rocks are generally divided into three categories: **igneous**, **sedimentary** and **metamorphic**. Most textbooks start out with igneous rocks first, so we we'll stick with tradition and begin with that category. The word **igneous** comes from the Latin word "ignis" meaning "fire." The fire these rocks came out of isn't made of open flames, but of hot magma—melted rock coming up out of the ground. How hot? The pot on your stove boils away at 100° C (212° F). Magma can be as hot as 1400° C (2400° F). At this temperature, all the atoms are just floating around, making a hot, atomic soup. When the magma begins cooling, the atoms will slow down and start to form bonds with each other, creating some of those patterns we saw in chapter 3: solid solutions, chains, sheets and crystals.

Sometimes magma comes all the way out of the ground. It spews out of the tops of volcanoes, or oozes out of cracks in the ground. After the magma comes out, we change its name to lava. Various things can happen to the lava: it can ooze along the ground, it can be hurled through the air, it can have air bubbles mixed with it, it can have water mixed with it, or it can be turned into a fine, powdery ash. When each of these things cools, they will look very different. The same lava can produce different rocks, depending on what happens to it as it exits. All types of rocks that are formed in these ways are called either **volcanic** or **extrusive**. (Extrusive rocks exit the ground.) People have been watching lava cool for hundreds of years. There are many places on earth where we know exactly when the magma erupted. We can study the rocks in these places and make accurate observations about extrusive rocks.

Lava is really, really hot—much hotter than any stove or oven, even an industrial oven.

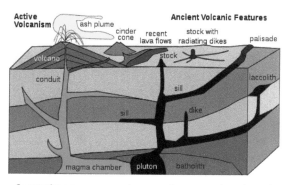

Sometimes magma stays underground and cools, forming a pluton (named after Roman god, Pluto).

Sometimes magma stays under the ground. When it cools while still under the ground, it is called either **plutonic** or **intrusive**. (Intrusive rocks stay in.) Unlike extrusive rock, no one has ever observed cooling magma deep underground. You'd have to drill a very deep hole and then get dangerously close to the hot magma. The magma would probably melt all of your scientific observation instruments since metal and glass melt at magma temperatures. There are places in the earth where we can dig up cold, hardened intrusive rocks, but no one was there to observe them when they were still hot. Much of the rock that forms the crust of the earth is technically classified as igneous, but there are serious problems with the theory that they were all once molten. The origin of intrusive rock is a bit of a mystery.

If you open a rock and mineral book and look at the igneous section, you will see many rocks with strange names that look very similar. There are just too many names to learn! We are going to trim the list to just the rocks that best represent large categories. And to help us organize the rocks, we'll use a handy chart that helps us to see the relationship between them.

Before we look at the chart, we need to go back to the chemistry we learned in chapter 3. Do you remember those iron and magnesium atoms that were part of the olivine stone soup? Iron and magnesium also showed up in the pyroxenes (the ones with the single chains) and the amphiboles (the ones with the double chains), helping the silicon chains to stick together. By the time we got to the framework minerals (feldspars) and quartz, the magnesium and iron had been completely used up. The feldspars contained potassium and sodium instead, and quartz crystals form without any help from extra metal atoms.

Magnesium and iron atoms turn out to be super important when classifying igneous rocks. Scientists decided to create a special word to describe rocks that contain a lot of Mg and Fe. They used "ma" for <u>ma</u>gnesium, and "f" for iron, <u>F</u>e, to make the word "**mafic**." If a rock (or magma) is mafic, it contains a lot of magnesium and iron. (Olivine has so much magnesium and iron that it is called ultramafic!) They also decided that rocks containing very little magnesium and iron also needed a special name. Since these rocks are high in <u>sili</u>con, they used the letters "sic," and to represent the <u>fel</u>dspars, they used "fel." They put the letters together to make the word "**felsic**." Felsic rocks (and magma) are high in silicon and very low in iron and magnesium. As a general rule, iron and magnesium cause a mineral or rock to have a darker color. Felsic rocks, since they contain little or no Fe or Mg, tend to be light in color. This will be one of the ways we figure out the chemistry of an igneous rock.

Hey, this isn't too hard- I think I can handle this!

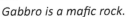

Gabbro is a mafic rock.   *Andesite is less mafic.*   *Rhyolite (RIE-o-lite) is felsic.*

Another important feature of igneous rocks is the size of the mineral crystals they are made of. Some rocks, such as granite, have very large crystals, easily seen without any magnification. Any rock that has crystals larger than 1 millimeter is considered to be "coarse-grained." (1 mm is about the size of the inside of this circle: o) **Granite** is a good example of a coarse-grained rock. Rocks that have crystals larger than about 2 centimeters (almost an inch) get a special name: **pegmatite**. This word comes from the Greek "pegma" meaning "joined together." Pegs are kind of like wooden nails that join things together. If you imagine these large, long crystals as being pegs, it might help you remember what a pegmatite is.

This is pink granite.   *This diorite is very coarse-grained.*   *This qualifies as a pegmatite.*   *pegs*

Fine-grained rocks don't have any crystals that you can see. They do have crystals, but they are microscopic in size. **Basalt** *(buh-SALT)* is an example of a fine-grained igneous rock. It is dark and dull—not much to look at. However, basalt is an important rock to know because not only do we find a lot of it on land, as evidence of past volcanic activity, we also find it at the bottom of the ocean. Basalt is what forms the crust of the earth below the oceans.

Another member of the fine-grained group is **obsidian**. It is the "extreme" member of this group, having really no grains at all. It looks like black glass. Since both obsidian and glass are made of SiO_2, obsidian really is black glass.

Basalt is fine-grained.   *Obsidian has no grains at all.*

Now we are ready for the chart. The columns represent textures and the rows represent chemistry, so each rock is associated with both a texture and a type of chemistry. The only texture that has an odd name is **pyroclastic**. "Pyro" means "fire" and "clastic" means "broken in pieces." Pyroclastic rocks have obvious bits and chunks in them, like a cookie loaded with nuts and chips. Basically, pyroclastic rocks are made of fragments that stuck together because they were cooked by the heat of a volcano. (The cookie analogy isn't far off.) "Frothy" means that it had lot of tiny air bubbles in it, like whipped cream.

The chemistry is very simple here, just whether the rock contains a lot of Fe and Mg (making it mafic) or not (felsic). Some rocks are in between, so there is also an intermediate category where the rocks have a medium amount of Fe and Mg. (Felsic rocks also usually contain potassium, sodium and calcium.)

I'm thinking I could make cookies or candies with all these textures!   Brilliant idea!

	coarse grains	fine grains	pyroclastic	frothy	glassy
Mafic	gabbro	basalt	tuff	scoria	none
Inter-mediate	diorite	andesite	tuff	pumice	none
Felsic	granite	rhyolite	tuff	pumice	obsidian

INTRUSIVE (plutonic) | EXTRUSIVE → (volcanic)

The first column is our coarse-grained, intrusive rocks that must have cyrstals large enough to see. Bear in mind that not all specimens look like these. That's what makes identifying rocks and minerals tricky. The piece you are holding may not look like the picture in the book. It takes a lot of experience looking at many samples to be able to positively identify a rock just by looking at it.

Gabbro was named after a small village in Italy. If gabbro is mafic, what minerals can we expect to find? From chapter 3, we learned that olivine, pyroxene and amphibole are three types of minerals that contain a lot of Fe and Mg. Sure enough, here in gabbro we see definite chunks of green which could be olivine. The black pieces are probably the pyroxenes and amphiboles. The light areas are likely some type of felsic rock (plagioclase feldspar) that got mixed in at some point.

Diorite is halfway between mafic and felsic. We see more light-colored feldspar here than in gabbro. Not all diorites look like this one, but black and white is very typical. It is not as mafic as gabbro, so we have more of the light-colored feldspars. The black spots are often biotite or hornblende. Diorite is not as common as gabbro or granite, but it is found worldwide, and was used extensivly by the Incans and Mayans as building stone.

Granite can have speckles of many different colors, but pink, white and black are the most common. Pink granite is very popular for monuments and kitchen counter tops. The pink is feldspar, the white is either feldspar or quartz, and the black is usually biotite or hornblende. Sometimes you also find granite that contains clear muscovite. The crust underneath all the continents is made of granite. Granite has been cut, carved, and polished since ancient times.

These fine-grained igneous rocks make it easy for us to guess their chemistry. Basalt is obviously dark and mafic. Rhyolite is very light and obviously felsic. Andesite belongs in the middle.

Basalt is a very important rock for us to know. It is the rock of which the ocean floor is made. Much of the lava that pours out of volcanoes is melted basalt. To be classified as a basalt, a rock must contain between 45-55% SiO_2, and feldspar that is high in sodium and calcium (plagioclase). Because basalt is mafic, it also contains pyroxene and amphibole minerals and maybe some olivine, as well.

Andesite was named after the Andes mountains in South America. For a rock to be classified as andesite, it must contain 55-65% SiO_2. You can't tell by looking whether a rock has exactly this amount of SiO_2, so it isn't easy to identify an andesite just by looking. Andesite is often found between areas of basalt and granite, which makes sense since it is an intermediate.

Rhyolite is a felsic rock, so it contains over 65% SiO_2. Like granite, which is also felsic, rhyolite can also contain small amounts of biotite, hornblende and other mafic minerals. When granite melts, it sometimes hardens into rhyolite. (Granite never hardens back into granite.) There are different types of rhyolite, so other samples may not look like this. Some look more white or tan.

Finally, we have rocks that were formed from cooling lava, mostly basaltic lava.

Tuff (or "tufo") is a very general word for rocks that were formed from volcanic ash. The ash got cemented together very tightly, making a hard, smooth rock that can be carved and used for building stones. There are many kinds of tuff, including basaltic, andesitic, rhyolitic and welded. The statues on Easter Island are made of dark, basaltic tuff.

Scoria is melted basalt that got blown out of a volcano in such a way that it got all frothed up with air bubbles from volcanic gases. The bubbles stayed in place while the lava cooled, resulting in a sponge-like texture. Scoria is also called "volcanic cinders." Sometimes scoria is red or brown. The "lava rocks" found in some gas grills are often scoria.

Pumice (PUM-iss or PUME-iss) is a lighter, more brittle version of scoria. Lava spewed out the top of a volcano in such a way that it mixed with the gases coming out of the volcano. It cooled very quickly and the air bubbles left holes, giving it a sponge-like texture. Pumice will float until all the air pockets have filled with water.

Obsidian is formed when felsic lava, rich in SiO_2, flows into water and cools very quickly. It cools so quickly that the silicon doesn't have time to form a single crystal. The texture is smooth and glassy. Most obsidian is black (despite being felsic) but some is red, and rarely it will have little gray shapes that look like snowflakes ("snowflake obsidian").

Now you really understand that chart! Yeah, you'll be explaining it to your friends in your college classes!

Of course, there are hundreds of other igneous rocks. I know. I sure can't keep track of all of them!

If you want to learn the names of more igneous rocks, you can pick up a good rock book. Yeah. But we're kind of done with igneous. My brain is full!

Before we leave the topic of igneous rocks, we need to mention magma and lava again. Since magma is melted rock, we can have **mafic magma** (basically melted basalt) or **felsic magma** (basically melted granite). These two types of magma will behave differently when they come out of the ground as lava. The high iron and magnesium content of mafic lava makes it runny. Mafic lava will spread out and flow quickly, kind of like melted butter. The ocean floors appear to be covered with hardened mafic lava that once spilled out and flowed in all directions. Mafic lava also comes out of volcanoes in places like Hawaii. Felsic lava is thicker and not as runny as mafic lava. It's more like pudding. The thick lava traps gases inside until the pressure builds up so much that the volcano explodes. Giant blobs of magma are thrown out like bombs. Lava can form all kinds of different shapes as it cools.

This is mafic lava in Hawaii flowing out onto a road. When lava flows like this, it is called **pahoehoe**.

This is also mafic pahoehoe lava. When it makes this shape they call it **rope lava**.

What a spectacular explosion! We might guess that this is felsic lava.

Sedimentary rocks are much different from igneous. **Sedimentary** means "made of sediments." Sediments are basically all the little particles in any kind of earthy, dirty slop—dirt particles in a mud puddle, sand particles washed about by waves and water currents, layers of silt left behind by flood waters, even tiny pebbles transported by streams. There is a special name for any rock or mineral particle that can be carried by water. We call them **clasts**. Clasts can be smaller than a grain of sand or as large as a boulder. (Boulders can be rolled and moved by water even though they are very large and may not actually rise in the water.) Geologists classify clasts by their size.

Not everyone agrees with these exact numbers. Other similar charts might give you different measures for sand and silt.

clay	silt	sand	pebbles	cobbles	boulders
less than .002 mm	from .002 mm to .06 mm	from .06 mm to 2 mm	from 2 mm to 6 cm	6-25 cm	over 25 cm
size of bacteria	size of cell	size of sand	size of pea/grape	size of apple	size of basketball

a cobblestone street

Any rock larger than 25 centimeters (about the size of a basketball) can be called a **boulder.** If it is less than 25 centimeters, it is a very large **cobble.** An average cobble is about the size of an apple. For centuries, cobbles have been used to pave streets. Anything smaller than 6 centimeters (about the size of a golf ball) is called a **pebble.** Mostly, we think of pebbles as being the size of peas. Pebbles can be as small as 2 millimeters, which is about the size of this circle: o Smaller than that, and we call the particles grains of **sand.** Sand grains can be as small as .06 millimeters (60 microns), which is invisible to our eyes. Microscopic single-cell protozoa in ponds and the cells in our bodies are the same size as the smallest grains of sand. When the grains get smaller than .06 millimeters, we start calling them **silt.** Silt particles can be as small as .002 millimeters (2 microns) which is about the same size as a bacteria. When the particles start getting smaller than a bacteria, they are called **clay.** When we think about clay, we think of something wet and sticky or mushy. Clay particles without any water make a very soft powder.

All of these clasts, from clay to boulders, can be found in sediments of various kinds. The clasts are carried along by moving water or wind until the motion stops and they fall out. Some particles, like boulders, never really get off the ground. They are pushed and rolled along by water currents. The tiny particles, like clay and silt, might take a very long time to fall out of the water. If you fill a glass with muddy water and watch it for a while, it might take a few days for the water to be clear and all the mud particles to settle to the bottom. Water has to be mov-

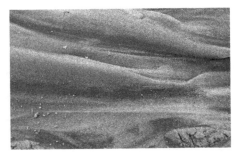

ing very quickly to be able to keep pebbles and sand from falling out. As the water slows down, it becomes less able to keep larger particles afloat. Slow-moving water can hold dissolved clay and perhaps some silt. A river can pick up slit and clay particles and transport them many miles, perhaps even hundreds of miles. Sometimes clasts will form a pattern as they fall out of the water, making stripes or ripples, as seen in this photo. Deposits of sediment that are very well sorted, having particles that are exactly the same size, are assumed to have been transported by water at one time, even if the water has now disappeared.

There are two kinds of sedimentary rocks: **clastic** and **chemical.** Clastic sedimentary rocks, as you can easily guess, are made out of these clasts we have been talking about. Clastic rocks started out as silty, muddy sludge that contained one or more types of clasts. In addition to water and clasts, the wet sediment had another key ingredient: something that would cement the clasts together. Sand doesn't stick very well once the water evaporates. Perhaps you've watched a sandcastle dry out and begin to crumble as it dries. For dry grains of sand to stick together permanently, a strong cementing agent must hold the grains together. The "glue" that cements clastic rocks is usually SiO_2 or $CaCO_3$.

Won't it be a shame when this castle crumbles?

Clastic rocks are classified by the size of the clasts they contain. When the clasts are clay particles, the rock will be called **shale**. Silt particles harden into **siltstone**. Sand makes **sandstone**.

Shale is made of clay particles and has the same smooth texture that clay does. Shale is usually a dark gray color, but can also be found in red or greenish-brown. Shale always comes in layers. It is a soft rock and can often be broken into thin sheets just using your bare hands. Fossils are commonly found in shale. The most famous shale formation is in Canada and is called the Burgess Shale. It is about 160 meters (520 feet) deep and contains tens of thousands of plant and animal fossils, including many bottom-dwelling ocean creatures.

Siltstone started out as silt. The texture is not quite as smooth as shale, but it is still very smooth. The surface of siltstone has a slightly rougher feel than shale, but not nearly as rough as sandstone. This sample of siltstone was cut out of a large slab. Siltstone is not as soft as shale and can be cut into blocks. However, even though it can be cut into blocks, it is still too soft to be used as a building stone. Siltstone can be many different colors, including white, tan, yellow, green, red, purple, orange, and black.

Sandstone has visible grains of sand, giving it a much rougher texture than shale or siltstone. Also, unlike shale and siltstone, sandstone is porous, meaning it can soak up water (slowly) because it has tiny spaces between the sand grains. Usually these spaces are not visible, but in this sample you can see some very small holes. Sandstone often has red or brown stripes in it, caused by extra minerals such as iron. Sandstone is harder than shale or siltstone and can be cut into building blocks. Fossils can be found in some sandstones.

Shale is extremely layered. You can break off chunks with your fingers.

This formation of siltsone looks a lot like shale. (Not all shale is black.)

Corona Arch, near Moab, Utah, is made of red sandstone.

Even professional geologists can find it hard to accurately identify siltstone. If they are out in the field and not near any laboratory, they have a quick and easy way to help them guess which is which. They break off a piece of rock and smash it gently until they have a pinch of loose particles that they can rub between their fingers. If they can feel the grains with their fingers they know the particles are large enough to qualify it as sandstone. If they can't feel the particles with their fingers, they put a pinch in their mouth. (This is NOT recommended for you to try! Let the professional do it.) If they can feel any particles with their tongue, they know it must be siltstone. If their tongue feels nothing but smooth texture, they classify it as a shale.

We still have some clasts left—the pebbles, cobbles and boulders. These three types of clasts are often grouped together and called **gravel**. Gravel can also be cemented together to form clastic rocks. (Of course, most often you will see this happen to pebbles rather than cobbles or boulders.) If the clasts are rounded and fairly smooth, as though they had been through a rock tumbler, the resulting clastic rock will be called **conglomerate**. If the particles are ragged and sharp, the rock will be called **breccia**. You'll hear the word breccia pronounced in different ways, even among geologists. You can choose the way you like best. 1) *BRETCH-ee-ah* 2) *BRETSH-ah*, 3) *BRESH-ee-ah* 4) *BRECK-ee-ah*. (You will need pronunciation number 3 for the joke below, but that does not mean it is the "correct" one. The pronunciation guides on the Internet tend to favor numbers 1 and 2.) The word "breccia" is Italian for "gravel," but it also means "broken."

Conglomerate looks a lot like a blob of the concrete they use on construction sites. It has rounded clasts (pebbles) in it.

This breccia is sedimentary, unlike the piece of volcanic breccia on the right. Both are called breccia because they have angular clasts.

This is volcanic breccia, so it would be classified as an igneous rock. The word breccia is not limited to sedimentary rocks.

When we look at a rock like breccia, we can very easily distinguish the particles from the stuff around them. Geologists decided that they needed a word for that stuff around them, and decided to use the word **matrix**, which is a common term in science and technology. You find "matrix not only in geology but also in biology, mathematics, and electronics, as well as in sculpting and manufacturing. It means "the surrounding stuff." In geology, the matrix is the solid stuff around the item(s) of interest. If you look back to page 11, that pyrite cube at the top of the page is embedded in a matrix of some kind of rock, possibly sandstone. In breccia, the rock around the particles is called the matrix.

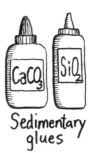

Sedimentary glues

The cementing agents in clastic rocks are very important. We have not mentioned them very much so far, but that does not mean they are not important. We met SiO_2 in chapter 3 as: 1) the recipe of the silicon framework of quartz, and 2) as part of a hot, liquid solution that filled in bubbles and cracks, then cooled to make geodes and agates. In this chapter we've seen how silicate minerals combine to form igneous rocks, and how SiO_2 can be the glue that holds clasts together to make a clastic rock. The focus will soon shift to $CaCO_3$ as we explore the chemistry of carbonate minerals and rocks.

Clastic sedimentary rocks are made of two things: 1) clasts, and 2) mineral "glue." If we get rid of the clasts and just use the "glue," we get a **chemical** sedimentary rock. (Imagine that instead of using a craft glue to stick two things together, you just squeeze out a blob of glue and let it dry by itself.) Since these glues are, in fact, minerals, we'll find a lot of overlap between rocks and minerals in this category.

Chemical sedimentary rocks are classified into several groups, but there doesn't seem to be any one correct format that everyone uses. Each book or website will have its own version of the classification system. In this book we will separate them into four categories: **the evaporites, the organics, the silicates and the carbonates.**

The **evaporites** can be formed by evaporation. You know what evaporation is. You've seen puddles dry up and disappear as the water in them evaporates into the air. Water molecules are light enough that they can leave liquid water and float up into the air. The temperature of the water controls how fast this happens. When water is frozen into ice, the molecules can't escape nearly as easily. When water is very hot, the molecules are energetic and can escape quickly. The steam that comes out of a pan of boiling water is made of water molecules leaving the water and going up into the air. Heavier molecules can't do this. For example, if salt or sugar molecules are dissolved into the water, only the water molecules will escape; the salt and sugar molecules will stay in the pan. If you let a pan of salt water evaporate, the salt will stay behind as dry crystals.

However, evaporite rocks can also be formed WITHOUT evaporation. All you need to do is create a situation where the atoms (Na, Ca, K, Cl, Mg, etc.) decide that it is better to stick together than to float around separately. Water molecules are able to pull on these atoms, keeping them apart—that is what "dissolving" means. But in certain situations, water's power of attraction will not be strong enough, and the mineral atoms will begin to attach to each other forming a solid rock or crystal. (If you've ever made rock candy, you've seen a crystal form while still under water.) Situations that favor crystal formation are when the water is "super saturated" (overly full) of these atoms, or when a certain temperature is reached (cold water can't dissolve things as well as hot water) or when other chemicals are present. If evaporation is the one and only way that these evaporates can form, it's very difficult to explain the formation of salt deposits that are thousands of feet thick. However, if a flood of mineral-rich water came in all at once, a very pure deposit could easily be the result.

Travertine is formed when salty minerals come up from natural hot springs. Hot water can hold huge amounts of dissolved minerals. After the hot water comes to the surface, it begins to cool down and can no longer hold that same load of mineral atoms. The atoms begin leaving the water and sticking to surfaces such as the rocks around them. Thin layers begin to build up year after year, and result in these astonishingly beautiful formations. This one is at Yellowstone National Park in Wyoming, USA.

Salt (halite, NaCl) is an important natural resource all over the world. This diorama (from the Deutsches Museum in Munich) shows us how salt mining was done before we had large machines to help us. Salt deposits can be hundreds of meters (thousands of feet) thick and are very pure. The city of Detroit, in Michigan, USA, has a huge salt mine underneath it. The mine even extends out and under Lake Erie. These huge deposits can't possibly have formed just by evaporation because anything sitting that long would have layers of dust and dirt. But the salt is pure and clean.

Photo from Wikipedia, thanks to user "High Contrast"

Gypsum can also be classified as an evaporite sedimentary rock. Do you remember gypsum from chapter 2? It is number 2 on the Mohs hardness scale. The chemical formula for gypsum is $CaSO_4 \cdot 2H_2O$. There's water in this rock! If you heat gypsum, the water molecules will evaporate out, leaving $CaSO_4$, a form of gypsum we call **anhydrite** (meaning "no water"). Anhydrite is used to make plaster and plasterboard.

Gypsum can take several forms. Sometimes it forms large crystals, but not always.

Here's another mineral we met in chapter 2. **Coal** can be classified as a chemical sedimentary rock, though it is often called "biochemical" because it came from plants, which are living ("bio"). It is also called **organic**, as we already noted in chapter 2.

There are different types of coal, based on how they burn. The best quality coal is called **anthracite**. It is hard, lightweight and very shiny. The next best is **bituminous** coal. It is softer and heavier than anthracite and doesn't produce quite as much heat when it is burned. The lowest quality is **lignite** (also called brown coal). Lignite is very soft and doesn't burn nearly as well as the other kinds because it still has water in it. You can't squeeze the water out of lignite—the water molecules are all locked up in its chemical structure. But when you burn it, those water molecules are not very helpful to the burning process!

Because anthracite is so much harder than bituminous coal, anthracite is often considered to be metamorphic.

Coal is found in layers called **seams**. A coal seam can be as thin as an inch (a few centimeters) or as thick as 100 feet (over 30 meters). Coal seams are often found layered between rocks of other kinds, such as shale and limestone. These seams can stretch for many miles under ground, or they can be above ground. The picture shows an 80-foot high seam in Wyoming, USA.

Bituminous coal and lignite look very much the same. You really need to examine them "in person" to be able to tell the difference.

Notice the truck down here.

Some coal is fairly near the surface and can be mined just by digging down into the ground a bit. Miners can simply strip the coal off as they dig down. This picture shows an "open pit" lignite mine. Does it mess up the landscape? Yes, for a while it makes a big mess. Then, when all the coal is gone, the mining company goes to a lot of trouble to make everything look nice again. They fill in all the holes, put good soil on the top, and plant grass and trees. Sometimes they even make the area into a park.

Unfortunately, not all coal is near the surface. A lot of the really good anthracite coal is deep underground. Mining for deep coal began well over a hundred years ago. It's amazing that without any bulldozers, excavators or dump trucks, people began digging far down into the earth. At first this coal was used to heat homes, but later it was in the train industry, both to power the steam engines and to run the factories that made the trains. Coal mining companies were so desperate for miners that they would even hire children, as shown in this photograph. Mining was not only dirty, it was dangerous. Miners had to risk explosions, cave-ins, suffocation and lung diseases.

Today, coal often goes to power plants that generate electricity. The coal is used to heat water and make steam. The steam then turns giant turbines that generate electricity. Unfortunately, coal also contains elements such as sulfur, which can create air pollution if they are released into the air. The steam that comes out of chimneys and cooling towers has had the sulfur and other harmful substances removed.

Scientists struggle to explain how coal was formed. We know that coal is made of plants because we can still see pieces of tree bark in some of it. If you look at the softer coals under a microscope you can see where the living cells used to be. No one disputes that coal is made of ancient plants. How this happened, though, is a question that is still under investigation. The traditional theory you are likely to read in an older textbook is that plants died and sank to the bottom of swamps. The swamps, or bogs, collected dead plants until a thick layer of **peat** had formed. The layers got thicker and thicker over time, and all that pressure turned the bottom layers into coal.

There are several problems with this theory. First, fossils of ocean creatures have been found in layers of shale above and below coal, and even in the coal itself. Ocean creatures don't live in fresh water bogs. Second, limestone is often found layered with coal, and limestone is generally associated with marine (ocean) environments. Third, the texture of the plant debris at the bottom of modern bogs is very different from coal, but they should be similar. Apparently, coal didn't come from bogs.

Peat is dark brown and very soft, but since it contains a lot of carbon, it can be burned as fuel. Because peat was less expensive than wood and coal, poor people in past centuries often had no other source of fuel for heating their homes.

Peat gatherers in England in the 1800s.

Another type of chemical sedimentary rock is called **chert**. The cherts are very similar to the cryptocrystalline silicates we learned about at the end of chapter 3. They have crystals so small that they don't seem crystalline at all; their texture looks very smooth. Their luster is "waxy" so they can have an almost creamy look, very much like the jaspers. In fact, they are so much like the jaspers that jaspers are considered by some geologists to be part of the chert family. Some books even categorize all the cryptocrystallines as cherts. (It's important to know that scientists don't always agree!) The geologists who want to keep the categories separate say that the jaspers and agates have a tiny amount of fibrous texture to their molecular structures, whereas cherts do not. That's certainly not anything we can observe directly. If it were not for very expensive, high-tech x-ray diffraction machines and electron microscopes, we'd never know this difference even existed. To the human eye, the texture of cherts and jaspers looks very much the same, if not identical.

Cherts have been very useful to humans throughout our history. The very first tools ever made were made from a type of chert called **flint**. Flint, like all cherts, has a natural tendency to chip off in wedge-shaped chunks that leave sharp edges. This type of fracture pattern is called **conchoidal** *(kon-KOYD-al)*. All the framework silicates, like quartz, also have this same fracture pattern. They do this because they have no internal geometry like the micas or the pyroxenes. That's one way to identify members of the quartz family. This chipping pattern was used to make sharp edges on axes, knives, and arrowheads. The act of shaping a piece of flint (or jasper or agate) into a cutting tool is called **knapping**. Another piece of rock was used to do the knapping. There are still people today who knap flint into arrowheads for bow hunting.

As technology progressed, flint became useful in a second way. Flint can produce a spark when struck with a piece of steel. (In fact, that is one way to identify flint.) Pistols and rifles of the 1600s and 1700s had a firing mechanism that held a piece of flint. When the trigger was pulled, a piece of steel snapped down onto the flint and produced a spark which would then ignite gunpowder.

An unusual type of chert, found in only two places in the world, is **oölitic chert** *(oh-oh-lit-ik)*. The word root "oö" *(oh-oh)* means "egg." (The little dots over the second "o" means that you don't combine the two o's, but say each one separately. Sometimes you will see it spelled without the dots, too, and people will say *ew-lit-ic.*) The "eggs" in this rock are the tiny black dots. No one knows what these dots are or how they got there. It's a mystery. Since chert is a silicate, it can be polished and used for jewelry. The two places that oölitic chert is found (so far) are Germany, and central Pennsylvania, USA. (The author of this book happens to live in central PA, so this type of chert just has to appear in this book!)

The last category of chemical sedimentary rocks is the **limestone** family. Limestone is so abundant and so significant that we're going to devote a whole chapter to it. The chemistry of limestone is completely different from the silicon-based rocks. The limestones are called **carbonates** and are based on the mineral calcite, $CaCO_3$. There are many types of limestone, and some of them can be classified as clastic sedimentary rocks because they contain clasts. (Remember, rocks don't know that they are supposed to cooperate with our categories!) Limestone is extremely important for the construction industry because it's what cement and concrete are made of. Limestone can contain fossils, which makes it extra interesting. However, let's go on to our third and last category of rocks, metamorphic, before learning more about limestone.

A lot of limestones look gray and boring.

But some have fossils!

Metamorphic rocks used to be igneous or sedimentary. ("Meta" means "change," and "morph" means "shape.") They started out as one thing, then changed into another. The two big categories that geologists like to sort these rocks into are **contact** and **regional**. These categories aren't about chemistry, just physical changes caused by intense heat and pressure.

Contact metamorphic rocks were changed by heat. Magma came up from below, perhaps through cracks, and made its way to the surface. All the rocks in and around that magma came into <u>contact</u> with the heat of the magma. They might not have come into contact with the magma itself, just the heat. How hot? Over 1200º C (2000º F)! This intense heat affected the molecules and crystals in the sedimentary and igneous rocks and caused visible physical changes.

The black stripes used to be tunnels of magma called "dykes."

In regional metamorphism, a huge area, or region, gets squeezed. It's obvious that at some point in the past, the earth's continents experienced some serious compression. Entire mountains and hillsides look like they have been squeezed, showing bends and folds. These are good places to look for metamorphic rocks. This picture shows a hillside made of slate, which used to be shale.

Rocks deep under the earth experience pressure from the huge amount of weight sitting on top of them. This would be another good place to find metamorphic rock. (Can you imagine having several miles of granite on top of you? Those poor metamorphic rocks!)

Some metamorphic rocks have stripes. Before they were heated or squeezed, they did not have stripes. The stripes appeared as their tiny crystals were forced to line up the same way. Imagine putting a pile of pencils and pens on a table and then putting your arms on either side. As you move your arms closer together, the pencils and pens are forced to become more parallel to your arms. By the time your arms are almost touching, they are pretty much all going in the same direction, parallel to your arms. In much the same way, the crystals in a rock can be forced to line up in the same direction. The rock that shows this most clearly is **gneiss**. *(nice)* Diorite and granite can become gneiss.

Gneiss always has black stripes.

Another type of striping found in metamorphic rock is called **foliation**. The "folia" part of the word means "leaf." They don't look like leaves, but they have layers as flat as leaves. Mica, a foliated mineral, is found in metamorphic rocks such as **schist**. *(schist)* Schist means "split." All schist rocks can be split into layers, but not quite as easily as mica. A schist that contains mica is **mica schist**. Schists have a scaly texture, with only small sheets peeling off. If you see shiny flakes in a rock, it's probably mica schist.

Slate started out as shale, which is made of the tiniest particles: clay. Clay is made of silicon. The silicon tetrahedrons in shale were compressed and began to organize themselves into microscopic sheets. If they had been compressed more, perhaps they would have formed larger sheets and turned into mica. But the sheets stayed microscopic, and you can't tell from looking that slate contains tiny sheets. The presence of these sheets causes slate to form in layers. The very first blackboards were made of large pieces of slate. Small slate tiles are often used on roofs.

We learned that sandstone is not cemented together as tightly as most other rocks. You might not be able to see the tiny spaces between the sand grains, but if you pour water on sandstone, at least some of it will be absorbed into those spaces. (That's one test for sandstone.) Sandstone makes a wonderful natural water filter. Rainwater trickles down through it and is clean by the time it reaches natural storage areas deep under ground. But what happens to sandstone if magma comes up through it? The heat melts the cement and presses the grains closer together. Some of the grains might stick together and form larger crystals, too. We end up with a harder rock called **quartzite**. Since

quartzite is much harder than sandstone, it makes durable building blocks and floor tiles. It is tough enough to be used as gravel under the rails on railroad beds. Ancient peoples made tools out of quartzite, because it fractures in the same way that flint does, leaving sharp chipped edges that are good for cutting (conchoidal fracture).

The last metamorphic rock that absolutely must be mentioned is **marble**. Marble is very different from most other metamorphic rocks because it does not contain silicon. Marble used to be limestone, $CaCO_3$. No Si atoms anywhere in sight! However, what happens to the $CaCO_3$ crystals is basically the same thing that happens to SiO_2 crystals. When limestone gets compressed or heated, the microscopic calcite crystals begin to join together to make larger crystals. Marble often has visible crystals. Like calcite, marble can come in different colors, or at least have swirls of other colors. Colors in rocks are always caused by other elements getting mixed in. These added elements are called "impurities" even if they make the rock look more beautiful. In marble, the impurities were probably very thin layers of clay or iron oxide that were present in the original limestone. As the limestone got heated and pressed, those layers began to melt and swirl, giving us the "marbled" look we like to see in our marbles.

Since marble is made of calcite, not silicon, this means it is much softer. Quartz is number 7 on the Mohs hardness scale and calcite is number 3. It is much easier to sculpt marble than it is to sculpt granite or any silicate rock. Marble has been a favorite material of sculptors for centuries. You can achieve fine details such as the outlines of muscles and hair. (Ancient Greeks would paint their sculptures, adding color for skin, eyes and hair. However, paint won't stick to stone for very long and the paint has long since washed off. We only know they did this because of tiny remnants of paint stuck in places like the corners of eyes.)

Limestone and marble are both found in some places in the world as huge desposits that can be mined. When removing limestone and marble, great care is taken not to damage the blocks. The blocks are cut using drills that are harder than marble. Is steel harder than marble?

Some blocks might go to sculptors, but most will go to businesses that sell marble for construction. Many famous buildings have been made of marble, such as the Taj Mahal in India, shown here.

If marbles are made from glass (SiO_2) then why do we call them marbles?

Yeah, shouldn't they be something like "glassies" or "silicies"?

Activity 4.1 Comprehension self-check

IGNEOUS

1) Which kind of igneous rock comes up out of the ground, intrusive or extrusive? _____

2) After magma comes out of the ground, we call it _____.

3) What is another name for intrusive? p_____ (Named after Roman god: _____)

4) Why might it be difficult to observe intrusive rocks being formed?

   a) They are deep under ground.   b) They are made of silicon.   c) They are hot.   d) They contain crystals.

5) Mafic rocks contain a lot of these two elements: _____ and _____.

6) What green mineral is called "ultramafic"? _____

7) Felsic rocks are the opposite of mafic. "Fel" stands for _____ and "sic" stands for _____.

8) TRUE or FALSE? Mafic rocks are darker in color than felsic rocks. ____

9) Which one of these is NOT a felsic rock? a) granite   b) basalt   c) rhyolite

10) Name a rock that is halfway between felsic and mafic. a_____ (Named after _____ mountains)

11) Which of these contains super large crystals? a) granite   b) diorite   c) pegmatite   d) gabbro

12) Which of these contains no grains at all? a) basalt   b) obsidian   c) rhyolite   d) tuff

13) Which of these is NOT a coarse-grained rock? a) granite   b) basalt   c) gabbro   d) diorite

14) Which mineral would you probably NOT find in granite? a) feldspar   b) quartz   c) biotite   d) calcite

15) Gabbro was named after: a) a village   b) a famous geologist   c) a mountain   d) a mineral

16) Which of these is NOT considered to be intrusive? a) granite   b) basalt   c) gabbro   d) diorite

17) Which of these has a "frothy" texture? a) granite   b) obsidian   c) pumice

18) Which of these is sometimes called "lava rock" and can stand the heat of a gas grill?

   a) tuff   b) scoria   c) obsidian   d) pumice

19) Which of these rocks will float? a) tuff   b) scoria   c) obsidian   d) pumice

20) Which of these is made from compacted volcanic ash? a) tuff   b) scoria   c) obsidian   d) pumice

21) Which is thin and runny and is less likely to explode out of a volcano? a) mafic lava   b) felsic lava

22) If you melt granite, you will get _____ lava, and if you melt basalt you will get _____ lava.

23) When felsic lava runs into water and cools so quickly that crystals do not have time to form, what is the result?

   a) granite   b) basalt   c) obsidian   d) pumice   e) tuff

24) The ocean floor is made of mostly: a) granite   b) basalt   c) gabbro   d) pumice

25) What does "clastic" mean? a) hot   b) volcanic   c) frothy   d) broken

SEDIMENTARY

26) Which is larger, a pebble or a cobble? _____

27) Which is smaller, clay or silt? _____

28) TRUE or FALSE? Both clay and silt particles are microscopic. ____

29) Technically, a boulder can be as small as a: a) baseball   b) basketball   c) beach ball

30) TRUE or FALSE? By definition, all clasts are made of silicon-based minerals. _____

31) TRUE or FALSE? Clay particles take a longer time to settle to the bottom of a pond than sand particles do. _____

32) When sand carries sediment a long distance the particles are usually (more/less) sorted than they normally are.

33) TRUE or FALSE? The "cement" in a clastic sedimentary rock is always a solution of SiO_2. ____

34) Clastic sedimentary rock made out of clay particles is called _____.

35) Clastic sedimentary rock made out of silt particles is called _____.

36) Clastic sedimentary rock made out of sand particles is called _____.

37) Which one of these rocks (from 9, 10, 11) is porous and can absorb water? _____

38) What do field geologists use to differentiate clay from silt? a) eyes b) fingers c) tongue d) nose

39) Conglomerate rocks have (angular/smooth) clasts, whereas breccia has (angular/smooth) clasts.

40) TRUE of FALSE? The word "breccia" can refer to either a sedimentary or igneous rock. _____

41) TRUE or FALSE? Both "breccia" and "clast" basically mean the same thing ("broken"). _____

42) The word "matrix" means: a) an object that is embedded in sandstone b) the surrounding stuff
   c) the bottom layer d) basic crystal structure

43) Which one of these is NOT a chemical sedimentary rock?

   a) sulfates b) silicates c) carbonates d) evaporites e) organics

44) TRUE or FALSE? The only way an evaporite mineral can form is by evaporation. _____

45) The correct mineral name for salt is _____.

46) Which one of these is NOT an evaporite? a) gypsum b) travertine c) salt d) coal

47) Yellowstone National Park in Wyoming, USA, is one place you can find:

   a) salt deposits b) seams of coal c) hot springs

48) TRUE or FALSE? The water that comes out of hot springs contains lots of silicon but very few other minerals.

49) Gypsum is a very soft mineral and is number ___ on the Mohs hardness scale.

50) If you were a coal miner, which kind of coal would you hope to find in your mine? (best quality)

   a) anthracite b) bituminous c) lignite

51) Which of these is sometimes brown? a) anthracite b) bituminous c) lignite

52) TRUE or FALSE? All coal mines are deep underground. _____

53) TRUE or FALSE? Power plants that burn coal have black smoke coming out of their chimneys. ___

54) TRUE or FALSE? All scientists agree that coal came from ancient plants. ___

55) TRUE or FALSE? All scientists agree about how coal formed. _____

56) What made cherts and flints so useful to ancient peoples? a) They could be ground into a fine powder.
   b) They contained metal ores. c) Their fracture pattern left sharp edges. d) They were shiny when polished.

57) What special trait does flint have? _____

58) What does "oö" mean? _____

59) TRUE or FALSE? Limestone is a type of rock that can have fossils. _____

60) Limestone is made of: a) $CaCO_3$ b) SiO_2 c) SiO_4 d) $CaSO_4$ e) CaF_2

METAMORPHIC

61) "Meta" means _____ and "morph" means _____.

62) This type of metamorphic rocks always has stripes: _____

63) This metamorphic rock has scaly layers due to the presence of very small sheets of mica: _____

64) Contact metamorphic rocks experienced _____.

65) Regional metamorphic rocks experienced both _____ and _____.

66) Gneiss used to be _____. Shale turns into _____.

67) Foliated rocks have: a) lots of thin layers b) fossilized leaves c) stripes d) quartz crystals

68) "Schist" means _____.

69) What causes the colored swirls in marble? a) impurities b) layers of clay or iron oxide
   c) elements other than silicon d) all of the above

70) TRUE or FALSE? Marble is softer than granite. _____

Activity 4.2 Videos

Go to the "Rocks and Dirt" playlist at: **www.YouTube.com/TheBasementWorkshop**, and watch the videos that correspond to the topics in this chapter. You can visit a hot springs, a salt mine, and some erupting volcanoes!

Activity 4.3   Igneous pairs
TIP: The chart on page 29 will be very helpful for this.

1) Both are coarse-grained, but one is mafic and one is felsic. _____ and _____
2) Both are felsic, but one is frothy and one is glassy. _____ and _____
3) Both are frothy, but one is mafic and one is felsic. _____ and _____
4) Both are fine-grained, but one is mafic and one is felsic _____ and _____
5) Both are felsic, but one is coarse-grained and one has no grains at all. _____ and _____
6) Both are intermediate (mafic/felsic) but one is coarse-grained and one is fine. _____ and _____
7) Both are extrusive but one is fine-grained and one is pyroclastic. _____ and _____
8) Both are mafic, but one is coarse-grained and one is fine-grained. _____ and _____
9) Both of these are the only example in their texture category. _____ and _____
10) Continents are made of _____ and ocean floors are made of _____.

Activity 4.4   Up-close igneous
Here are some magnified views of igneous rocks. See if you can figure out which is which. Do the easy ones first, then use the process of elimination to figure out the rest. (Use information you learned in the chapter!)

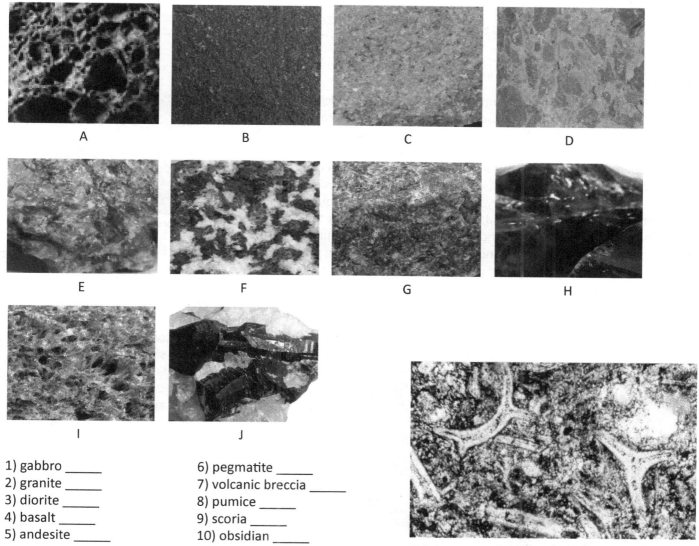

A   B   C   D

E   F   G   H

I   J

1) gabbro _____
2) granite _____
3) diorite _____
4) basalt _____
5) andesite _____

6) pegmatite _____
7) volcanic breccia _____
8) pumice _____
9) scoria _____
10) obsidian _____

This is a microscopic view of one type of tuff. The angular pieces are shards of SiO_2 glass.

NOTE: In "real life," igneous rock might not be so easy to distinguish. Granite can be black and white, for example, and look like diorite.

Activity 4.5 What's the use?

Decide which rock you would use for each task. (Choose from the list at the bottom.)

1) You need fuel to burn. _____
2) You want to make a piece of jewelry. _____
3) You want make roof tiles. _____
4) You want to make a road without any asphalt. _____
5) You want to make plaster. _____
6) You want to make an arrowhead. _____
7) You want to melt the ice on your sidewalk. _____
8) You want to make a finely detailed sculpture. _____
9) You want to make concrete. _____
10) You want to make an outdoor monument. _____
11) You want crushed rock to fill a railroad bed. _____
12) You need rocks to absorb very high heat in a gas grill. _____
13) You want to make a set of stone coasters that will absorb the water that drips off the glasses. _____
14) You want to make a lightweight, mildly abrasive rubbing stone to take off skin calluses. _____

limestone, gypsum, granite, pumice, quartzite, coal, slate, marble, chert, flint, scoria, cobbles, halite, sandstone

Which rocks in this chapter do you think are the least useful? _____

Activity 4.6 Shale

Did shale get your vote for one of the least useful rocks? It is far too soft to be any good for building walls or roofs, like sandstone and slate. You can't burn it like coal. Its chemistry isn't helpful for any manufacturing processes, like gypsum is. It just sits there are takes up space in the landscape.

Part of the Marcellus Shale in Pennsylvania

However, shale that is buried deep underground has a companion that is very useful: natural gas (methane). Deep shale has natural gas, and sometimes even oil, in and around it. Recently, energy companies have found ways to drill down into these shale deposits and break up the shale so that the gas can leak out and come to the surface. The method they use is called hydraulic fracturing because it pumps water (along with sand and chemicals) down into the shale to break it up. "Fracking" has become very controversial because of reports of groundwater and wells that appear to have been contaminated by this process. The energy companies assure us that what they are doing is safe, and if nothing goes wrong, maybe it is. However, people being people, mistakes will be made, and have already been made, especially at the surface where fracking trucks have been known to dump their dirty water in places they are not supposed to.

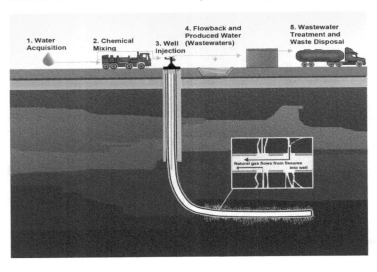

If you would like to see some animations showing how they drill down into shale, there are some posted on the YouTube playlist. Nothing controversial has been posted, just explanations of the technology. People who are pro-fracking point out that traditional sources of oil and gas are beginning to dwindle and we need to find new ones. We like having energy resources that allow us to travel as much as we want to and to have air conditioning in our homes, but these won't be possible unless we find new sources of energy. So there are valid points on both sides of the argument. The energy companies need to find a way to extract the oil and gas without polluting the environment. The playlist represents both sides of the issue.

Activity 4.7   The Rock Song

You can find the audio for this song in two places. One is on the YouTube playlist, where they appear as a music video, with the words printed on the screen. If you want just the audio without the video, you can go to www.ellenjmchenry.com/rocks-and-dirt.

The Rock Song

Sedimentary, sedimentary, sedimentary rocks were slop.
They once started out as silt and sludge and particles of rock.
Then pressure came, and packed them, packed them down.
That's how limestone, shale and coal got into the ground.

Metamorphic, metamorphic, metamorphic rocks have changed.
They once started out as other kinds of rock, though that seems strange.
Then pressure came, and squeezed them, squeezed them tight.
Marble used to be limestone, gneiss was once granite. *(say "nice" and "gran-ITE")*

I-G-N-E-O-U-S spells igneous, yes igneous.
They once started out as magma, piping hot and dangerous.
Then cooling came, either fast or very slow.
Granite and obsidian are rocks that you should know.*

*Granite is an example of coarse-grained igneous, and obsidian is the opposite with no grains at all.

Activity 4.8   Nice pictures

These pictures wouldn't fit into the text, but they are really worth a look.

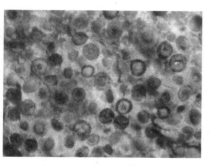

I turned this obsidian into the light so that the "conchoidal" fracture pattern would show up very well. Anything made of SiO_2 will fracture like this.

Here is a very close-up view of oolitic chert. The "eggs" don't all look the same when you get this close, even though from a distance they just look like black dots. You can see why some people love this kind of chert!

Flint is often found in nodules like this, with an outer coating of chalk. Chalk is made of $CaCO_3$, not SiO_2, so it is strange that these two minerals should be found together.

This is a really nice, and rather unusual, specimen of gneiss. It looks like it used to be pink granite. It is very easy to imagine this rock having been under stress!

This "pillow lava" was formed when blobs of lava plopped into the ocean.

This is a mosque built out of salt blocks at the bottom of a salt mine in Pakistan.

Few people know the word "diorite," but the steps of St. Paul's in London are made of it.

Code of Hammurabi on diorite (Louvre)

45

Activity 4.9 Lithologic patterns

Geologists have patterns that they use to indicate different types of rock. Use their patterns, below, to fill in the sections of this landscape. Four of them don't have official patterns so you can make up your own, or use colors such as red, brown, gray and black. Notice that all the metamorphic rocks have some kind of wavy pattern, and also notice the contact metamorphism going on. (Amphibolite is metamorphic basalt.) Put two layers of each sedimentary rock in the bands to the right of the magma, then figure out where to put their metamorphics. What will the lava in the water turn into?

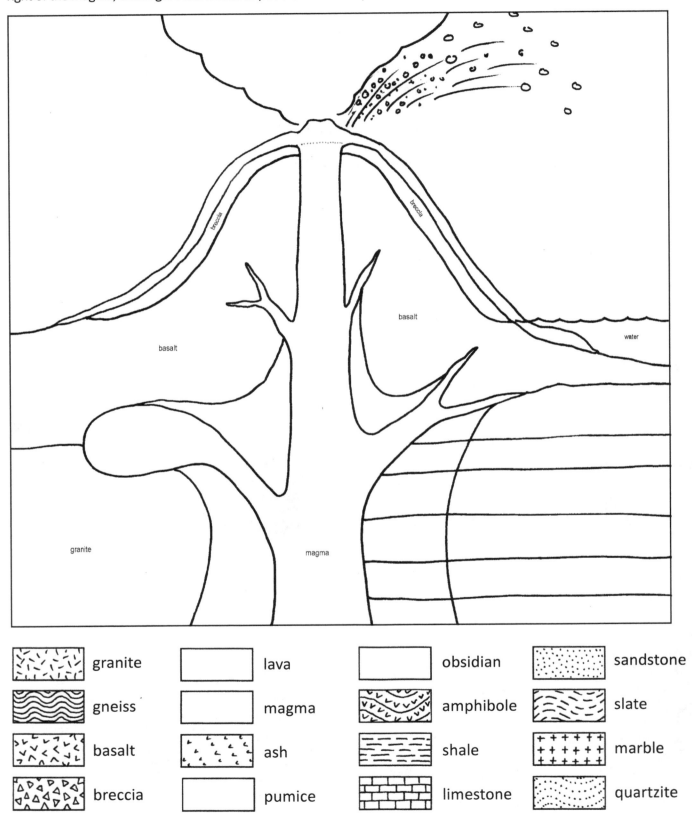

	granite		lava		obsidian		sandstone
	gneiss		magma		amphibole		slate
	basalt		ash		shale		marble
	breccia		pumice		limestone		quartzite

CHAPTER 5: LIMESTONE

Since limestone gets its own chapter, it must be pretty important. In fact, it is one of the handiest rocks on the planet. It can be used in almost every industry, from construction and manufacturing to agriculture and food science. To understand its many uses, we need to study its chemistry, so we'll be doing a lot of chemistry in this chapter. But first, let's look at the many different kinds of limestone and where they are located.

The word "limestone" is a very general word meaning any rock that contains over 50% calcium carbonate, $CaCO_3$. We've already met $CaCO_3$ as the mineral **calcite**. Actually, we've also met $CaCO_3$ as the mineral **aragonite**. How can these minerals look so different and yet be made of the same stuff? (And neither of these look anything like limestone.) Think about graphite and diamond. Both are made of nothing but carbon atoms, yet one is black and soft and the other is clear and hard. The arrangement of the at-

Calcite and aragonite are both made of $CaCO_3$.

oms is what makes the difference. The atoms in graphite are arranged into flat sheets of hexagons. The atoms in diamond are locked into a strong framework. In calcite and aragonite, the $CaCO_3$ crystals have been squeezed into different crystal shapes. We looked at some basic crystal shapes back on page 10. Calcite and aragonite have different crystal shapes while still being made of the same stuff. They formed under different circumstances in different environments. Calcite can take other shapes, too, especially if there are some stray metal atoms in the area. Blue, orange and brown calcite don't look like clear calcite or aragonite.

Calcite and aragonite are very pure examples of calcium carbonate. There might be small amounts of other atoms mixed in, but the amount is so small that we can basically ignore it. In limestone, however, large amounts of other stuff can be present. A rock can be called limestone as long as it is more than half calcium carbonate. This means that there are many types of limestone and they can look quite different. Here are some notable examples of limestone formations around the world.

"James Bond Island" in Thailand is made of limestone. (It became famous after being featured in a James Bond movie.)

Limestone landscapes like this are at various places around the world, including central China.

The Grand Canyon has layers of limestone. One of the layers is called the Redwall Limestone.

This is a limestone quarry, where blocks of limestone are being cut for use as building stones.

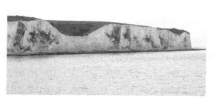

The famous white cliffs of Dover, on the coast of England, are made of chalk, a type of limestone.

Caves are made of limestone. This is Bears Cave in Transylvania, Romania.

Here is another limestone landscape. (This is just down the road from where the author lives.) It's a much more typical limestone formation—not spectacular at all. Most of the eastern USA has limestone like this. It might look boring on the surface, but limestone landscapes are often full of interesting caves and caverns underneath. Limestone areas with caves are called **karst** landscapes.

Limestone can be white, gray, brown, tan, pink or even kind of bluish or yellowish. Some limestones are soft, like chalk. Other limestones are very hard and make excellent building stones. Some limestones have fossils, others do not. Oölitic limestone has tiny dots or spheres, similar to what we saw in oölitic chert.

Here are some samples that show the wide variety in limestones.

This is that boring gray limestone in the picture above. Not much to look at, but very useful to many industries.

This is **chalk**, the stuff that those white cliffs of Dover are made of. Some chalks are harder than others.

This limestone is called **coquina**, using the Latin word for seashell. This is basically a cemented lump of tiny seashells.

This is **oölitic limestone**. Up close, it looks like it has tiny spherical egg shapes embedded in it.

This is **fossiliferous** (fossil-containing) limestone. Not all specimens are as loaded with fossils as this one.

Travertine is considered to be a type of limestone. In chapter 4, we learned that it forms around hot springs.

If a limestone is gray or brown, that means that it has a lot of impurities in it, such as clay, iron, magnesium, and sulfur. If a limestone has a significant amount of magnesium, the name of the stone changes from limestone to **dolomite**. In dolomite, the magnesium atoms actually go in and replace some of the calcium atoms. Some of the molecules switch from being $CaCO_3$ to $MgCO_3$. This substitution is possible because calcium and magnesium atoms both have the same electron arrangement in their outer shells.

To study limestone further, we need to learn a little more chemistry. But that doesn't mean we have to be boring. Let's make it fun! We'll represent the atoms as little cartoon characters with a certain number of hands. The hands will represent the number of bonds that the atoms would like to make. For example, we learned that both carbon and silicon have 4 electrons in their outer shell but would like to have 8. That means that they would like to make 4 bonds, to fill those empty places. Oxygen has 6 electrons in its outer shell and would like to have 8, so it would like to make 2 bonds. Calcium and magnesium both have 2 electrons in their outer shell and would like to have 8, but they know it is probably unrealistic to hope for 6 other atoms to bond with. It is better for them to just share the 2 electrons they have. That means that Ca and Mg are looking to make 2 bonds, just like oxygen is. Hydrogen is a tiny atom and can only ever make 1 bond. So here is our cast of characters:

HYDROGEN   OXYGEN   CARBON   CALCIUM   MAGNESIUM

These guys can hold hands in various ways to make different molecules. First, let's meet **carbonate, CO_3**. This group of atoms is extremely common in chemistry. (Most high school chemistry classes require you to memorize it.) It is made of 1 carbon atom attached to 3 oxygens. Carbon is very flexible about bonding, and it reaches out two hands to one of the oxygens, making that oxygen very happy. This is called double bonding. The double-bonded oxygen has both hands occupied, so it is content. The other oxygens are attached with only one hand so they still have one hand free and can bond with something else. If a hydrogen atom and a sodium atom happen to come by, they will each pair up with an oxygen, making $NaHCO_3$, which you know as baking soda, the white powder that you put into muffins and cookies to make them puff up in the oven. If a calcium atom happens to float past a carbonate molecule, it will want to hold hands with both oxygen atoms, making $CaCO_3$, calcite.

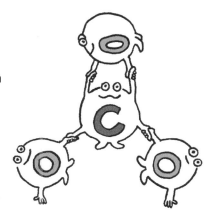

The mineral calcite is made of trillions upon trillions of these $CaCO_3$ molecules. The way they stack together makes a rhombohedral pattern, which gives calcite its rhombohedral crystal shape. If other atoms are present, it messes up this crystal pattern. If limestone was made of pure $CaCO_3$, it would look like a piece of calcite (shown on page 47). Since limestone does not have a distinct crystal shape, we know it must have other things mixed in. Geologists all know that limestone is not pure, so they go ahead and use the formula $CaCO_3$, and just assume that everyone knows that limestone is not pure.

As mentioned on the previous page, one thing that can get mixed into limestone is the element magnesium, Mg. If an Mg atom floats past a $CaCO_3$ molecule, it is able to kick the Ca out of the way and take its place. (Pure $MgCO_3$ is called magnesium carbonate and is a white powder, like baking soda.) When limestone has a lot of Mg in it, we change its name to **dolomite**. Its formula has both Ca and Mg inside parentheses, telling us it could be either one.

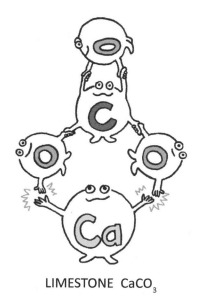

LIMESTONE $CaCO_3$

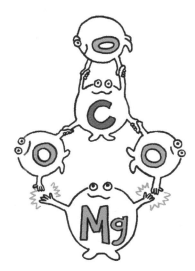

DOLOMITE $(Ca,Mg)CO_3$

You may have noticed that in both the $CaCO_3$ and $MgCO_3$ molecules there are purple zigzag lines right where Ca and Mg are holding hands with the oxygens. This is an important feature. The hand-holding that is going on between the calcium and the oxygens is not the same type as between the carbon and the oxygens. The bond between the carbon and the oxygens is called **covalent** bonding. We saw this type of bonding inside the silicon tetrahedron, SiO_4. We are showing covalent bonds as hands joined together. The calcium and magnesium atoms are forming **ionic** bonds, not covalent bonds. We are showing ionic bonds as hands *almost* touching, with zigzaggy purple lines around them.

Even though covalent and ionic bonds are strong enough to hold rocks together, they are not impossible to break. In fact, let's break some right now! We'll need to apply heat. We'll use a propane gas torch that can reach about 2000° C (3000° F). Considering that water boils at 100° C (212° F) that's some serious heat! Let's torch a piece of limestone, $CaCO_3$. (Yes, we know it is not pure $CaCO_3$.)

It looks like the heat managed to break two bonds, giving us two smaller molecules, CaO and CO_2. But wait—isn't CO_2 just carbon dioxide? Yep, this little molecule floats away and is gone! CaO is a solid, though, so it can't float away. The scientific name for CaO is **calcium oxide**, but it is often called by its informal name, **quicklime**. In this case, the word "quick" doesn't mean "fast," but "alive." This name was given several hundred years ago by people who witnessed water being poured over CaO. The CaO sputtered and spit and fizzled as it melted, acting very lively, so they called it quicklime, meaning "living lime."

If you go to the YouTube playlist, there is a video in the lime section, made by some professional chemists at the Royal Institution in England. They heated large blocks of limestone in a kiln in order to drive out all the carbon dioxide. The blocks didn't look like they had changed very much, but when they took them outside and sprayed them with a hose... well, you can watch it for yourself. This "quick" reaction produces a lot of heat, too. In fact, calcium oxide can be used as an instant source of heat. Just add water!

Calcium oxide, CaO, turns out to be very useful stuff. As we just mentioned, when you add water to CaO it produces heat. (A chemical reaction that produces heat is called **exothermic**. "Exo" means "out" and "thermo" means "heat.") CaO has been used as a warming device at the bottom of self-heating soup cans. (It sits in a separate canister, apart from the food.) Besides heat, it also generates light. Before the days of electricity, cylinders of calcium oxide were heated with natural gas burners to produce a very bright, white light in theaters. We still use the term "in the limelight," meaning something that is getting a lot of attention.

a actor standing in the limelights

Calcium oxide is important in other ways, too. It is vitally important in the manufacturing of steel, glass, paper, porcelain and cement. It can be used to neutralize acids and improve drainage of clay soils. CaO can also be turned into a safer form, $Ca(OH)_2$, that can be used to put stripes on football or baseball fields. A long time ago, powdered lime was even used as a weapon. Soldiers would launch bags of lime powder onto their enemies. The lime would create a huge, white dust cloud and the enemy soldiers would get lime in their eyes. Eyes are watery, so the lime gets activated and burns your eyeballs. Sports fields are a much nicer use of lime.

Let's look at what happens when we add water to CaO. Both the Ca and the O atoms have a hand free and would like to find someone to hold hands with. Along comes a water molecule, H_2O. The O looks over and thinks that he could try to steal one of the hydrogen atoms off the water molecule. Hydrogen atoms are relatively easy to steal, so this is what happens. That leaves water as just an OH. The calcium is glad to bond to the O in that OH. The final result is that the calcium looks like it is holding on to 2 OH's. The cartoon makes it easy to understand but is hard to draw. If we use just letters to represent this arrangement, we write it like this: $Ca(OH)_2$

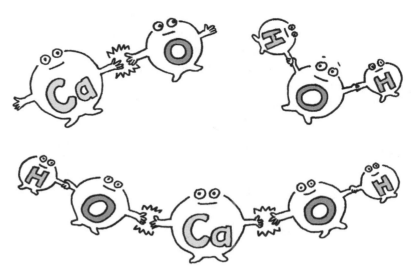

$Ca(OH)_2$ is properly called **calcium hydroxide**, but is informally called **slaked lime**. (It is also called pickling lime, builders' lime, or caustic lime.) In its pure form, $Ca(OH)_2$ is a white powder, but usually plenty of water has been added to make it, so often it looks like a thick, white pudding. If a lot of water is added, it is called limewater.

Calcium hydroxide is very useful in the food industry. Since it is sometimes called pickling lime, obviously it can be used to make pickles. It is also used to process sugar beets or sugar cane, and it is added to fruit juices to fortify them with calcium. It is very helpful when making corn tortillas because it helps the corn flour to stick together. At the other end of the food spectrum (meaning when food has turned to waste), calcium hydroxide can be used by sewage treatment plants to help take particles out of the waste water.

How would all this chemistry look if we used just letters and numbers, not cartoon characters?

The first step, where we heated the limestone, would look like this: $CaCO_3 \rightarrow CaO + CO_2$ *(This used heat.)*
The second step, where we added water, would look like this: $CaO + H_2O \rightarrow Ca(OH)_2$ *(This made heat.)*

We can add a third and final step to this series. $Ca(OH)_2 + CO_2 \rightarrow$ _____ + _____

What molecules will be on those blank lines? Here are the cartoon molecules on the left side of the arrow. Before turning the page, you might want to study this situation and take a guess as to how these atoms could rearrange themselves to make two different molecules. (Hint: This series of equations is called the limestone cycle. In a cycle, you get back around to where you started.)

This is how these molecules rearrange. We have limestone again! Adding carbon dioxide to lime-water produces one calcium carbonate molecule and one water molecule. The water will evaporate, leaving just the limestone. Here is how we write it.
(The "aq" stands for "aqueous" which means "in water.")

$$Ca(OH)_2 \, (aq) + CO_2 \rightarrow CaCO_3 + H_2O$$

This is how the atoms reshuffled:

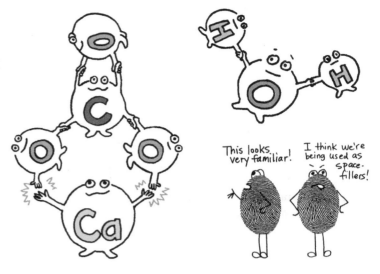

So now we are back at the beginning of what is called the **limestone cycle**. Not all chemical reactions are as easy to study as this one. Many reactions are "one way" and you can't ever get back to where you started. Also, in this cycle the math works out very nicely. In a lot of reactions you need 2 of one thing and 3 of another in order to get the molecules rearranged correctly.

The last part of this cycle is something you can observe for yourself. You can dissolve a few spoonfuls of pickling lime (or agri-cultural lime) into a glass of water. Let the water sit overnight. In the morning the water will be clear but there will be some white powder at the bottom of the glass. Pour just the clear water into another glass. (This clear water is "limewater" and still has lime in it.) Then use a straw to blow bubbles into the water. The carbon dioxide in your breath will be enough to allow some tiny particles of $CaCO_3$ to form.

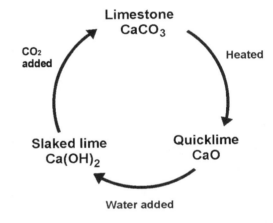

Now, for the grand finale! Just one more chemical equation for this chapter. This equation is also key to under-standing limestone. This reaction isn't a cycle, but is just as interesting because it works both forwards and backwards. (Don't worry about the tiny 2+ and 2- signs. Just look at the letters.)

$$H_2O + CO_2 + CaCO_3 \leftrightarrow Ca^{2+} + 2H^+ + 2CO_3^{2-}$$

Carbon dioxide is a gas, but it can be dissolved into water. The fizzy bubbles in carbonated beverages are made of carbon dioxide. The bubbles appear as the CO_2 comes out of the solution and goes back into the air. On the left side of this equation we see CO_2 dissolved in H_2O. There is also limestone on this side of the equation. (Imagine seltzer water being poured over a piece of limestone.)

On the right side we see what happens. The calcium atom comes off the limestone and floats around by itself. The H_2O and CO_2 combine with the remaining CO_3 to form 2 hydrogen atoms (ions) and 2 molecules of carbonate. Everything on the right side is dissolved in water.

We can reverse this process, and go from the right side back to the left side. If we have water that contains calcium atoms (ions) and molecules of carbonate, and hydrogen ions, these can react together to form one molecule of water, one molecule of carbon dioxide, and one molecule of solid limestone. That means that our watery solution (on the right) will produce one molecule of a gas, CO_2, and one molecule of a solid, $CaCO_3$. A gas and a solid will come right out of the water!

When a solid comes out of a liquid, we call this **precipitation**. This is what happens when you blow CO_2 bub-bles into limewater— tiny white particles appear seemingly out of nowhere. The atoms were there all along, but were dissolved in the water so they were not solids. The addition of carbon dioxide made them form a solid.

Now here's where we get to take a side track off into biology. Many sea creatures can take these molecules out of the sea water. It's not called precipitation when they do it, however. It's just called making a shell. Scientists still don't completely understand how these animals do this. They do know that the part of their body responsible for this task is the **mantle**. The mantle is just one section of their soft, squishy bodies. The mantle is somehow able to extract the calcium and carbonate out of the water and use it to fill a very thin protein framework that they've built. (Your body does something similar when it builds bone, but it uses calcium, phosphorus and magnesium.) A mineralogist would classify a seashell as **organic** $CaCO_3$.

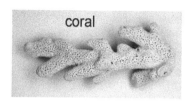
coral

Geologists have wondered how much of earth's limestone was formed by sea creatures. There are places in the world where we can observe limestones being formed from shells and corals. When the creatures die, their shells fall to the bottom and begin to collect. The shells can dissolve over time, and, if conditions are right, the tiny bits of shell can be cemented together to form a hard rock. But was all limestone in the world made this way?

If we look at limestone under a microscope, we see tiny clasts. These clasts might be made of shells, rocks, crystals, or even microscopic poop particles. If these microscopic clasts get coated with thin layers of calcite, they will turn into oölitic limestone (shown on page 48). If the clasts happen to be very large shells, you get coquina. The clasts are cemented together with either calcite or aragonite crystals, both of which are made of $CaCO_3$. There are also places in the world where we can observe $CaCO_3$ precipitating directly out of the water without the help of shell-making sea creatures. Apparently, limestone can come from either organic or inorganic processes. Could the vast amount of limestone covering large areas of the world have been made by organic processes alone? Let's use that equation again, along with some data about how much carbon dioxide limestones have absorbed.

First, we need to count all the carbon atoms in the world and find out how many of them are in limestone.

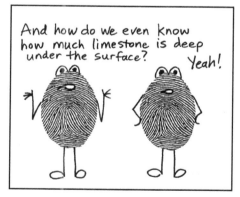

Remember, 3/4 of the earth's surface is covered with water. Is most of the carbon in the oceans, then?

Good questions! Of course, we can't really count atoms. Our numbers will be rough estimates, not exact counts, and we'll have to count **grams**, not individual atoms. (One gram is about the weight of a paper clip. There are trillions upon trillions of atoms in one gram.) Fortunately, several people have already done these calculations for us. (They are listed in the bibliography in the limestone section.) They've estimated figures like the volume of the oceans, the number of plants and animals on the earth, the amount of limestone that we know about, and the approximate amount of carbon in a gram of each substance.

The number of grams of carbon in the atmosphere is about 720,000,000,000,000,000. Wow, that's too many zeros! We can make this number easier to look at by putting all of those zeros into scientific notation: 10^{16}. The tiny number, in this case, 16, tells you how many zeros we have removed. So we can write 72×10^{16}, and that means 72 with 16 zeros after it. Ready for the rest of the results?

ATMOSPHERE: **72** x 10^{16}
PLANTS and ANIMALS: **200** x 10^{16}
COAL, OIL and GAS: **413** x 10^{16}
OCEANS: **3740** x 10^{16}
LIMESTONE: **6,000,000** x 10^{16}

Plants take in CO_2 and use it for photosynthesis. Animals eat the carbon-containing plant sugars.

The amount of carbon locked up in limestone is thousands of times more than in the rest of the earth combined! The math doesn't add up. But there's more...

Let's look at this equation again: $H_2O + CO_2 + CaCO_3 \leftrightarrow Ca^{2+} + 2H^+ + 2CO_3^{2-}$

The right side would represent the ocean environment with calcium and carbonate floating around in the water. As $CaCO_3$ is created by either organic or inorganic processes, we move to the left side of the equation. On the left we see a molecule of CO_2. That CO_2 would be released either into the water or into the air. So for every molecule of limestone that is formed, a molecule of CO_2 goes into the environment. Look at how many grams of limestone we have on planet earth. Do you see a problem here? Environmentalists are already concerned that there is too much CO_2 in our atmosphere right now. They call it a "greenhouse gas" and claim that it is raising global temperatures. So if 72 x 10^{16} is too much CO_2, then what would $6,000,000$ x 10^{16} have done to the planet?! But there's more...

A researcher back in the 1990s looked at some imestone he knew was old, and compared it to recently formed limestone. If limestone has always formed the same way, then all limestones should be similar. He went to the Grand Canyon and looked at one of its well-known layers of limestone. Here is a summary of his research:

Grand Canyon limestone	modern limestone ("limemud")
$CaCO_3$ looks like calcite	$CaCO_3$ looks like aragonite
clast size is 1-4 microns	clast size is about 20 microns
quartz sand grains mixed in	no quartz sand grains
fossils pointing in same general direction	fossils not in same direction
fossils of delicate, soft-bodied animals	no fossils of soft animals

The layers of limestone in the Grand Canyon have names. The most famous layers are the Kaibab Limestone, at the top, and the Redwall Limestone, a layer stained by iron oxide.

The quartz sand grains and the fossils pointing in the same direction suggests that the water was moving in one direction at the time of formation. Modern lime muds form in very calm, shallow areas. The fact that soft-bodied animals, even as soft as a jellyfish, are found as fossils in old limestone suggests that fossilization occurred very rapidly. Soft-bodied sea creatures disintegrate quickly, in a matter of weeks, if not days. But there's more...

No matter how hard I try, I just can't seem to use magnesium in my shell!

Clams don't use magnesium.

Have you ever heard of the "dolomite problem"? Of course not, because you aren't a professional geologist. Professionals of any kind don't print their problems in books for the general public to read; you have to go to their professional conferences or read their professional journals. There are quite a few problems that stump geologists; the dolomite problem is just one of them. We met dolomite on page 49. Dolomite is basically limestone in which about half of the calcium atoms have been replaced by magnesium atoms. Why is this a problem? Because most shell-making animals don't use $MgCO_3$. A few do, but not enough to account for all the magnesiusum in dolomite. Did water containing magnesium somehow get into the limestone after it was formed? There isn't any other evidence of water getting in, and the deposits are far too thick and too uniform to support this theory.

So how could so much limestone have formed without poisoning the planet with CO_2? Is there a good explanation for dolomite? And what about fossils? We'll investigate several ideas in chapter 7.

Let's go back to that equation one last time. $$H_2O + CO_2 + CaCO_3 \leftrightarrow Ca^{2+} + 2H^+ + 2CO_3^{2-}$$

This equation can also help us to understand how caves form. The left side tells us that if water containing dissolved carbon dioxide is poured over limestone, the limestone will dissolve into the molecules on the right. Rain contains dissolved CO_2. If rainwater gets into cracks in limestone, it can gradually eat away at the rock and create holes. More rainwater begins washing into these holes and they get bigger. Gradually, the holes get larger and larger until they make empty areas we call **caves** (or caverns).

Once a cave has opened up, the water seeping in can then start making formations called **speleothems**. (Any word starting with "speleo" will have something to do with caves.) The rainwater coming into a cave has dissolved limestone in it. As the water evaporates inside the cave, the $CaCO_3$ precipitates back out again, but often in very interesting ways. The precipitated limestone can form stalactites (on the ceiling) and stalagmites (on floor), or flowstone, dripstone, and drapery. Some caves have streams or pools at the bottom.

The thin lines are "straws" that can break easily.

Cave formations are often named after what they look like. You hardly need a picture to know what these look like: shelfstone, soda straws, bubbles, pearls, bacon, popcorn, fried eggs, butterflies, twists, and corals. Sometimes other minerals are present, too, not just calcite. Gypsum ($CaSO_4$) can form beautiful crystals. Some of the formations will break easily, so cavers must be careful as they explore.

Landscapes made of limestone are called **karst** landscapes. Karst is the German name for a limestone region in Slovenia. Karst is found all over the world, though, not just in Slovenia. Anywhere there is limestone there can be a karst landscape. These pictures show two karst areas that have caves.

Karst landscapes are known not only for their caves and caverns, but also for their **sinkholes**. Sinkholes are places where the "roof" of a cave has collapsed, turning the cave into a deep hole in the ground. Sinkholes can be very small, only a few meters wide, or they can be as large as a small field.

Photo by Luis Fernández García - Own work, CC BY-SA 4.0, https://commons.wikimedia.org/w/index.php?curid=45303372

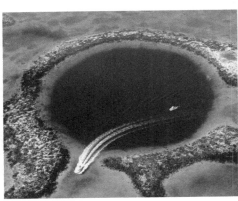

Sinkholes in water are called blue holes. They are not whirlpools and they don't really pose any danger to boats. (It's just the idea of this deep dark hole underneath you...) This blue hole is off the coast of Belize in Central America.

Some sinkholes are very old, such as this deep sinkhole in Mexico. This hole has been around as long as anyone can remember. Since it has a pool of water at the bottom, the local folks have cleverly turned it into quite a tourist attraction. This sinkhole presents little danger, since it collapsed a long time ago and appears to be very stable now. Geologists probably examined the walls of the hole before allowing it to be used for recreation. They will continue to examine it from time to time to assure everyone's safety.

Some sinkholes are quite new. The one in the top picture is in Spain and the hole is in gypsum, not limestone. The people walking along the road give you an idea of how large this sinkhole is.

The bottom picture is a very fresh sinkhole that opened up overnight. One day there was grass, and the next day there was a hole several meters wide.

What if your house was sitting on a future sinkhole and you didn't know it?

That's actually happened! It's scary! There are videos about it on the YouTube playlist.

56

Activity 5.1   Comprehension self-check

1) True or False? Calcite and aragonite are both made of $CaCO_3$. _____

2) If the chemistry of a rock is 55 percent $CaCO_3$, can it be considered a limestone? _____

3) What makes calcite and aragonite look different? _____

4) Which one of these is NOT considered to be a type of limestone? a) chalk   b) gypsum   c) coquina   d) travertine

5) Which color is limestone least likely to be? a) white   b) black   c) gray   d) tan   e) pink

6) Why is magnesium able to replace calcium in $CaCO_3$?

   a) Because it has the same number of electrons in its outer shell.   b) Because it has the same number of protons.

   c) Because it has the same number of electrons.   d) Because the atoms are the same size.

7) What does coquina contain? a) quartz   b) iron impurities   c) fish   d) shells

8) Oxygen atoms would like to make what number of bonds? a) 1   b) 2   c) 3   d) 4

9) Carbon atoms would like to make what number of bonds? a) 1   b) 2   c) 3   d) 4

10) Calcium atoms would like to make what number of bonds? a)1   b) 2   c) 3   d) 4

11) True or false? Carbon makes a double bond in $CaCO_3$. ____

12) Can you draw a picture of how a molecule of $NaHCO_3$ might look? You can use this box:

    (You don't have to draw the cartoon guys! Just use letters and lines.)

13) Is the bond between carbon and oxygen covalent or ionic? _____

14) Is the bond between oxygen and calcium covalent or ionic? _____

15) What does "quick" mean in the word "quicklime"? _____

16) True or false? Heat can break ionic bonds but not covalent. _____

17) What is the formula for calcium oxide? a) CaO   b) $CaCO_3$ c) Ca(OH)   d) CaO_2

18) Why can't calcium oxide float away?

    a) It is a liquid.   b) It is a solid.   c) It is bonded to other atoms.   d) It is broken.

19) Which of these will "melt" when water is poured on it? a) CaO   b) Ca(OH)   c) $CaCO_3$ d) CO_2

20) True or false? This reaction, $CaO + H_2O \rightarrow Ca(OH)_2$, is exothermic (produces heat). _____

21) Name at least four things that $Ca(OH)_2$ can be used for: _____

_____

22) True or false? Gases can dissolve into liquids. _____

23) True or false? Solids can dissolve into liquids. _____

24) How many double bonds are in CO_2? _____

25) What does "aq" mean in a chemical equation? a) solid   b) gaseous   c) liquid   d) in water

26) When a dissolved solid comes out of a solution we say that it: a) consolidated   b) evaporated   c) precipitated

27) True or false? Limestone can come from both organic and inorganic processes. _____

28) Which of these contains the most carbon? a) atmosphere   b) oceans   c) limestone   d) coal and oil

29) What body part of a shell-making sea creature can take calcium out of sea water?

    a) the shell   b) the mantle   c) the gills   d) we don't know

30) True or false? The limestone in the Grand Canyon is identical to modern limestones. _____

31) True or false? Sea creatures can use either $CaCO_3$ or $MgCO_3$ to build their shells because Ca and Mg have the same number of electrons in their outer shells. ____

32) If a word begins with "speleo" it will have something to do with _____.

33) True or false? Gypsum is never found in limestone caves. ____

34) True or false? Karst landscapes usually contain caves or sinkholes. _____

35) Sinkholes in water are called _____ _____.

Activity 5.2 Videos galore!

The "Rocks and Dirt" playlist has lots of great videos about limestone, quarries, caves and sinkholes. The sinkhole videos are slightly scary, because people, cars and buildings really do disappear into sinkholes. (However, the videos don't actually show anyone getting hurt.)

Activity 5.3 Decoding cave facts

Find the answers to these cave questions by using the secret code strip at the bottom. You can either cut off the letter strip or make a quick copy on another piece of paper. (If you want to make a copy, lay the edge of a piece of paper right below the letter strip. Trace the dividing lines down onto the paper, then fill in the letters.)

Next to each answer, you will see one of the shapes on the number strip. Slide the letter strip so that shape is directly above the question mark on the letter strip. That will make the letters and numbers line up correctly to spell out the answer. If you use the wrong shape key, the answer will not make sense.

1) The longest cave in the world is called Mammoth Cave. It is located in: ● __ __ __ __ __ __ __ __
            10 4 13 19 20 2 10 24

2) The deepest cave in the world at 2197 meters (7208 ft), is in the country of: ◆ __ __ __ __ __ __ __
           7 5 15 18 7 9 1

3) The fish you find in caves are often missing this anatomical part: ● __ __ __ __
        4 24 4 18

4) Guano beetles in caves eat guano. What is guano? ◆ __ __ __ __ __ __ __
        2 1 20 16 15 15 16

5) Many small cave creatures eat this stuff. ◆ __ __ __ __ __ __
        3 18 11 4 18 16

6) Carlsbad Cavern in New Mexico has a huge cavern called the: ● __ __ __ __ __ __ __
        1 8 6 17 14 14 12

7) Species of this amphibian have adapted to live in the dark. ◆ __ __ __ __ __ __ __ __ __ __
        19 1 12 1 13 1 14 4 5 18

8) Bats use this to navigate in total darkness. ◆ __ __ __ __ __
        16 12 18 11 1

9) This cave creature has poisonous pincers. ■ __ __ __ __ __ __ __ __ __
        1 3 12 18 7 14 3 2 3

10) This cave creature eats small insects that come into the cave. ■ __ __ __ __ __ __
        17 14 7 2 3 16

11) Most crickets are black. Cave crickets are: ● __ __ __ __ __
        22 7 8 19 4

12) Bats drop these into cave pools where they are eaten by fish. ◆ __ __ __ __ __
        6 12 5 1 19

13) The Waitomo caves in New Zealand have these on the walls. ◆ __ __ __ __ __ __ __ __ __
        4 9 12 20 20 12 15 10 16

14) Many streams and pools in caves have these 10-legged creatures. ◆ __ __ __ __ __ __ __ __
        3 18 1 25 6 9 19 8

15) This one-eyed microscopic creature lives in cave streams and pools. ■ __ __ __ __ __ __ __
        1 13 14 3 14 13 2

16) Caves are usually made of calcite, but a few have some: ◆ __ __ __ __ __ __
        7 25 16 19 21 13

| ◆ | ■ | ● | ◆ | 1 | 2 | 3 | 4 | 5 | 6 | 7 | 8 | 9 | 10 | 11 | 12 | 13 | 14 | 15 | 16 | 17 | 18 | 19 | 20 | 21 | 22 | 23 | 24 | 25 | 26 |

| ? | A | B | C | D | E | F | G | H | I | J | K | L | M | N | O | P | Q | R | S | T | U | V | W | X | Y | Z |

Activity 5.4   Learn about coccolithophores

Chalk used to be something that every child was very familiar with because it was used on blackboards in almost every classroom. Blackboards (originally made of slate) are now being replaced by "white boards" so modern students don't use chalk very often. If you've used chalk, it was probably in the form of sidewalk chalk.

Originally, blackboard chalk came from natural chalk deposits. The White Cliffs of Dover on the southern shore of England are the most famous chalk deposits in the world. Chalk is also found on the other side of the English Channel, on the coast of northern France. Denmark also has large chalk deposits, and so does Poland, Russia, Ukraine, the USA and Canada. Since the supply of natural chalk is limited, manufacturers have found other substances that can be used to make chalk. Most blackboard chalk sold today is actually made from gypsum, $CaSO_4$.

white cliffs of Dover, England

white cliffs of France

white cliffs of Denmark

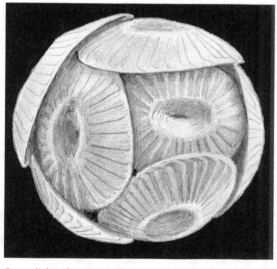

Coccolithophores are living cells, even though they might look like someone's art project. There is a living cell underneath all those plates. The cell is similar to a plant cell in that it can use sunlight and carbon dioxide to make its own food. Like a plant cell, it will also produce oxygen as a "waste."

Chalk is pretty interesting stuff if you look at it under a microscope. You will see broken bits of shell from microscopic animals called **coccolithophores** *(kok-oh-LITH-o-fores)*. Coccolithophores are made of just a single cell, like a paramecium or an amoeba. Like sea shells and corals, they can take calcium and carbonate out of the water and put them together to make calcite, which they use to build a protective shell. They make little round plates that they stick all over the outside of their cell. Coccolithophores use sunlight to do photosynthesis, just like plants do. They make their own food from sunlight. Calcite is clear, so these plates don't block the light from reaching the inside of the cell where photosynthesis takes place. There aren't any males or females; to reproduce they just split in half.

Large masses of coccolithophores occur in the Southern Ocean, around Antarctica. Scientists are tracking them and measuring how fast the populations grow. They predicted that an increased level of CO_2 in the ocean would cause the population to shrink, but it turned out that the opposite was true—the population grew quite quickly. When the population of a microscopic creature grows to immense proportions, we call it a "bloom." Perhaps an ancient bloom help to create the chalky cliffs.

You might remember from the chapter text that flint is strongly connected to chalk deposits. No one knows why this is true. Flint is found as smooth, round nodules right in the midst of the chalk. Very odd. Some of the fossils found in chalk cliffs are made of flint. Water that was full of silicon tetrahedrons would have entered the dead animal's body and replaced all the cells with liquid silica that then hardened. Chalk contains many fossils, but most of them are made of limestone, not flint. All of the fossils are of sea creatures, from tiny coccolithophores, to full-size fish, seashells, sea stars, and lobsters.

Use Google image search (or another search engine) with key words, "fossils found in chalk," and you'll see some fabulous pictures of chalk fossils.

Activity 5.5   Learn about sulfur's role in cave formation

This picture shows how we think limestone caves form. Rain water containing dissolved CO_2 gets into tiny cracks. (If the cave is in a cold climate, water can freeze and expand in the cracks, making this process go faster.) The water continually eats away at the rock until hollow areas start to form. If the rock above the cave collapses, the cave will turn into a sinkhole.

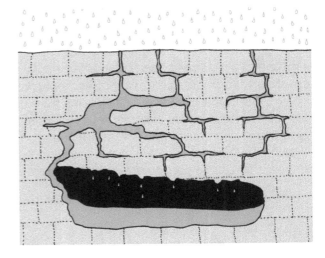

How long does this take? Cave experts used to think it took eons of time for this to happen. However, recent research indicates that this can happen far more quickly than anyone had ever thought possible, especially if other elements, such as sulfur, are involved. We don't know if cave formations have always grown at the rate we see today. Conditions in the past may have been different.

The role of sulfur in cave formation was discovered in the Carlsbad Caverns of New Mexico, USA. The cavern shown here is Lechuguilla Cave, one of the most spectacular caves in the world. Lechuguilla has many gypsum formations. Remember, gypsum is $CaSO_4$, similar to limestone but having sulfur instead of carbon. What difference does this make?

When carbon gets into water it forms carbonic acid, H_2CO_3. When sulfur gets into water it forms sulfuric acid, H_2SO_4. Carbonic acid is a weak acid, but sulfuric acid is very strong. Sulfuric acid is one of the ingredients in drain cleaners—the stuff that promises to dissolve absolutely anything clogging your drain. That's strong stuff! If sulfuric acid started eating away at rock, you might be able to carve out a cave in a few years, especially if there happened to be a few super heavy downpours of rainwater to flush out loose particles.

Another unexpected find in the Carlsbad Caverns is a form of life that loves the toxic sulfuric acid. There are types of bacteria that can use sulfur as a source of energy. Because of this, they are much less reliant on the sun and can live in total darkness. (Sulfur-loving bacteria are also found near hot water vents at the bottom of the ocean. Many of these vents pour sulfur into the water.) It is thought that the presence of these bacteria has influenced the shape and structure of some of the rock formations in the Carlsbad Caverns.

Photo by Dave Bunnell, CC BY-SA 2.5, https://commons.wikimedia.org/w/index.php?curid=2027473

Do you remember John Dalton from chapter 1? He was the person who came up with the theory that atoms are the smallest particles of matter. Here is a diagram from one of Dalton's books (1808), showing his idea of what a sulfuric acid molecule might look like. The atom in the center, with the plus sign, is sulfur. The other circles represent oxygen atoms.

The red and yellow model is a modern sulfuric acid model. The yellow is sulfur, the reds are oxygens and the whites are hydrogen. How many bonds is sulfur making? Sulfur, like oxygen, has 6 electrons in its outer shell and would like to have 8. But instead of making 2 bonds, it is making 6! Those hydrogens will not stay attached. They will go wandering off as H+ ions.

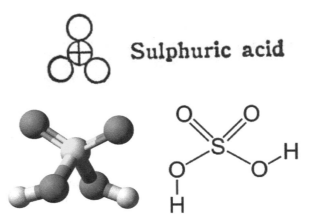

Sulphuric acid

CHAPTER 6: "DIRT"

The word "dirt" isn't a science word. We should really call this chapter: SOIL. So what is dirt? What does it mean when we say that something is "dirty"? With laundry, dirt can mean soil from the garden or yard, but it also might be sweat or food stains. On our shoes, though, dirt is almost always soil that has gotten out of place. Soil is not a problem as long as it stays where we want it to be, in gardens and under the grass. It's when soil is removed from those places and gets onto our clothes or tracked in the house that it becomes dirt. Because the word "dirt" can refer to things other than soil, it's not a very useful word when we are studying soil science. Since this is a science book, not a manual on how to clean your house, let's switch over to using the word "soil."

Soil is something you are familiar with. Even if your family doesn't have a garden, you certainly have patches of dirt and grass somewhere near your home. If you live in an apartment in a big city, perhaps you have potted plants in the window, or a grassy park nearby. We all know that soil's most important job is to grow plants. Without soil, not only would we have no fruits or vegetables, we'd also have no eggs or meat or milk, because animals eat plants. (Fish are less dependent on plants, but certainly benefit from the filtering action that soils provides before rainwater enters streams and rivers.) Most of us don't think about soil very much, but our lives depend on it!

How much soil is on planet earth? About 2/3 of the earth is covered with oceans. If we add in all the lakes, river, and streams, we can estimate that as much as 3/4 of the earth is covered with water. Of the 1/4 of the earth that is land, half of that land can't support much plant life. The Sahara desert covers a substantial portion of Africa. Antarctica, Greenland, northern Siberia and northern Canada are covered with snow most or all of the year. So we are down to only about 1/8 of the planet. How much of that 1/8 can be used to grow plants that we can eat for food? About half is too rocky, too steep or too shallow to support food crops. That leaves only 1/16 of the planet. Some of that 1/16 has been taken over by cities, highways, neighborhoods and shopping malls. Another sizable portion has been set aside for national or state parks. That leaves less than about 1/20 of the planet capable of growing food. The soil on that fraction of land might only be a meter or two thick. We need to take care of every precious inch of soil we have. Later in this chapter, we'll find out what happens when soil is washed or blown away.

How do you study soil? Look at it under a microscope?

Don't ask me- I didn't know dirt was something you could study!

Perhaps the best way to start studying soil is simply to begin digging and see what's down there. If we dig until we hit solid rock, we will be able to see the **soil profile**. The profile is a cross section where you can see all the various layers. To a soil scientist, soil is not just the brown stuff on the top. Soil goes way down, often to a depth of 2 meters (6 feet) or more. The layers are different colors and textures. All soils have similar layers, and soil scientists have assigned letters to these layers. They call the layers **horizons**. (The horizon of a landscape is the long stripe of land right where the sun comes up or goes down. Soil horizons are long stripes of land, but certainly never see the sun!) In this picture, the horizons are different colors, so it is easy to tell them apart. Not all soils have this much color variation.

The **O horizon** is the very top of the soil. "O" stands for "organic," which means living. The O horizon is about how far you can push your finger down into the dirt. It includes grass, leaves, sticks, insects, animal poop, etc.

The **A horizon** is the dark brown layer near the surface, and is often called the **surface soil**. This layer also has a lot of organic matter in it. The organic things can be alive, like roots, bugs, worms and bacteria, or they can be dead things that have decomposed (rotted). If you have a compost pile, you can witness dead plants turning into dirt. Bacteria and fungi love to eat dead plants, and by eating them they turn them into fresh nutrients that can be taken up and used by new plants. The nutrients are recycled over and over again, thanks to the helpful soil microbes. The A horizon also contains some minerals, some water and some air.

The **B horizon** is also known as the **subsoil** and is often a red or gray color. ("Sub" means "under.") You still might find a little bit of organic matter here, such as very long roots, but mostly you will find particles of sand, silt and clay, and also elements such as iron, aluminum, calcium, magnesium and potassium. These elements probably came from the rock underneath.

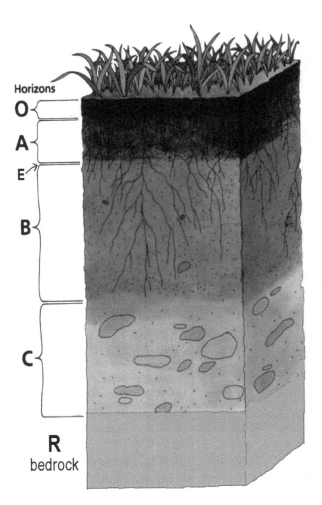

Horizons
O
A
E
B
C
R
bedrock

The **C horizon** is made of **parent rock**. If we have parent rock, where are the children rocks? The kids are upstairs, in the B and A horizons. The parent rock is believed to be the source of the sand, silt and clay particles as well as the elements in those top two horizons. However, the name "parent rock" can be misleading because then we imagine all the mineral particles having come up from below, fighting against both gravity and rain. (More about this on page 66.)

Underneath the C horizon, we find **bedrock**. Sometimes bedrock is called the **R horizon**. Bedrock is solid rock that is not in the process of being broken down. You'll definitely break your shovel when you hit this horizon!

Some soils look like they have an extra layer between the A and B horizons. This extra layer will be lighter in color than the A or B horizons because this is where a lot of the minerals have washed out and gone down into the colorful B horizon. This layer is the **E horizon**. It would be nice if E stood for "extra," but it actually represents a difficult word that you should not even be bothered with. The word means "the minerals washed out." Not all soil profiles have an E horizon.

Here is a "pie chart" showing the main ingredient of soil and how much of each there is. Half of soil is simply air and water. The air and water turn out to be very important, though. The other half of soil is mostly minerals, but with small portion of organic matter. The organic portion includes all types of living things, from bacteria and fungi to plants and animals. Let's look at that 45% mineral part.

The minerals in a soil are almost always the same as the minerals found in the parent rock down below, so it is likely that the parent rock has contributed to, or even created, the mineral part of the soil. The word "minerals" includes both clasts (sand, silt and clay) and elements such as iron (Fe), calcium (Ca), magnesium (Mg), and aluminum (Al).

The elements exist as individual atoms, in their ionic form where they have a positive or negative electrical charge. We saw the calcium ion, Ca^{2+}, all the time in the last chapter. When we learned about the silicates, we saw ions such as Fe^{2+} and Mg^{2+} holding the tetrahedrons together.

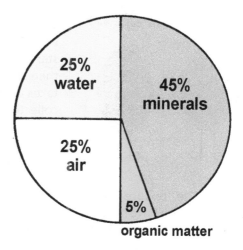

The clasts in soil are the same ones we met when we learned about sedimentary rocks: sand, silt and clay. Sand is visible without magnification, but silt and clay particles are too small to see. Silt can be as small as a bacteria, but clay particles are even smaller. Clay particles can't be seen under an ordinary microscope—you need a very expensive electron microscope. Some soils contain just one of these clasts, but most contain a mixture.

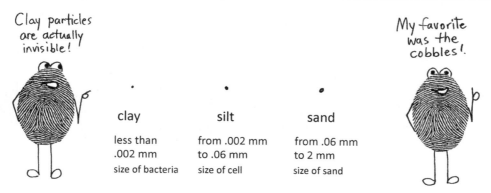

clay
less than .002 mm
size of bacteria

silt
from .002 mm to .06 mm
size of cell

sand
from .06 mm to 2 mm
size of sand

When a soil has a lot of sand, it is called sandy soil. When it has mostly clay particles, it is called clay. (So far this isn't rocket science!) When soil contains a mixture of all three, it is called a **loam**. This word comes from Old English, hundreds of years ago, and back then it meant "clay." Now that we have the word "clay," the word loam is used to describe a type of soil that is made from clay mixed with sand or silt. Loam is the kind of soil you want in your garden, as it is very good for growing plants.

Soil scientists like to be very accurate when describing the texture of a soil. They have many phrases they can use, such as "sandy, silty loam," or "clay loam," or "silty clay." These phrases sound so much the same that they get confusing, even for soil scientists. So they made a chart in the form of a triangle where each side represents a clast size: sand, silt or clay. The number scale on each side goes from 1 to 100, representing the percentage of each particle. To figure out what kind of soil you have, you really only need to know two of these numbers. The ideal is to hit the center of the triangle, in the loam area. Can you find percentage numbers of clay and sand that would make loam? You put your finger on a number, then slide it along the dotted lines that go out from that number. For example, if you have 20 percent clay, the line for 20 percent goes to the right, with the pink "sandy clay loam" over it, and the lavender "sandy loam" under it.

A simple way to get a rough idea of how much of each particle you have is to take a jar of soil and add a few cups of water, then stir very well. Don't put it in the blender or anything, but do stir well. Then let the jar sit. Sand particles are big and heavy and will start falling out of the water immediately. Wait one minute and then measure how much sand has collected at the bottom. Silt takes longer to settle out. Wait one hour, then measure what has collected on top of the sand. Clay takes the longest time to settle out. Wait at least one day, then measure how much clay has collected on top of the silt. Measure each amount with a ruler and then calculate the percentage of each (not including the water).

The "texture triangle" gives you a way to describe the texture. What other characteristics does a soil have? Color is important. Soils that are red contain a lot of iron; yellow soils also contain iron, but in a different form. If a clay soil is gray, that might mean it has been getting too much water. White soil often has a lot of calcium from calcite. Volcanic islands can have black soils that came from mafic lava. Dark brown or black soils have lots of organic material, mostly from decayed leaves. (Dark brown loam is considered to be the perfect garden soil.)

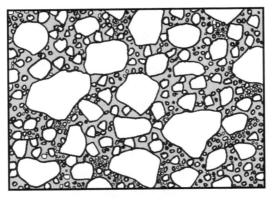

Another characteristic of soil is its **porosity**. Porosity describes the pores, or holes, in the soil. Holes come in large, medium and small, with the tiniest ones being microscopic. In this drawing, the round things are the grains of sand, silt and clay. The empty spaces around them are the pores. Pores are not round—they are the odd leftover spaces between the particles.

Each pore, no matter how large or small, can hold either water or air. After a heavy rain, every pore will be completely filled with water. After a few hours, the large pores begin to drain and air will come in. This is good because plant roots need some air, too, not just water. The middle-sized pores can hold water for several days, or maybe even a week. During this time, the plants send tiny root hairs into these pores to soak up the water. Root hairs are too large to be able to get into the tiniest pores, so once the water is gone from the middle-sized pores, the plants will begin to wilt.

This photograph shows soil that doesn't look particularly wet, but still has water in the smallest pores and probably even in some middle-sized pores. Water will stay in micro-pores a long time.

Sandy soils have many large pores and drain quickly. Clay soils have mostly micro-pores and therefore stay wet for a long time.

Soils have another important property: **structure**. You've probably noticed that soil tends to form clumps. Soil scientists have two names for soil clumps: **aggregates** or **peds** *(AG-greh-gates)*. The word "aggregate" is a very general word and is used to describe chunks of rock, too. Quarries that process limestone usually have piles of aggregates of various sizes. Medium-sized limestone aggregates are often called gravel and are used to make driveways and roads. Smaller pieces are often called "chips" and are used under patios and brick sidewalks. The word "ped" is more specific to soils. It is an abbrevation of the word pedolith. You've seen lot of soil peds but you just didn't have a name for them until now.

Limestone aggregates

Not all soils clump in the same way. Good garden soils tend to form roundish aggregates of various sizes. Round peds are called **crumbs**. This photograph shows some medium-sized soil crumbs. Very small crumbs are called **grains**. Some soils have peds that are not round, but are shaped more like **plates** or **blocks**. Some even have long **columns**. (There are also soils that have no structure at all.)

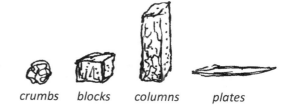

crumbs   blocks   columns   plates

These structures are caused by the interaction of many processes including wetting and drying, freezing and thawing, and by creatures that live in the soil, such as fungi and worms. The same soil might have a different structure in each horizon layer. Platy structures are found in the A horizons of forest soils, or around lakes. Blocky and columnar structures are usually found in the B horizons of clay soils. Blocky structures can mean that there is always a little bit too much water for these clay soils to handle. Blocky structures in the A horizons of farming fields might indicate that tractors have compacted (pressed down) the soil too much. Plowing or tilling a field too much can also destroy soil structure. Farmers interesting in preserving soil structure have stopped plowing their fields every spring. They use a "no-till" method of farming, instead.

In general, structure is good. Any structure is better than no strucure at all. Structure helps soils to hold on to water so that plants won't wilt so easily. During summers with low rainfall, good soil structure can mean the difference between saving or losing an entire crop. Structure also provides drainage so that soils don't become too wet. Wet soils are more likely to grow harmful molds that will destroy plant roots.

NO-TILL FARMING

Soil scientists are helping farmers to improve their soils so their crops can grow better. One way to improve soil is to stop plowing all the time. Plowing destroys soil structure, and plants grow better in structured soils.

No-till farmers plant "cover" crops after they harvest their main crops. Cover crops are usually plants that have roots that are able to put nutrients back into the soil. When the cover crops are mature, a special mowing machine flattens them, making them into mulch that will help to prevent weeds. Planting seeds in a no-till system requires a specialized piece of equipment that can deal with the layer of flattened cover crop. The cover crop plants will eventually rot and add more nutrients to the soil.

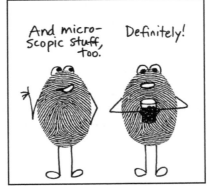

Before we move on to the living part of the soil (the 5%), there are two more aspects of the mineral part (the 45%) that really should be mentioned. First, we need to think about the mineral elements (calcium, potassium, magnesium, etc.) as part of soil formation. Geology books tell us that the parent rock contributed most of the minerals. Soils in limestone areas should be particularly high in calcium, soils above mafic rock would be high in iron and magnesium, soils above feldspar will have high levels of potassium and aluminum, and soils near gypsum might have an unusually high level of sulfur.

Descriptions of soil formation can easily make us think that element ions are going upwards, up from the parent rock. The diagram on the left shows what most of us probably envision in our minds as we read about soil formation. However, this can't be correct. Gravity and rain work together to constantly pull things down. When rain flushes minerals into lower horizons, we call this **leaching**. Leaching of minerals into lower horizons is a big problem if the soils are being used for agriculture. The minerals drain down to depths that plant roots can't reach. The plants them-

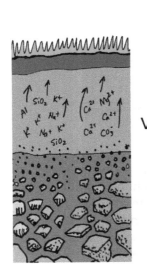

VS

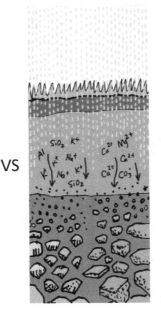

selves make the problem even worse by using up most of the minerals that are near the top. The soil becomes mineral deficient. Plants grown in mineral deficient soils will then be mineral deficient. Animals and humans that rely on these plants for food might then become mineral deficient. For example, people who live in areas where the element iodine is not found in the soil can develop thyroid problems. (The thyroid is a gland in your neck.) Plants, can also get sick if they don't get all the minerals they need. The no-till system of agriculture is one way to put minerals back into the soil. The minerals in the cover crop plants go back into the soil as they decompose. Farmers will also put mineral fertilizers on their fields, containing elements such as nitrogen, phosphorus, potassium and iron.

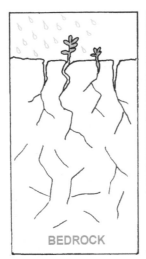
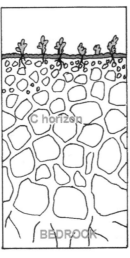
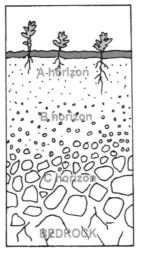

This diagram makes more sense. The parent rock down below isn't constantly sending up minerals. Where there is soil now, there used to be rock. The minerals have washed down over the years.

However, there *is* actually a way for chunks of rock to defy gravity and rise. When wet soil freezes, the water in it will turn to ice and expand. This expansion can push things upward, a process geologists call **heave**. Farmers often find new rocks appearing in their fields in the spring. These rocks probably came up because of heave during the winter.

Even though that diagram makes more sense, there are still other ways that soils can form. Large scale flooding can bring in vast amount of dirt and silt. Many geological features on the earth indicate that water played a significant role in the formation of the landscapes we see today. Large amounts of water would greatly speed up the breakdown of rocks (erosion). We are limited in our study of rocks and soil because we don't have a time machine that can take us back to when things were forming. Our knowledge of the past is very limited. We can only observe what is going on today. Soil formation can be observed in places like Surtsey Island.

SURTSEY ISLAND: A place where we can watch soil formation

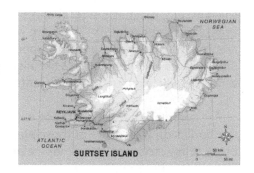

Surtsey Island was born in 1963. There was a huge volcanic eruption off the southern coast of Iceland. The volcano produced so much lava that it bubbled up out of the ocean and cooled to form an island. The black and white photograph was taken when the eruption was just starting. Before the volcano erupted, it was 130 meters (425 feet) below the surface of the water.

The lava was made of melted basalt and olivine. When it cooled, it turned into mafic scoria with chunks of olivine scattered through it. Some of the lava ran down into the ocean and cooled quickly, turning into a mafic form of obsidian. The massive amount of ash that came out of the volcano turned into tuff. Over the years, this tuff has combined with the oxygen in the air and changed its chemistry, turning into a harder, more durable rock. In a few places, hollow tubes formed in the lava as it cooled, and then minerals such as calcite, gypsum and opal, precipitated into cracks.

As soon as the magma cooled, the forces of water and wind began to erode the island. Waves crashed onto the shore and before long the chunks of scoria became smooth and round, as though they had been tumbled in a rock tumbler. Rain fell on the island and cut channels into the cliffs. Anything loose was swept away. Wind swept over the soft rock hills and eroded them into a striped texture.

Less than a year after the eruption, driftwood and floating plants began washing up on the shores. The plants were dead, but their seeds were alive. Small plants began growing in areas that had loose sand that could hold moisture. The roots began breaking up the surface of the rocks. Before long, lichens were growing on some of the rocks. Lichens make acid that can slowly dissolve rocks. Soon birds came and began leaving their droppings on the rocks, which added a source of nitrogen, an essential mineral for plants. The bird droppings also contain more seeds. In only a few years, the black lava sand turned into a thick carpet of grass.

Scientists were shocked at how fast Surtsey was colonized by plants and animals. Processes that they thought took thousands of years were happening right before their eyes. Surtsey is now home to seals, birds, insects, spiders, mites, butterflies, worms, and many species of plants.

lichens

The second important aspect of mineral elements in soils is how plants use them. Certain elements are more necessary for plant growth than others. Some elements, like phosphorus, are essential nutrients, meaning plants can't live without them. Other elements, like silicon, are not taken up by plants, but they don't do any harm. A few elements, such as sodium, are harmful if there is too much.

Which elements on the Periodic Table are essential to plants? Besides carbon (C), oxygen (O), and hydrogen (H), plants need about 14 elements. (Find C, O, and H on the poster on the next page. Oxygen is red, just like in our diagrams. Plants get these elements from the oxygen and carbon dioxide in the air.) The three mineral elements that plants need the most are nitrogen (N), phosphorus (P), and potassium (K). (Find these on the poster.) Tying for second place are calcium (Ca), magnesium (Mg) and sulfur (S). (Where are these?) Those six elements are the biggies—the ones that plants need a lot of. There are also elements that plants need only a small amount of. These "micronutrients" are iron (Fe), manganese (Mn), copper (Cu), zinc (Zn), molybdenum (Mo), boron (B), sodium (Na) and chlorine (Cl). Some plants also need metals like nickel or aluminum, but not all plants do. Some lists include silicon, also, though it is not usually taken up by most plants. Silicon in the soil helps plants grow in other ways.

Humans need all of these elements, too, as well as small amounts of many others. The poster tells you what our bodies do with some of these elements. You can see how important it is to eat plants! Plants provide us with other things, too, not just mineral elements. Plants make sugars, starches, and fats that we "burn" for energy, and proteins that our bodies can use to build muscles and make enzymes. The poster was created to make people aware of how important soil is to human health.

So how do plants take in these minerals? We can guess that they do it with their roots because that's the part of the plant that is down in the soil where the minerals are. But how do they actually pull them out of the soil? The answer turns out to be... boring old clay. Yes, that sticky mud that we don't want on our shoes has a vitally important job!

Clay particles are made of tiny sheets of silicon tetrahedrons. They are sort of like tiny pieces of mica. Their flat shape is what makes clay feel slippery. The sheets slide around on top of each other.

Clay is like small sheets of mica. This is what clay looks like under a microscope.

The silicon tetrahedron has four oxygen atoms that each have a bond site open. When they grab an electron, that makes their overall charge negative.

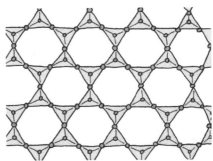

In this diagram of mica, we see tetrahedons bonded to each other to form sheets of hexagons. These hexagons are too small to see, but they do give mica and clay the flat, hexagonal structure that we can see (sketch on left).

This poster is printed here as "fair use for educational purposes." It is a free download on the FOA website. Other posters are available, too.

69

The tan things in this picture are clay particles, though they are not particularly hexagonal looking. (Not all clay particles look as hexagonal as the ones shown in the sketch on the last page.) The clay particles are carrying a negative electrical charge partly because of those oxygens in the tetrahedrons. They can also be negative because of aluminum atoms that have gone in and take the place of some silicons. Aluminum only bonds in 3 places, not 4 like silicon, so this changes the molecular structure of the clay, often adding to the negativity.

*A root hair is exchanging ions with a clay particle. The ions (circles with letters) are WAY too large! They have been enlarged so that we can see what is going on. When things stick on the edges like this, we call it **ad**sorption.*

Since the clay is negative, it will attract things that are positively charged, such as ions of calcium, potassium, iron, and magnesium. These ions will stick to the clay particle but won't form an ionic bond with it. We saw ionic bonds forming in olivine, pyroxenes, mica, quartz, etc. Here, a thin film of water molecules surrounds the ions and keeps them from sticking too tightly. This allows the ions to fall off and go into the plant roots. When molecules stick to the outside of something, it is called **adsorption** (not absorption). These mineral ions have been **adsorbed** by the clay.

The clay particles are not going to give up their ions too easily. To get the clay particle to release an ion, you must do a fair exchange and give it an equivalent ion. Fortunately, the cells that form the tiny root hairs are capable of making ions. The rectangular things on the left are plant cells. The blobs inside are cell parts such as nucleus, Golgi bodies and mitochondria. Notice how one cell is specialized so that it forms a long "hair." (This hair would be too small to see.) There are little "machines" embedded in the walls of this cell that can make hydrogen ions. These hydrogen ions are nothing but a proton, having been stripped of its electron. This gives them a positive charge, since protons are positive. The plant cell now has something to trade with!

Hydrogen ions are protons.

The hydrogen ions, H^+, float over and trade places with a mineral ion. The math has to work, though. Since hydrogen ions are (1+), they can be exactly exchanged for a potassium ion, K, which is also (1+). Calcium ions are (2+), so it will take 2 hydrogen ions to equal one calcium. Magnesium is also (2+). Iron, Fe, can be either (2+) or (3+), but here on the clay is it usually (3+), which makes it cling more tightly to the clay. It will take 3 hydrogen ions to release an iron, Fe. Once the mineral ion has fallen off the clay particle, the root hair cell can bring in inside. Bringing something inside is called **absorption**. When something is **absorbed**, it is taken inside. (To learn more about both "sorptions" watch the sorption video on the YouTube playlist.)

What would happen if a soil did not contain clay particles? What would catch and hold the mineral ions so that plants could get them? Soils that don't have any clay tend to be low in minerals and are, therefore, not so good for growing crops. When it rains, minerals wash right out. This is called **leaching**. Clay particles help to prevent leaching.

Rainforests provide us with an example of soils that don't really contain any clay. Nutrients are recycled right at the surface by fungi, bacteria and worms. Recycled nutrients are called **humus**, a substance we'll meet on the next page. Humus can function in a similar way to clay particles, and hold mineral ions, but the minerals don't get deep down into the soil.

Did you hear the word "worms"? It's about time!

So now we are down to that 5% that is, or was, alive. Organic matter includes decayed plants and animals as well as living ones. Let's look at the decayed stuff first. We'll focus what happens to a dead leaf.

When a leaf falls off a tree and drops to the ground, it is already covered with bacteria and mold spores. While the leaf was alive, it was actively fighting against these microorganisms. Now that the leaf is dead, the bacteria and fungi (molds) go to work digesting the leaf. Of course, bacteria and fungi don't have stomachs, but they do make digestive chemicals similar to the ones found in our own stomachs. The bacteria and fungi release digestive chemicals called **enzymes**. The enzymes act like little scissors and they cut apart the leaf's cells. With millions of bacteria working on the leaf, it is soon reduced to brown dirt.

External digestion seems very strange to us. If humans could do external digestion, we'd have digestive chemicals oozing out of our skin. We'd be able to digest our food simply by touching it! The food would turn into a soup of nutritious chemicals that we could absorb through our skin, right into our bloodstream. Now, think of how much more efficient it would be if we sat in a bathtub of water, with our food floating around us, and let our digestive chemicals fill the water. We would be able to digest a lot of food all at once. Bacteria and fungi like to have water around them, too. If it is too dry, they have trouble digesting. Water is an essential part of their digestion processes. Dry leaves will not decompose. They will break into little pieces, but those pieces won't decompose.

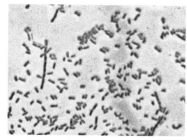

soil bacteria under a microscope

During decomposition, bacteria can release gases that smell unpleasant. Gases are given off in many chemical reactions. We saw carbon dioxide being released from limestone in the first part of the limestone cycle. Carbon dioxide is a gas that does not have an odor. Many gases are odorless, but some smell pretty bad. Decomposition processes often involve the release of gases like hydrogen sulfide, H_2S, which smells like rotten eggs. If animal remains are being digested, nasty-smelling liquid acids are also created as part of the break-down process. While bacteria are actively digesting, bad smells are to be expected. However, when they are done, the smells stop. No digestion, no smells.

When the microorganisms have finished, and nothing is left but molecules, we call it **humus**. *(HUE-miss)*. The word "humus" is Latin for "dirt." (Don't confuse the word "hu_mus_" with the word "hu_mm_us" *(HUM-iss)*. Hummus is a food made from beans and spices.) Humus does not smell bad. If you've ever picked up a handful of garden soil and smelled it, you know that soil doesn't smell bad. In fact, many people say that soil smells good. Humus contributes to the sweet smell of a forest in mid-summer.

humus   hummus

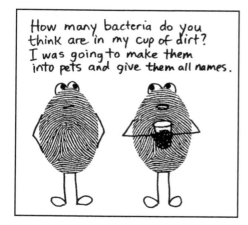

How many bacteria do you think are in my cup of dirt? I was going to make them into pets and give them all names.

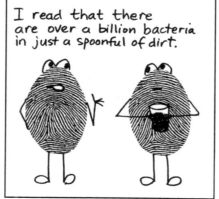

I read that there are over a billion bacteria in just a spoonful of dirt.

That's a lot of names!

Yeah- maybe I'll just name the worm Wiggly.

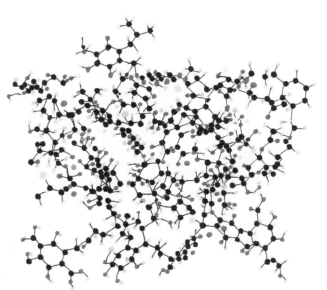

Exactly what is humus? Even soil scientists have a hard time defining it and describing it. This picture shows what soil researchers think a molecule of humus might look like. The black dots represent carbon atoms, the reds are oxygens, the blues are hydrogens, and the other colors are various other atoms. Molecules this large are called macromolecules.

A humus macromolecule is like a clump of broken molecules all tangled together. Some of the broken molecules are things you are familiar with, such as sugars, fats, proteins and starches. Others are less familiar, such as cellulose and lignin from plants cells. These broken bits are "leftovers" from the digestive process of bacteria and fungi. It's like taking the leftovers in a cafeteria and scraping them all into one big pot. You would have bits of vegetable, fruit, bread, meat, and dessert all mixed together. How would you describe this mix? It's not a salad or a main dish or a dessert. It's a... bunch of stuff. That's what humus is—a bunch of stuff—only at a molecular level. Chemists can look at humus and say, "I see some carbon chains that look like

This might be a large molecule, but it is far too small to see, even under a microscope. Visible humus is made of billions of these.

they used to be part of a starch molecule," or "That carbon ring looks like it came from a plant hormone molecule." But here they are all jumbled together, like scraps in a garbage can. Every humus molecule is different, being a random assortment of whatever the bacteria and fungi didn't eat.

We know for sure that humus is produced by organisms in the soil that decompose dead plants, but we don't know exactly how they do it. In fact, there's a lot we don't know about humus. What we *do* know about humus is that even though plants don't use it directly, it still benefits them greatly. Soils high in humus are likely to produce healthy and productive plants. Here are some ways that humus helps plants:

1) Humus acts like a sponge, helping the soil to not dry out so fast. Plants need a lot of water. Lack of water is the number one thing that can slow down plant growth. During times of very low rainfall, the quality of the soil is vitally important. If there is a lot of humus in the soil, the soil will be much better at holding on to water. Humus can hold 7 times its own weight in water!

2) Humus can hold on to mineral ions the same way that clay particles can. Some of the atoms that are hanging off the edges of the humus molecule have a negative charge and can attract positive ions such as sodium, magnesium, potassium and calcium. Just like clay particles do, the humus molecules hold these ions tightly enough so that they don't get washed away, but loosely enough so that root cells will be able to pull them off using that "ion exchange" process (trading H^+ ions for mineral ions).

3) Humus is still edible, and can be used as an energy source by bacteria, fungi, and worms. There is energy locked up in carbon-to-carbon bonds. Humus has a lot of carbon atoms bonded to other carbon atoms—just look at all those black dots. Most cells have "machinery" that can break these bonds and release the energy. (We have to say "most" because there are some really weird types of bacteria that get their energy from sulfur and iron.) **Bacteria, fungi and worms perform useful functions for plants.** Some bacteria can put nitrogen back into the soil. Fungi produce chemicals that plants can take in and use as pesticides (bug killers). Both bacteria and fungi produce sticky sugar molecules that help the soil to form the crumb structure we learned about on page 65. This type of soil structure is the very best for plants. Those crumbs provide the perfect amount of water and air. Worms also help soil to have good structure. Their burrows provide channels for rain water. Worms also eat dead plants on the surface of the soil, bringing these nutrients down into the deeper layers.

4) Very small humus molecules can help seeds to germinate (begin to grow). Certain types of humus molecules, called humic acids, seem to be very helpful to seeds. The humic acid molecules go into the seeds and cause them to sprout, but exactly what is happening is still a mystery.

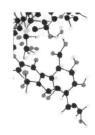

5) **Humus can take toxins out of the soil**. Many of the atoms on the edges of the humus molecule are able to "grab" molecules that are passing by, and form permanent bonds with them. Carbon atoms love to bond with each other to form chains or rings. (Notice the carbon rings in the humus molecule.) Toxins that contain carbon atoms can often be permanently added to the humus molecule. The humus molecule is already a hodge-podge of miscellaneous carbon-based molecules, so it's not a big deal to add one or two more! Once the harmful molecules are added to the humus molecule, they are no longer harmful. Thus, humus helps to keep the environment clean.

Humus molecules mix together not only with other humus molecules, but with clay and minerals, too. Soil is a mixture of humus, clay, sand, silt and loose mineral ions. Separating out the humus is extremely difficult. Often, harsh chemicals must be used, causing some soil scientists to question the results of tests on extracted humus. Can we be sure that the strong chemicals did not change the humus in some way while it was being extracted? No, we can't. Thus, soil scientists can have different theories on how humus is formed.

Organic material, both living and decomposed, makes up only 5% of the soil. Most of that 5% is humus. Less than 1% is living things that we can see. However, this is part of soil that many of us find most interesting. Here are some of the most common creatures you will find living in soil.

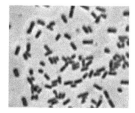

BACTERIA
Rod-shaped bacteria of many species, most of them harmless to us.

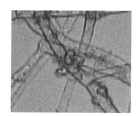

FUNGI
Microscopic fungi most often look like jointed sticks.

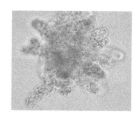

PROTOZOANS
This is an amoeba. Only certain types are found in soils.

SNAILS
Soil snails can be very small. You really have to look for them.

SLUGS
Slugs hide in the soil all day and come out at night to eat plants.

NEMATODES
These are also called roundwords. Most are microscopic.

EARTHWORMS
There are a few different types, but all of them are beneficial to soil.

CENTIPEDES
Centipedes are predators. Don't confuse them with millipedes.

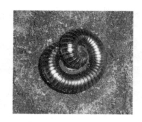

MILLIPEDES
These are often found coiled up in a circle. They eat dead plants.

WOODLICE
Also called the pill bug, sow bug, potato bug, or roly poly.

ANTS
Some species dig deep tunnels underground.

MITES
Mites have 8 legs, like spiders. All are tiny, some are microscopic.

SPIDERS
Not all spiders spin webs. Some live in soil. They are predators and have fangs.

SPRINGTAILS
Have 6 legs and used to be called insects but now have their own group.

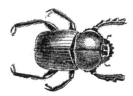

BEETLES
Beetles are insects and have 6 legs. Some beetles love soil (and even dung).

Enough bugs, but we need to go back to bacteria for a minute and see another way they benefit soil and plants. Plants use a lot of nitrogen. That's the nutrient that plants are most likely to run out of. That's surprising, if you consider that 80% of air is nitrogen. If soil is 25% air, then there should be plenty of nitrogen in it. The problem is that plants can't use nitrogen as it is found in the air, as a molecule of N_2. Plants must get their nitrogen from either ammonium, (NH_4^+), or nitrate (NO_3^-). Bacteria are required to convert N_2 to these other forms. Follow along on the chart below.

The black rectangle in the middle shows that bacteria and fungi decompose plants and animals and turn them into humus. One of result of this process (red arrow number 1) is the production of ammonium, NH_4^+. That's fortunate for the plants because that is one of the forms of nitrogen that they can use. More bacterial action can turn the ammonium into nitrite (arrow 2), and then nitrite gets turned into nitrate (arrow 3). That's good, too, because nitrate is the other form of nitrogen that plants can use. Follow that arrow up to number 6. That's where the plant roots will take in the nitrates. Now go back to arrow 4. A different type of bacteria will take some of those nitrates and turn them back into N_2, which will go into the air. At arrow 7, we see plants dying and being decomposed by bacteria and fungi. We are back where we started.

At arrow 8 we see a rabbit who has been eating the plants. The rabbit's body now contains nitrogen. First, the rabbit might urinate and put urea, which contains nitrogen, into the soil. Then the rabbit will die and be decomposed by bacteria. We're back at the decomposers again. Now look at arrows 9 and 10. Those arrows started out up at the top, coming from the atmospheric nitrogen. Air gets down into the soil and is picked by "nitrogen-fixing" bacteria, who convert the N_2 into ammonium, NH_4^+. And now we're back at number 1 again.

THE NITROGEN CYCLE

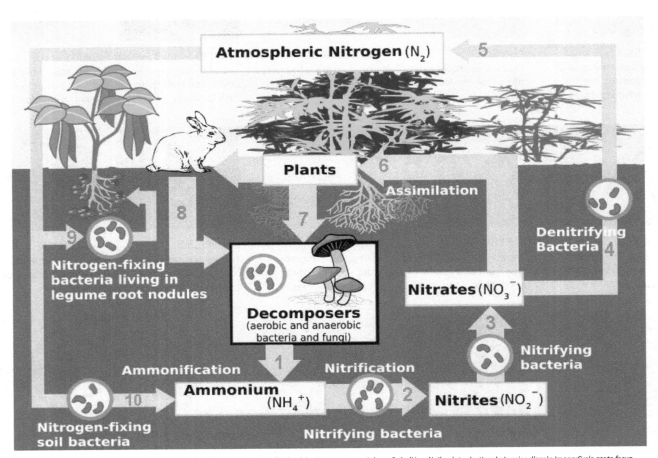

Arrows have been added by E. McHenry. Original file by Cicle_del_nitrogen_de.svg: *Cicle_del_nitrogen_ca.svg: Johann Dréo (User:Nojhan), traduction de Joanjoc d'après Image:Cycle azote fr.svg. derivative work: Burkhard (talk)Nitrogen_Cycle.jpg: Environmental Protection Agencyderivative work: Raeky (talk) - Cicle_del_nitrogen_de.svgNitrogen_Cycle.jpg, CC BY-SA 3.0, https://commons.wikimedia. org/w/index.php?curid=7905386

What the nitrogen cycle chart did not show is that some plant roots have a special adaption to help out the nitrogen-fixing bacteria. Some bacteria can live by themselves in the soil, but a lot of them need help from plants called **legumes**. Legumes include beans, peas, soybeans, peanuts, lentils, clover and alfalfa. These plants love to have nitrogen-fixing bacteria living right in their roots. Legumes produce seeds that are high in protein, which contains nitrogen. To assure a constart supply of nitrogen, they cooperate with the bacteria. Those bumps on the roots are called **nodules** and they're sort of like houses for the bacteria. The bacteria get a nice place to live, and in return they produce nitrogen for the legumes. Fortunately, there is nitrogen to spare and some of it will go back into the soil. Also, if you use legumes as cover crops, you can plow them back into the soil and return the nitrogen that way.

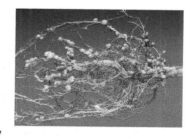

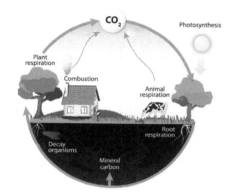

Carbon, also, has a cycle. Remember back in the limestone chapter we saw an estimate of how much carbon is contained in the atmosphere, oceans, coal, plants and animals, and limestone. The immense amount of carbon locked up in limestone isn't going anywhere anytime soon. The carbon in other areas, however, moves along quite frequently. The carbon atoms in CO_2 are taken in by plants and turned into sugars and starches. Animals eat these sugars and starches and exhale them back into the air as CO_2. Then the animals eventually die and the carbon goes into the soil. The carbon in the soil might get washed into the ocean by rain water, or it might go back into the air as CO_2 produced by the process of bacterial decomposition. Occasionally, something really interesting might happen to the carbon. Carbon in trees can end up in furniture, or as firewood. Charcoal (from burned logs) was once used to make things like ink and soap.

Soil does so many things, as this poster shows. We need to take care of our soils and preserve them. Soils take a long time to form, but can be washed or blown away almost overnight. One of the most famous soil disasters in history was called the Dust Bowl. It took place in the western U.S. in the 1930s, the decade of the Great Depression.

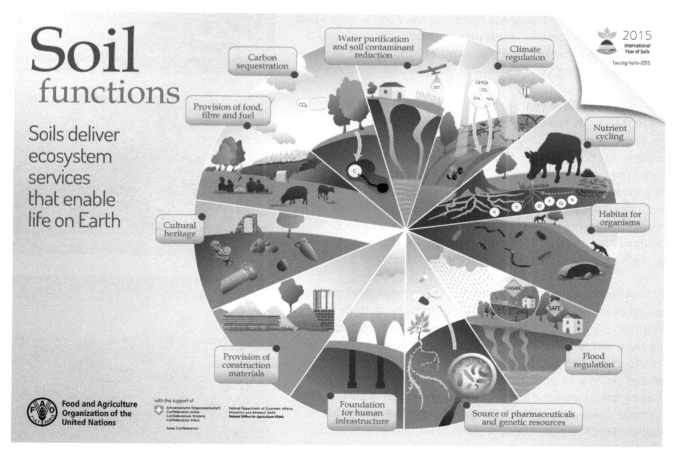

THE DUST BOWL: A painful lesson in the importance of soil conservation

The 1920s were very good years for farming in the American states of Kansas, Oklahoma, Texas, Colorado and New Mexico. This decade was unusually wet. Rain poured down on the crops and they grew like crazy. Farmers knew that this area had historically been dry, but they thought that the climate had changed for the better and assumed it would stay this way. The size of farms increased from 160 acres to 640 acres. The natural deep-rooted grasses that had grown on these prairies for centuries were uprooted and the land was plowed. During the winters, the fields lay empty, without any cover crops. This was fine as long as the weather cooperated. However, the climate reverted back to its dry old self during the 1930s.

1934 was perhaps the worst year. It hardly rained at all. Wind storms would sweep across the plains and pick up all the dirt that the farmers had plowed up, creating huge dust storms. People who remember these dust storms say the dust was so thick that you could only see about a meter (three feet) ahead of you. The storms would be so severe that the dirt would blow not just across the state, but across the country. The dust clouds blew all the way up to Chicago, where they dumped 12 million pounds of dust. Cities as far away as Washington, D.C. were covered with dust. Winter dust storms made red snow in New York and Boston.

The worst storm of the decade occurred on April 14, 1935. This day became known as Black Sunday because the dust clouds were so thick that they blocked out the sun. This storm was also larger than any other and stretched from Canada to Texas. Rich topsoil that had taken hundreds of years to form was stripped off a large portion of the Great Plains area and scattered all over the northern hemisphere. Hundreds of homes were so damaged that they had to be torn down. Thousands of people decided they'd had enough. They packed their bags and headed further west, many of them ending up in California.

DUST STORM APPROACHING SPEARMAN, TEXAS. APRIL 14. 1935

The Dust Bowl years brought much suffering to the people of these states. By the end of the decade, 75% of the topsoil was gone. Without crops to sell, farming families went hungry. And to make matters worse, the 1930s also happened to be the era we call "The Great Depression." The entire country was experiencing great economic hardship. California continued to receive thousands of people fleeing the Dust Bowl. Not all of these were farmers; many were owners of small businesses that had gone bankrupt. About 1/8 of California's population today are the descendants of these immigrants from the Dust Bowl.

After the drought was over, the U.S. government put vast amounts of money and human resources into trying to fix the problem and making sure it never happened again. 200 million trees were planted from the Canadian border down to Texas, to try to stop the wind and to hold water in the soil. Farmers were educated on how to prevent soil erosion by planting cover crops and using better plowing and planting techniques. Today, these areas have deep wells drilled into them, providing water during droughts. Fertilizers are used, in place of natural soil minerals.

Activity 6.1 Comprehension self-check

Questions about water, air and minerals in soil:

1) "Dirt" is soil that is: a) very dark in color   b) out of place   c) wet   d) good at growing plants

2) How deep is a typical soil profile? a) a few inches   b) 3 to 6 feet (a meter or two)   c) 50 to 100 ft (20-30 m)

3) Which horizon contains the most organic matter? a) A   b) B   c) C   d) O   e) E

4) Which horizon is usually more red?   a) A   b) B   c) C   d) O   e) E

5) Which horizon is right above the bedrock?   a) A   b) B   c) C   d) O   e) E

6) Which horizon is called the subsoil?   a) A   b) B   c) C   d) O   e) E

7) Which horizon is a place where minerals have washed out, leaving it lighter in color? a) A   b) B   c) C   d) O   e) E

8) Which horizon is called topsoil?   a) A   b) B   c) C   d) O   e) E

9) Which horizon is least likely to have roots? a) A   b) B   c) C   d) O   e) E

10) Which horizon is a good place to look for grasshoppers and crickets? a) A   b) B   c) C   d) O   e) E

11) True or false? Good soil doesn't have any air in it.

12) True or false? The word "loam" used to mean "clay."

13) Look at the texture triangle. If a soil is 20% clay, 20% sand and 60% silt, it is called a _____.

14) If a soil is 30% clay, 30% sand and 40% silt, it is a _____.

15) If a soil is 20% sand and 80% clay, it is a _____.

16) True or false? Sandy soils have many large pores. ___

17) True or false? Clay soils have few large pores. ___

18) True or false? Plants are okay as long as there is still water in the smallest pores. ___

19) True or false? The word "aggregate" is only used to describe soil structure. ___

20) Which one of these is NOT a type of soil structure? a) crumbs   b) blocks   c) columns   d) plates   e) tubes

21) Which of these has the LEAST effect on the structure of a soil?   a) what season of the year it is

   b) amount of moisture   c) presence of worms and roots   d) plowing and tilling

22) True or false? In gardening, there's no such thing as too much water. ___

23) If you stop plowing your fields, you'd better: a) plant a cover crop   b) put on extra weed killer

   c) hope the rabbits don't move in   d) do all your weeding by hand

24) If a soil was made from parent rock that contained a lot of limestone, it should be high in:

   a) iron   b) sulfur   c) calcium   d) potassium

25) When gravity works together with rain to flush minerals out of the soil, we call this _____.

26) The lack of what element causes thyroid problems in humans? a) nitrogen   b) chlorine   c) iodine   d) iron

27) When dirt freezes and thaws and freezes and thaws, this can slowly push rocks to the surface. This is called:

   a) heave   b) erosion   c) weathering   d) adsorption   e) gravity

28) True or false? Plants use up minerals in the soil.

29) When was Surtsey Island born?   a) 1893   b) 1903   c) 1963   d) 1993

30) Which one of these is NOT found on Surtsey? a) olivine   b) basalt   c) scoria   d) gypsum   e) tuff   f) granite

31) True or false? Scientists were surprised at how slowly soil was forming on Surtsey. ___

32) What is the electrical charge of clay? a) positive   b) negative   c) neutral

33) What does a root cell make for the purpose of ion exchange? _____

34) What is it called when a molecule sticks to the outside of something?   a) adsorption   b) absorption

35) Where do you find soils that contain little or no clay? _____

36) Which of these is NOT considered to be a macronutrient? a) nitrogen   b) phosphorus   c) potassium d) iron

37) Particles of clay are most like: a) spheres   b) plates   c) blocks   d) ropes

Questions about the organic part of the soil:

38) True or false? Bacteria and fungi make digestive chemicals similar to the ones humans make. _____

39) How important is water to decomposing bacteria? a) not at all   b) a little   c) essential

40) What gas is given off when limestone dissolves? a) O_2 b) H_2S c) CO_2 d) N_2 e) H_2

41) What gas smells like rotten eggs? a) O_2 b) H_2S c) CO_2 d) N_2 e) H_2

42) True or false? Rotting things never stop smelling bad. _____

43) True or false? Humus stinks. _____

44) True or false? There is an exact chemical formula for humus. _____

45) True or false? Humus can hold on to mineral ions the same way clay particles can. _____

46) What kitchen item does humus act like? a) knife   b) sponge   c) soap   d) blender

47) Which one of these would NOT eat humus? a) rabbit   b) bacteria   c) fungi   d) worms

48) Where can you find energy in the humus molecule?   a) in the mineral ions that hang on to the edges
    b) between the carbon atoms   c) in the oxygen atoms   d) nowhere

49) What do bacteria and fungi produce that helps soil stick together and form round crumbs?
    a) mineral ions   b) humic acid   c) proteins   d) sugars

50) True or false? If worms would stop burrowing, the soil quality would improve. ___

51) Fungi produce chemicals that plants can use as: a) fertilizer   b) pesticide   c) vitamins   d) food source

52) True or false? Humus is of no benefit to seeds, only to fully grown plants. _____

53) True or false? All soil scientists agree about how humus is formed. _____

54) What percentage of the soil is made of organic things? _____

55) What percentage of the soil is made of minerals? _____

56) Which one of these can be seen without a microscope? a) protozoans   b) nematodes   c) springtails

57) Which two soil creatures have 8 legs? _____ and _____

58) Which soil creatures have more than 8 legs? _____, _____, and _____

59) Which one of these is a predator? a) slug   b) millipede   c) centipede

60) If you had to do a science project on one of these creatures, which would you choose? _____

61) Which one of these is "ammonium"? a) H_2S b) NO_3^- c) NH_4^+ d) NO_2^- e) CH_4

62) Which of these can bacteria NOT do?   a) make ammonium   b) turn ammonium into nitrate
    c) take nitrogen out of the soil   d) put nitrogen back into the air   e) turn oxygen into nitrogen

63) Where do nitrogen-fixing bacteria like to live? a) roots   b) leaves   c) atmosphere   d) rotten logs

64) Which one of these is NOT a legume? a) peas   b) beans   c) corn   d) lentils   e) peanuts   f) clover

65) Which one of these is NOT a part of the carbon cycle? a) ocean   b) granite   c) atmosphere   d) plants

66) Which one of these needs to take in CO_2? a) plants   b) animals   c) neither

67) What was the weather like in the western states during the 1920s? _____

68) Which of these caused the Dust Bowl disaster? a) the weather   b) increase in farm size
    c) the uprooting of all the natural grasses   d) fields lying bare during the winter months   e) all of these

69) Many people left these states during the 1930s. Where did most of them go? _____

70) What was done to try to fix the Dust Bowl problem? _____

Activity 6.2 Videos

The "Rocks and Dirt" playlist has lots of video about soils. You can learn from soil scientists as they take you out in the fields to see soil profiles, textures and structures. Also, there are videos on the nitrogen and carbon cycles, microorganisms in the soil, and the Dust Bowl.

Activity 6.3 Soil crumb word search

Wow, what beautiful soil crumbs! It's the weirdest soil in the world, though, because each crumb has a letter embedded in it. The clues below give you hints about words you can spell out with these letters. The letters of each word are connected, crumb to crumb, and generally go towards the right or towards the bottom, but some of the words do curve around a bit. The first one is done for you. (TIP: Use a colored pencil to mark the words.)

CLUES:

1) When soil gets in the wrong place, we call it: _____. (pg 61)
2) If a soil contains a lot of calcium, the parent rock is probably this rock: _____. (Figure it out!)
3) The O, A, E, B, and C horizons make up a soil's _____. (pg 62)
4) Soil that contains sand, silt and clay is called a _____. (pg 63)
5) The ability of soil to hold water in tiny pores is called its _____. (pg64)
6) A very short word for aggregate is _____. (pg 65)
7) In no-till farming, a farmer will plant a _____ crop at the beginning of winter. (pg 65)
8) When rocks work their way up to the surface with the help of freezing and thawing, this is called _____. (pg 66)
9) When minerals wash out of soil, this is called _____. (pg 66
10) Surtsey Island is off the coast of what island nation? _____ (pg 67)
11) What colorful plant-like organism lives on rocks and gradually dissolves them? _____ (pg 67)
12) The three most essential elements to plants are nitrogen, phosphorus and _____. (pg 69)
13) Plant roots make _____ ions so that they can exchange them for minerals hanging on clay particles. (pg 70)
14) When an ion sticks to the outside of something, this is called _____. (pg 70)
15) Digestive chemicals are called _____. (pg 71)
16) Bacteria need this in order to do their external digesting: _____ (pg 71)
17) This is a food that is easy to confuse with humus: _____ (pg 71)
18) A very large molecule can be called a _____. (pg 72)
19) This element forms the skeleton of all organic molecules. _____ (pg 72)
20) Some types of bacteria can put this element back into the soil. _____ (pg 74)
21) Humic acid can help seeds to _____. (pg 72)
22) A microscopic roundworm: _____ (pg 73)
23) This soil creature is nocturnal (comes out at night). _____ (pg 73)
24) NH_4^+ is _____. (pg 74)
25) Many people migrated to this state during the Dust Bowl decade. _____ (pg 76)

79

Activity 6.4 FOCUS ON FUNGI (We really haven't given them their due!)

Fungi (the mushroom family) are a very important part of most soils, especially in forests. They hold soil together and prevent erosion, they decompose wood and leaves and turn it into humus, they eat away at rocks (thereby helping to create new soil), and they can get rid of toxins that come into the soil.

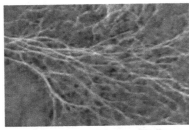

Fungi have a very simple anatomy. The main part of their "body" is called the **mycelium** *(mi-SEEL-ee-um)*. The mycelium looks like a bunch of white roots. But you can't call them roots because fungi are not plants. The individual fibers (the stringy things) of the mycelium are called **hyphae** *(HI-fi, or HI-fee)*. Just one is a hypha *(HI-fuh)*. The mycelium stays under the soil all the time and grows very quickly. When it is time to reproduce, the hyphae near the surface put up **fruiting bodies**. We are familiar with fruiting bodies because mushrooms are a common type of fruiting body. The fruiting bodies produce **spores** that act like seeds. Wherever a spore lands, it can begin growing into a new mycelium. You've probably seen spores on the underside of a mushroom cap. (The actual spores themselves are microscopic, but we can see the structures that contain them.)

These fruiting bodies grow on trees.   *These are microscopic fruiting bodies.*   *Puff balls are a type of mushroom.*

A mycelium can live for a long time and become very large. The largest known mycelium is in the U.S. state of Oregon. It covers an area of 2,400 acres. This is the largest living organism on earth! Scientists guess that it might have taken 2,000 years to get this large.

A mycelium can also be very small, even microscopic. The picture here on the right shows a microscopic view of a mycelium that is smaller than a pin head.

Fungi are amazing. Not only do they hold the soil together, they can do lots of other things as well.

1) Fungi seem to cooperate with plants, especially trees, providing them with nutrients that they need. They grow around all the tiny root hairs, making a cozy environment for helpful soil bacteria to live in. Crops grow better when certain species of fungi are present in the soil.

2) Fungi seem to be able to help plants communicate. The mycelium is made of living cells so it responds to changes in the environment. Trees roots may be able to pick up on changes in a mycelium and somehow respond. Perhaps a mycelium is like an underground "Internet" for plants?

3) Fungi have been found to be one of the best cures for oil spills. Fungi use their external digestion to cut apart those carbon-to-carbon bonds in oil molecules and harvest them for energy. That basically means they can "eat" oil! They can turn oil-soaked soil back into a healthy underground community of life.

4) Fungi produce chemicals that are very useful as medicines for people. The first antibiotic, penicillin, came from a mold, and since then hundreds of other medicines have been made from various species of fungi.

5) Mycelia can be shaped and baked as an "eco-friendly" alternative to disposable plastics.

6) Fungi might even be used to make an alternative fuel, better than the plant-based ethanol that is being made from corn right now.

Activity 6.5   FUNGI WORD PUZZLE

We recognize fungi by their fruiting bodies. Remember, the main part of the fungus is that mycelium, under the ground. Most of the time the mycelium isn't fruiting. It would be very difficult to positively identify a mycelium, so we really rely on those fruiting bodies for identifying them. Learn the names of some well-known fruiting bodies by filling in the letters after you have solved the clues below.

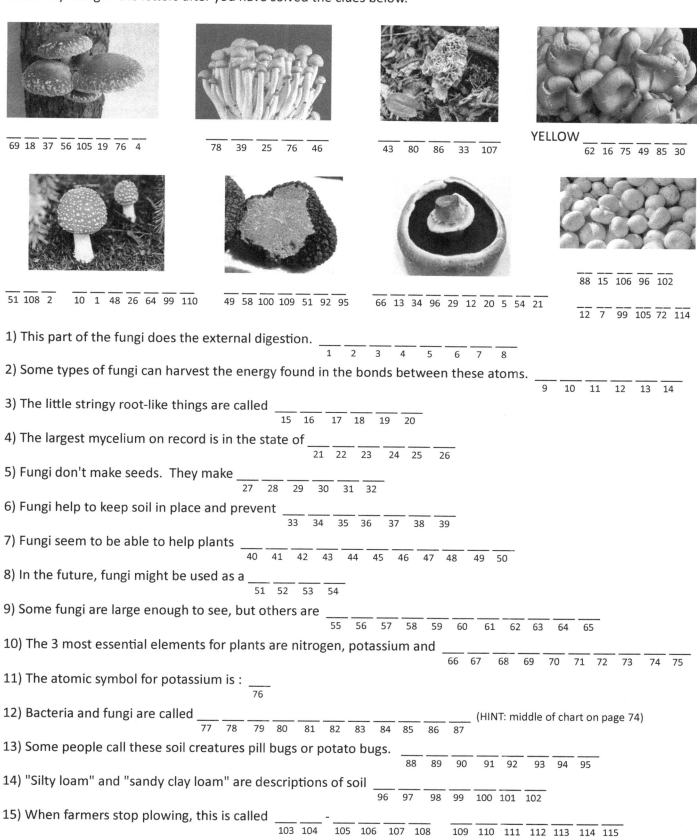

___ ___ ___ ___ ___ ___ ___ ___
69 18 37 56 105 19 76 4

___ ___ ___ ___ ___
78 39 25 76 46

___ ___ ___ ___ ___
43 80 86 33 107

YELLOW ___ ___ ___ ___ ___ ___
62 16 75 49 85 30

___ ___ ___
51 108 2

___ ___ ___ ___ ___ ___ ___
10 1 48 26 64 99 110

___ ___ ___ ___ ___ ___
49 58 100 109 51 92 95

___ ___ ___ ___ ___ ___ ___ ___ ___ ___
66 13 34 96 29 12 20 5 54 21

___ ___ ___ ___ ___
88 15 106 96 102

___ ___ ___ ___ ___ ___
12 7 99 105 72 114

1) This part of the fungi does the external digestion. ___ ___ ___ ___ ___ ___ ___ ___
1   2   3   4   5   6   7   8

2) Some types of fungi can harvest the energy found in the bonds between these atoms. ___ ___ ___ ___ ___ ___
9   10 11 12 13 14

3) The little stringy root-like things are called ___ ___ ___ ___ ___ ___
15 16 17 18 19 20

4) The largest mycelium on record is in the state of ___ ___ ___ ___ ___ ___
21 22 23 24 25 26

5) Fungi don't make seeds. They make ___ ___ ___ ___ ___ ___
27 28 29 30 31 32

6) Fungi help to keep soil in place and prevent ___ ___ ___ ___ ___ ___ ___
33 34 35 36 37 38 39

7) Fungi seem to be able to help plants ___ ___ ___ ___ ___ ___ ___ ___ ___ ___ ___
40 41 42 43 44 45 46 47 48 49 50

8) In the future, fungi might be used as a ___ ___ ___ ___
51 52 53 54

9) Some fungi are large enough to see, but others are ___ ___ ___ ___ ___ ___ ___ ___ ___ ___ ___
55 56 57 58 59 60 61 62 63 64 65

10) The 3 most essential elements for plants are nitrogen, potassium and ___ ___ ___ ___ ___ ___ ___ ___ ___ ___
66 67 68 69 70 71 72 73 74 75

11) The atomic symbol for potassium is : ___
76

12) Bacteria and fungi are called ___ ___ ___ ___ ___ ___ ___ ___ ___ ___ ___ (HINT: middle of chart on page 74)
77 78 79 80 81 82 83 84 85 86 87

13) Some people call these soil creatures pill bugs or potato bugs. ___ ___ ___ ___ ___ ___ ___ ___
88 89 90 91 92 93 94 95

14) "Silty loam" and "sandy clay loam" are descriptions of soil ___ ___ ___ ___ ___ ___ ___
96 97 98 99 100 101 102

15) When farmers stop plowing, this is called ___ ___ - ___ ___ ___ ___ ___ ___ ___
103 104   105 106 107 108   109 110 111 112 113 114 115

81

Activity 6.6 Learn about a very strange dirt: diatomaceous earth

There are hundreds of different types of microscopic cells that make hard shells. Coccolithophores are actually one of the least well-known of this group; most people have never heard of them. Probably the best known of this group are the **diatoms**.

Diatoms are made of a single cell and they use sunlight for photosynthesis. They are extremely abundant in both the ocean and in fresh waters, and they are responsible for putting more oxygen into the air than all plants on earth combined. Diatoms are sort of like the grass and weeds of the ocean. They are at the bottom of the ocean food chain, and without them, every ocean creature would starve. Microscopic animals eat diatoms, then the microscopic animals get eaten by larger animals, which get eaten by larger animals, until you get up to fish, sharks, and whales.

Diatoms are microscopic, so you can't see them without a microscope. One drop of ocean water has thousands, or even millions, of diatoms in it. Unlike coccolithophores, they don't take calcium and carbon out of the water. Instead, they take silicon and oxygen out of the water and make shells out of SiO_2. Yes, these guys live in glass houses! (Or maybe quartz houses.) Their shells are clear, so the sunlight pours right in. Since they need sunlight, they have to float near the top. They have various ways to avoid sinking, but eventually that's what happens—their silicate shells fall to the bottom. When a whole lot of them collect they form a thick layer called **silaceous** (sill-ay-shus) **ooze.**

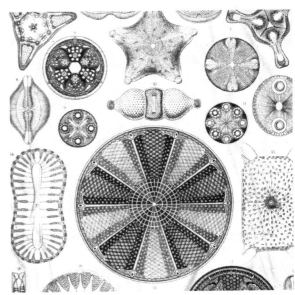

Diatoms come in weird and wonderful shapes. Their shells have tiny holes. Some diatoms can move around, but they are slow.

At some time in the past, certain areas of the world seem to have had very large numbers of diatoms that got buried and turned into soft sedimentary rock similar to chalk. (In some places, the deposits can be up to several hundred meters thick.) However, it isn't chalk because it is made of SiO_2, not $CaCO_3$. This soft, white rock is called **diatomaceous earth.** *(di-ah-tom-AY-shus),* often abbreviated as DE.

DE is mined, like chalk is, and sold as a white powder. Commercial industries use it for things like stabilizing dynamite, filtering food products like wines and syrups, filtering water, polishing metal, as insulation, as a abrasive in toothpastes, and as a clean alternative to dirt in hydroponic greenhouses.

The photo on the right shows how DE looks under a microscope. On a microscopic level, it isn't a white power; it's more like broken glass. These sharp bits of glass are not harmful to humans or animals, but they are lethal to insects. The sharp pieces cut them, as well as absorbing all their protective natural oils so that they die of dehydration (being too dry). DE also works on parasitic worms in animals. The worms feed on whatever the animal eats. If diatomaceous earth comes its way, the worm eats it. Because the worms are small, the sharp diatoms cut their intestines to pieces.

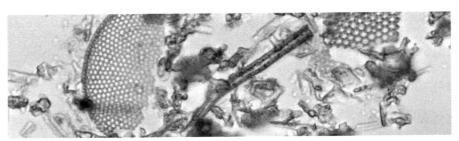

Activity 6.7   TIME TO REVIEW!

How much can you still remember from past chapters? If you have trouble with this activity, you are welcome to take a peek in those previous chapters and find the answers.

PART 1: Who am I? (minerals)

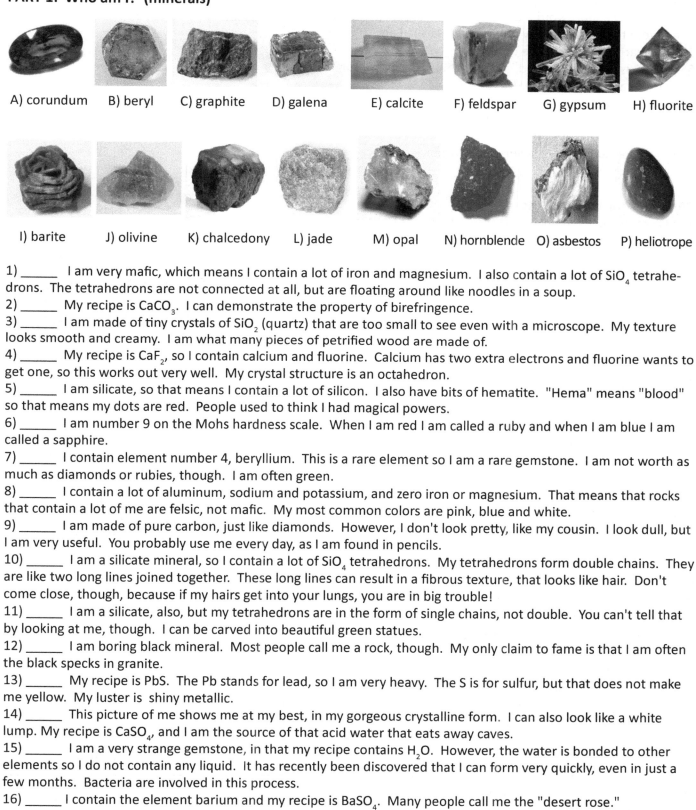

A) corundum   B) beryl   C) graphite   D) galena   E) calcite   F) feldspar   G) gypsum   H) fluorite

I) barite   J) olivine   K) chalcedony   L) jade   M) opal   N) hornblende   O) asbestos   P) heliotrope

1) _____ I am very mafic, which means I contain a lot of iron and magnesium. I also contain a lot of SiO_4 tetrahedrons. The tetrahedrons are not connected at all, but are floating around like noodles in a soup.

2) _____ My recipe is $CaCO_3$. I can demonstrate the property of birefringence.

3) _____ I am made of tiny crystals of SiO_2 (quartz) that are too small to see even with a microscope. My texture looks smooth and creamy. I am what many pieces of petrified wood are made of.

4) _____ My recipe is CaF_2, so I contain calcium and fluorine. Calcium has two extra electrons and fluorine wants to get one, so this works out very well. My crystal structure is an octahedron.

5) _____ I am silicate, so that means I contain a lot of silicon. I also have bits of hematite. "Hema" means "blood" so that means my dots are red. People used to think I had magical powers.

6) _____ I am number 9 on the Mohs hardness scale. When I am red I am called a ruby and when I am blue I am called a sapphire.

7) _____ I contain element number 4, beryllium. This is a rare element so I am a rare gemstone. I am not worth as much as diamonds or rubies, though. I am often green.

8) _____ I contain a lot of aluminum, sodium and potassium, and zero iron or magnesium. That means that rocks that contain a lot of me are felsic, not mafic. My most common colors are pink, blue and white.

9) _____ I am made of pure carbon, just like diamonds. However, I don't look pretty, like my cousin. I look dull, but I am very useful. You probably use me every day, as I am found in pencils.

10) _____ I am a silicate mineral, so I contain a lot of SiO_4 tetrahedrons. My tetrahedrons form double chains. They are like two long lines joined together. These long lines can result in a fibrous texture, that looks like hair. Don't come close, though, because if my hairs get into your lungs, you are in big trouble!

11) _____ I am a silicate, also, but my tetrahedrons are in the form of single chains, not double. You can't tell that by looking at me, though. I can be carved into beautiful green statues.

12) _____ I am boring black mineral. Most people call me a rock, though. My only claim to fame is that I am often the black specks in granite.

13) _____ My recipe is PbS. The Pb stands for lead, so I am very heavy. The S is for sulfur, but that does not make me yellow. My luster is shiny metallic.

14) _____ This picture of me shows me at my best, in my gorgeous crystalline form. I can also look like a white lump. My recipe is $CaSO_4$, and I am the source of that acid water that eats away caves.

15) _____ I am a very strange gemstone, in that my recipe contains H_2O. However, the water is bonded to other elements so I do not contain any liquid. It has recently been discovered that I can form very quickly, even in just a few months. Bacteria are involved in this process.

16) _____ I contain the element barium and my recipe is $BaSO_4$. Many people call me the "desert rose."

PART 2: Three-letter words
The answer to each of these questions is a three-letter word.

1) Alchemists found that if they heated certain rocks, metals would come out.
   The rock that contains a certain mineral is called its ___ ___ ___.
2) Soil bacteria are able to pull nitrogen out of the air and put into the soil. We say that they ___ ___ ___ nitrogen.
3) This is a very short word that means an aggregate (lump) of soil. ___ ___ ___ (HINT: page 65)
4) An atom that is electrically unbalanced is called an ___ ___ ___.
5) One quarter of the volume of soil is ___ ___ ___.
6) "Hema" is Greek for "blood," so hematite is often this color. ___ ___ ___
7) Sn is the atomic symbol for the element ___ ___ ___.
8) Felsic rocks are ___ ___ ___ in iron and magnesium, but high in silicon.
9) Even though all substances can exist as solid, liquid or ___ ___ ___, water is the only one where we can observe all three in the natural world.
10) Hydrogen sulfide, H_2S, smells like a rotten ___ ___ ___.

PART 3: Four-letter words
The limestone pattern below has the first letters of all the answers to these clues. Can you finish the words?

1) The crystal shape of pyrite
2) NaCl
3) The softest mineral on the hardness scale
4) Au
5) Fe
6) Pb
7) A mineral that contains chemical H_2O.
8) When volcanic ash turns to rock
9) A green silicate mineral that has single chains of tetrahedrons.
10) Shale used to be this.
11) Clasts that are very small but you can still see them.
12) Clasts you can't see but can feel with your tongue.
13) Anthracite and bituminous are types of this.
14) This Greek word means "change."
15) This feature is found in karst landscapes.
16) This is a combination of water, air, minerals and organics.
17) Mafic rocks tend to be _____ in color.
18) When magma cools it turns into solid ____.
19) This is considered to be the perfect texture for soil.
20) This mineral comes in flat sheets, either biotite or muscovite.
21) Coccolithophores are alive! They are made of a single, living ____.
22) Ca(OH)$_2$ is better known as slaked ____.

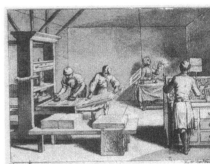

ANOTHER USE FOR LIMESTONE:
Very fine-grained (smooth) limestone is used by artists to make prints called lithographs. ("Litho" means "stone.") The artist polishes the stone slab until it is ultra-smooth, then he/she draws on it with a waxy pencil. Ink will stick to the wax, but not to the wet stone. Paper is put on top of the inked stone, then it is run through a press that pushes the paper firmly onto the stone. The image comes off onto the paper.

c		g		c		o		s		m	
	l		d		c		i		c		
j		m		s		t		s		l	
	s		t		r		l		c		

CHAPTER 7: DEEP DOWN UNDER

When you were younger, did you ever think about trying to dig a hole to the other side of the earth? Small children often wonder if this is possible. Children in North America imagine coming up in China, though Australia is more accurate. Even adults have fun imagining what it would be like. In 1864, author Jules Verne wrote a novel called *Journey to the Center of the Earth*. In the book, three men find an inactive volcano in Iceland that has a secret passage down into the earth. After many days of traveling, they come to a huge cave, large enough to hold a small ocean, gigantic mushrooms, 12-foot tall humans, and a population of dinosaurs! When they finally get back to the surface, they find themselves coming out of a volcano in Italy.

the underground ocean

We know, of course, that Jules Verne's story is unscientific. Even in 1864, people knew that it was complete fiction. Most of us assume that scientists are very careful to keep facts separate from fiction. When it comes to studying the inside of the earth, however, we've got very few facts to go on. The inside of the earth isn't a place you can actually observe. Scientists end up taking educated guesses a lot of the time. Unfortunately, these guesses are often presented as though they were facts. In this book, we will be very careful not to make guesses look like facts.

We know we can't dig a hole to the center of the earth, but how deep can we go? In 1970, Russia set up a digging project way up north in the Kola Peninsula. (The peninsula is marked in red. Russia is white.) Though this peninsula is farther north than Alaska or Norway, the climate is a little less cold. The surrounding waters are full of fish, and the land contains many natural resources such as copper, nickel, iron, and diatomaceous earth. Thousands of people live on the peninsula and many of them are involved in mining operations. (Unfortunately, the Russian government also set up military manufacturing here and polluted some places with toxins and nuclear waste.)

A very tall building was constructed to cover the very tall drilling rig. The actual hole was only 9 inches (23 cm) wide. (The picture on the right shows the top of the hole after they sealed it shut.) They kept drilling until the year 1992. When they got down to 12.26 km (7.6 miles) they had to stop because the temperature at the bottom of the hole was just too high (180° C/350° F).

What did they find at the bottom of the hole? Most geologists predicted that they'd find basalt, but this turned out to be wrong. Only one scientist correctly predicted what they found: **crushed granite and hot water**.

The top of the hole is now welded shut.

There were other surprises, too. Gases also came up from the Kola Superdeep Borehole: hydrogen, helium, nitrogen and carbon dioxide. And in very deep layers, they found what appeared to be fossils of single-cell microscopic organisms.

To help us understand how deep this hole is, we can imagine shrinking the diameter of the hole to the size of a drinking straw. How many drinking straws would we have to connect end-to-end to make our model borehole the correct length? Over 1,000! Since Kola, longer boreholes have been drilled, but they all go off to the side a bit, not straight down. Kola still holds the world record for deepest hole.

Now, if imagining 1,000 straws didn't "boggle" your mind, this certainly will. If you imagine that the earth is an apple, the outer part of the earth that we call the "crust" would be like the skin on the apple. The Kola borehole (at 12.26 km/7.6 mi.) didn't even get close to the bottom of the crust. If you were to add the Kola borehole to our apple model, it would be a shallow pin-prick that doesn't even penetrate the skin! That means that no human has ever seen any part of the big interior of the earth, represented here by the white part of the apple.

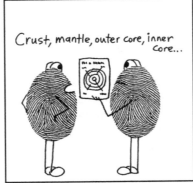

All those nice pictures in the books are just guesses about what might be inside the earth. It's important to remember that no one has ever been able to directly study rocks that are deeper than the Kola borehole. We've never seen what's down there. Our guesses might turn out to be correct, or they might not. The drillers at the Kola borehole put this statement on their official website: "We realized at once how little we still know about the structure of our planet." Other scientists have made statements like this: "We know more about the outer edges of the solar system than we do about the inside of the earth." So then how do we even begin to guess what is down there?

The only way we can study the inside of the earth is to observe how earthquake waves travel through it. The top picture shows a machine called a **seismograph**. You are probably already familiar with this kind of machine. It has a little needle that moves back and forth, recording vibrations. The needle would be attached to a sensor in the ground, and the machine would be on a surface that is protected from nearby bumps and bangs, like footsteps and closing doors. The needle rests on a roll of paper that slowly turns. These machines are sensitive enough to detect earthquakes that occur on other continents.

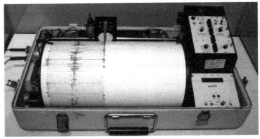

By Yamaguchi at the English language Wikipedia, CC BY-SA 3.0, https://commons.wikimedia.org/w/index.php?curid=1089235

The picture on the bottom shows a **seismometer**. We can't call this a seismograph, since "graph" means "writing." This machine does not have a needle that writes. Earthquake detection has gone digital, just like everything else. There are other types of seismometers, also, that look different from this one, including one designed to sit on the ocean floor. However, they all work on basically the same principle. They have something inside that can stay still while the earth shakes around it.

Imagine holding a heavy ball that is suspended by a spring. If you move your hand back and forth quickly a few times, the ball will

By Apple2000 - Own work, CC BY-SA 4.0, https://commons.wikimedia.org/w/index.php?curid=44406976

stay in the same place. In physics terms, the heavy ball has **inertia** *(in-ER-shuh)* and wants to stay put. A few quick motions of your hand aren't enough to get it moving. Seismometers have a part inside that is able to stay perfectly still even if the earth quakes and moves underneath it. The machine has been designed to compare the stillness of this part to the motions surrounding it. All these shakes are called **seismic waves**. "Seismos" is Greek for "shaking." Seismometers are also equipped with extremely accurate clocks so that they can record the exact time that certain seismic waves were detected. The timing of seismic waves is the one and only clue we have as to what might be inside the earth.

The study of seismic waves can be a bit confusing. There are many types of waves and sometimes it is hard to remember which is which. The nice thing, though, is that much of what you learn about earthquake waves can be applied to other types of waves, too, such as sound waves or water waves. Waves all behave in basically the same way.

Earthquake waves can be divided into two categories: 1) "**body waves**" that go through the body of the earth, and 2) "**surface waves**" that stay on the surface. Let's look at the body waves first.

There are two kinds of body waves: 1) **P waves** and 2) **S waves**. The letter P represents two words, both beginning with the letter P: **Primary**, and **Pressure**. (It's just a bit of luck that both these words begin with P.) Primary can mean "first." P waves are the first waves that the seismometer detects when there is an earthquake because they are the fastest waves. P waves can also be called "pressure waves" (or "compression waves") because of how they work.

The best way to learn about P waves is to play with a Slinky. If you don't have a Slinky and don't know anyone who does, just watch the video on the YouTube channel. However, if you can do this demo yourself, you will remember it much better. We'll let our volunteers demonstrate.

Stretch out the Slinky to a length of at least 2 meters (6 feet). Be careful not to over-stretch it. (If your Slinky already has a few bends and kinks in it, this demo won't look quite so pretty, but it should still basically work.) If you have someone to hold the other end, that is best, but you could also tape it to a chair or table leg. Let the Slinky lie perfectly still, and have one person give their end a tap in the direction of the other end. Watch the "wave" travel along the length of the Slinky till it gets to the other side. Sometimes the wave will even start to come back again. You can do this as many times as you want to. It's really fun to watch.

What you've done is apply PRESSURE to one end of the slinky. The pressure (tap) your hand applied to the first link was then transferred to the second link. The pressure on the second link was then transferred to the third link. From the third link it went to the fourth link, and so on, all the way down the line. Overall, the Slinky stayed in place. You just had this "bunched" area that traveled along it. (The YouTube video shows a slo-mo view of this.)

P waves can move through any medium: solid, liquid or gas. When P waves travel through the air (gas) we call them sound waves. P waves can travel through water, too, although our ears don't do so well at hearing them under water. P waves travel the fastest through a solid because the molecules are right next to each other, and are therefore easily bumped. P waves travel the slowest in air because the air molecules move around so much.

S waves | P waves

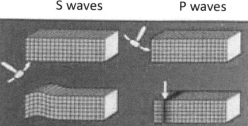

S waves are the other type of body wave. The letter S stands for both **Secondary** and **Shear**. (Lucky again! Good thing "shear" and "secondary" both start with S.) In an earthquake, the S waves are the second type of wave to hit the seismometer. They are called "shear" waves because of the way that they behave. Sometimes the word "shear" means "to cut" (like shearing a sheep), but in this case it means "a shift from side to side."

You can make an S wave with your Slinky. Just have one person (not both!) give their end of the Slinky a shake back and forth. You will see an S-shaped curve travel down the Slinky.

S waves go back and forth, quite unlike the P waves. Pressure waves don't have any wiggling. (That's an easy way to remember S waves. S waves look wiggly, like the letter S does.)

It's really important to understand these waves before you go on. If at all possible, stop reading and go watch the videos on the YouTube channel that demonstrate P and S waves.

Going back to our introduction, we said that there are two kinds of waves: **body waves** and **surface waves**. The body waves are the P and S waves. We still need to see how these waves are used to investigate the inside of the earth, but before we do that, let's finish our introduction to waves and learn what surface waves are. There are two types of surface waves: **Love waves** and **Rayleigh waves**.

Love waves were discovered in the early 1900s by a scientist named Augustus Edward Hough Love. The definition of a Love wave is something like this: "The interference of many shear waves guided by an elastic layer, which is welded to an elastic half-space on one side while bordering a vacuum on the other side." Understanding the physics of Love waves is definitely outside the scope of this book!

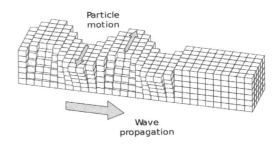

All we really need to know about Love waves is that they occur only on the surface of the earth, they travel more slowly than P or S waves, and they are VERY destructive despite their lovely name. Love waves are the ones that take down buildings and tear apart roads. The best way to learn about Love waves is just to watch some video animations showing how they roll along the ground. Check out the playlist.

The second kind of surface wave is the **Rayleigh wave** *(RAY-lee)*. As you might guess, this wave is also named after a scientist. ("Lord Rayleigh" was his title. His real name was John Strutt.) These waves are produced by earthquakes, but they are also made by scientists who want to use them to learn about deep layers of the earth where there might be gas and oil deposits. Geologists can set off explosions that send out vibrations into areas they want to explore. These small "quakes" don't do any harm and they provide helpful information.

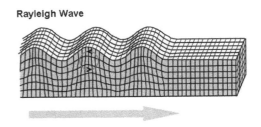

Love and Rayleigh waves cause great damage to buildings and roads.

Rayleigh waves are the slowest kind of wave, traveling about 3 km/sec. That's still pretty fast! (10 times the speed of sound) Like Love waves, they result from the interaction of P and S waves when they are on the surface. Both Love and Rayleigh waves can travel great distances, sometimes circling the earth several times before they stop. Diagrams often show little circles under the rolling lumps of Rayleigh waves. This shows the circular motion of the energy in the wave. (This motion is very much like an ocean wave.) It's important to remember that the rocks and dirt are not moving all that much. The energy of the wave moves *through* them.

Now back to P and S waves. Both of these waves can travel through solids, but only P waves can travel through liquids and gases. This is an important fact that helps to give clues about the inside of the earth. To see why this is true, we need to understand what P and S waves are doing to the atoms.

In these diagrams, the blue circles represent atoms. The red arrows represent the motion of the seismic waves. In diagram (1), we have a line of atoms that are all bonded together to form a solid. The bond lines bewteen them are a bit flexible, more like strings than toothpicks, but the atoms are solidly connected. You can easily imagine that if the red arrow gives a push on the first circle, all the rest will get pushed, too. In diagram (2), the circles are not bonded together, but are free to move around. This represents a liquid or gas. Imagine that these are balls on a table. What will happen if the red arrow gives a push? We'll get the "domino effect," won't we? The first ball will bump the second one, then the second one will bump the third one, and so on down the line. The circle at the end of the line in diagram (1) will get a harder push than the circle at the end of the line in diagram (2), but in both cases, all the balls will move. Thus, P waves will go through both solids and liquids.

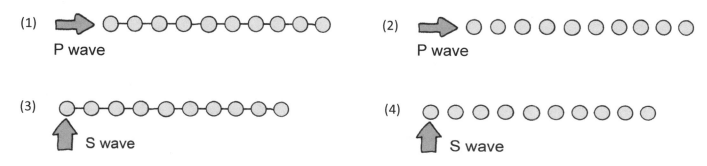

In diagram (3), we have a solid again, because all the circles are bonded together. Now we will start an S waves instead of a P wave. Remember, S waves are called "shear" waves because they go side to side (or up and down). To make an S wave, we need to push on the first circle from the side. If we push the first ball in diagram (3), will the other balls move? Yes, of course they will, because they are bonded, right? Now in diagram (4) we have the circles representing a liquid or gas because they are not bonded. If the red arrow pushes the first ball, will the others be affected? No, not at all. The first circle might go rolling off, but the others will just sit there.

NOTE: There are some nice video demonstrations of this on the YouTube playlist. One of them uses people instead of balls, and it's something you can do if you get some friends or family together.

The fact that S waves can't go through liquids is one of the keys to understanding why we think the center of the earth has several cores. When there is an earthquake, the seismic waves will travel across the globe in a matter of minutes. Seismometers all over the wrold will sense P waves from the quake. The diagram on the left shows P waves traveling to every part of the earth. With S waves, however, something strange happens. There is a section directly opposite the quake that does not receive any S waves. As you can see in the diagram on the right, this area is called the **shadow zone**.

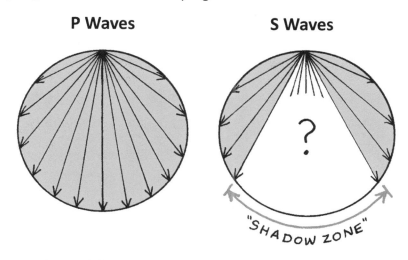

So what's going on here? Since S waves can only travel through solids, we can guess that there is an area in the center of the earth that is not solid. Is it liquid, then, or gaseous? Most scientists guess that it is a liquid because at that depth there is an immense amount of pressure. Gases turn into liquids under that kind of pressure. However, the S waves can't tell us what the liquid is. You'd get a shadow zone if the center of the earth was filled with milk or orange juice. Okay, so those are silly guesses. What would make more sense? The element mercury is a liquid. Can the earth be filled with mercury? Or maybe a soft metal that melts easily, like gold?

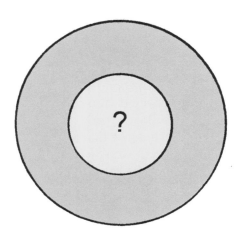

What would this liquid core be made of? We know quite a bit about the elements on the Periodic Table. We can study how they behave when they are heated or put under pressure. We can observe where they located on the surface and how abundant they are. Elements that are rare on the surface (like gold and silver) are likely to be rare (or non-existent) deep inside, too. This core is probably made of the more common elements in the crust. We also observe that the earth has a magnetic field, and that makes us think that there is probably a lot of iron inside. Most books and websites about the earth will tell you quite factually that the center of the earth is made of **iron** and **nickel**. This could be exactly right. But can we ever <u>prove</u> it? No. Some scientists think sulfur and oxygen are also present in this core. One physicist has even suggested there is helium. The only hard <u>fact</u> we have to go on is that S waves don't go through the center of the earth.

The next clues about the center of the earth came from looking carefully at P waves. First, they noticed that there was a shadow zone for P waves, too (the white areas). Second, P waves that went straight through the earth were not arriving on the other side at the expected times and in the expected places. Something inside the earth seemed to be bending the waves. Waves 1, 2, 3 and 8, 9, 10 are going through as expected. Waves 4 and 7 are being bent once, by the liquid core. Waves 5 and 6 are being bent not once, but twice!

How and why do waves bend? Light is an energy wave we can observe easily. Light waves travel at different speeds through water and air. If you put a pencil into a clear glass filled with water, the pencil will look as though it is bent right at the water line. This is because light travels more slowly through water than it does through air. When light slows down, the direction of the waves changes just slightly. This bending is called **refraction**. The substances in the center of the earth seem to refract seismic waves differently. P waves go faster through more dense things. This makes sense when you remember that the atoms are bumping into each other. When atoms are packed close together, the energy is transferred very efficiently from one atom to the next. In waves 4 and 7, the speed increases in the yellow area. This means that the yellow area is more dense. With waves 5 and 6, the speed increases again in the very center (the circular area inside the dots). This means that the very center is even more dense. Applying what we know about materials, we can guess that the center is probably solid. This central **inner core** might even have a core, too! Recent data suggests that there is another core inside the inner core.

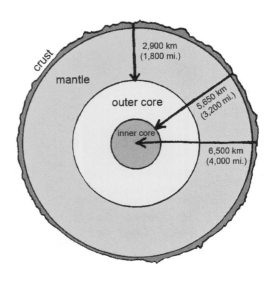

Changes in wave behavior can also be used to find out how thick the earth's outer crust is. Seismologists out in a boat over the ocean notice a difference in how waves travel at about a depth of 5-10 km (3-6 mi.). On land, the distance is more like 30-50 km (20-30 mi.). This top layer is what we call the **crust** of the earth. The crust is made of basalt in some places, like the ocean floor, and granite in other places, like on the continents.

The layer right under the crust is called the **mantle**. S waves can travel through the mantle, suggesting that the mantle is solid. The arrows in this diagram show the distance from the surface down to the outer core, the inner core, and the very center of the earth. The diameter of the earth (total distance across) is double the distance down to the center: 17,000 km (8,000 miles). These numbers have all been "rounded off" and are not exact figures.

So now we've been to the center of the earth! It's not nearly as interesting as the surface, is it? The very inner core is a hot solid, probably made of iron and nickel and maybe a few other elements. The outer core is a hot liquid, probably made of the same things as the outer core: iron and nickel and maybe a few other elements. The cores are very likely made of the same stuff, but the extreme pressure inside the inner core makes it a solid.

As we make our way back to the surface, we come to the **mantle**. This is a very thick layer and makes up most of the volume of the planet. As we noted in the picture to the left, S waves went right through it. (1, 2, 3, 8, 9, and 10.) When S waves go through a material, this is means it is a solid. Therefore, **the mantle is solid**. This is an important fact to remember. Another important fact is that no one has ever been able to drill down far enough to collect an actual sample of the mantle. Every drilling project attempting to do this, so far, has failed. (The first attempt was in the early 1960s and the latest attempt was in 2007.) The only provable facts about the mantle are how P and S waves travel through it. We know where its upper and lower boundaries are, and there seems to be some kind of change in the middle, causing some scientists to define an "upper mantle" and a "lower mantle."

What might the mantle be made of? The speed at which P waves travel through it is the exact same speed that they travel through basalt, so basalt is a good guess. The minerals of which basalt is made would also be likely candidates for mantle ingredients. These include high-calcium and high-sodium feldspars, olivine and pyroxenes. (We met olivine and pyroxenes on pages 18 and 19.) Basalt is basically a mixture of feldspars and olivine. A rock called "peridotite" is also often listed as a likely ingredient of the mantle. Peridotite is just a blend of olivine and pyroxenes, so it really isn't anything new.

One way that scientists can test these guesses about the mantle is to collect surface samples of these rocks and put them into fancy testing machines that can simulate the great heat and pressure these rocks would be feeling way down in the mantle or the core. The container that holds the rock sample must be amazingly strong, so it is made of the strongest substance on earth—diamond. The container is called a **diamond anvil**. It can hold up under unbelievable heat and pressure. While the samples are "cooking" in this diamond anvil, they can be tested with various techniques, mostly using energy waves, like x-rays. (The yellow lines in this picture represent energy waves going through.) This allows scientists to find out what happens when rocks and minerals are put under great pressure and heat. Sometimes the changes are predictable, other times they are a bit surprising.

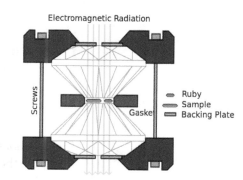

By Tobias1984 - CC BY-SA 3.0, https://commons.wikimedia.org/w/index.php?curid=19419201

The results of these tests with basalt, olivine and peridotite have suggested that our guesses about the mantle are probably right. Therefore, whenever a chunk of olivine or peridotite is discovered in a lava flow (as shown in this picture), it is assumed that it is a piece of the mantle that didn't melt. Can we be sure? No, because no one was down there to see what really happened.

Scientific, "educated guesses" like this are called **inferences**. An inference isn't a wild guess. It is based on as much data as possible. An inference might be entirely correct, especially if it is based on several pieces of data, not just one. However, sometimes inferences turn out to be wrong, too.

Yes, scientists can be wrong. This is more likely to happen when data is limited. If we are studying something that we can actually see, we can measure it, weigh it, photograph it, and perform tests on it. We'll have lots of information that comes from direct observation. When something can't be directly observed, it gets trickier. An inference can be a very good guess indeed, but still turn out to be wrong. (Example: bottom of Kola borehole)

This diamond anvil technology was used to try to find out more about the mantle. In 1998, scientists at Okayama University in Japan put mafic rocks like basalt and olivine into their diamond capsule and heated it to temperatures that would simulate the heat and pressure in the mantle and the outer core. The rocks melted into magma inside the capsule, and then were observed as pressure increased. When the pressure inside the capsule became as high as the pressure that would be at about 300 km (200 mi.) deep, something unexpected happened. The magma "shrank" to half its regular volume! In other words, it took up half the space it used to. This would be like taking a piece of cake the size of a tennis ball and squeezing it until it was the size of a golf ball. Cake compresses easily, so that's not too hard to imagine. Magma acts the same way, though. The atoms and molecules move closer together as they are squeezed, and take up less space. The result is that the compressed magma becomes very dense.

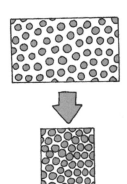

Do you remember reading about density in chapter 2? Hopefully, you did that experiment with materials of various densities and watched them sink or float in water and oil. No matter how many times you shook the jar, the objects always ended up at the same levels. Density is one of those laws of physics, like gravity or the laws of motion. These "laws" are the foundations of science because they always work the same way. Always. Things that are less dense always rise when surrounded by things that are more dense (unless they are prevented from doing so). Things that are more dense will sink. This is always true.

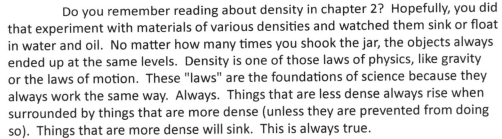

The dividing line between sinking and floating is called the **crossover depth**. Magma at this depth (300 km) suddenly becomes very dense, shrinks to half its volume, and will sink toward the center of the earth, never to rise again. It can't possibly go back up. The laws of physics say so. This wedge-shaped diagram shows where the crossover depth is located in the mantle. The mantle is about 2,900 km (1,800 mi.) thick, so the crossover depth is very close to the top.

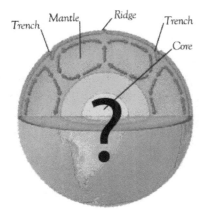

The picture on the right is from Wikipedia, but there are similar pictures on many other web pages and in many books. It's a nice picture, but how scientifically accurate is it? Can the mantle circulate around like this? Where would the crossover depth line be on this globe? So what's up with these pictures? We'll answer this eventually, but first we need to learn about the crust of the earth.

Before we move up to the crust of the earth, let's review what we know about the mantle.
1) It's huge. *(Most of planet earth is mantle.)*
2) It's solid. *(We know this because S waves travel through it.)*
3) It's mafic. *(Mostly olivine, pyroxenes, and basalt. Basalt is a mixture of feldspars and olivine.)*
4) It has a crossover depth at which magma sinks and never comes back up.

Mohorovičić on a commemorative stamp.

In 1909, a seismologist in Croatia, named Andrija Mohorovičić (don't worry about pronunciation), discovered that seismic waves were speeding up when they got to a certain point under the crust. He began tracing these waves and discovered that there was a very definite line below which the waves went much faster. He did not know why they went faster, but he understood that this line was some kind of major dividing line. Today we know this line as the boundary between the crust and the mantle. Its correct name is the Mohorovičić Discontinuity, but everyone just calls it **the Moho**. The Moho is the place where the crust stops and the mantle begins.

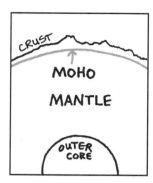

There are some other terms you can learn, too, if you are interested. The crust and the top of the mantle are called the **lithosphere**. ("Litho" is Greek for "stone.") There is a "weak" place under the lithosphere, called the **asthenosphere**. ("Astheno" means "weak.") We don't need to say a whole lot about them. If you want to know more about these, you can look them up on the Internet.

The crust of the earth is made of basalt and granite, with layers of sedimentary rock on top of them. Basalt is what most of the ocean floors are made of, beneath the layers of sediments. The continents are basically made of granite, with layers of sediment covering them. This diagram does not show the sediments.

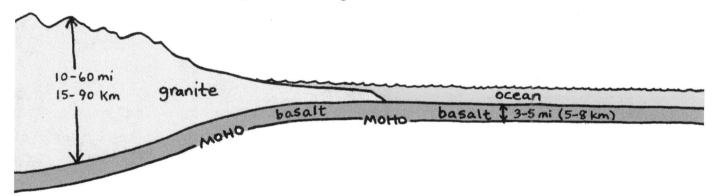

Basalt is the most common rock on earth. We know it is under the oceans, and we think it is probably under the continents, too. Underneath the basalt is the mantle. The top of the mantle can have hot, active magma. When the magma erupts up into the ocean floor it can create underwater volcanoes. The Pacific Ocean has thousands of underwater mountains

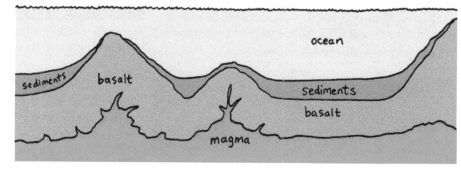

that are actually extinct volcanoes that never made it to the surface. When an underwater volcano does reach the surface, it creates an island. Surtsey Island (in chapter 6) was made in this way.

Some volcanic islands are now **dormant**, meaning they have not erupted for a very long time. One of Hawaii's dormant volcanoes, Mauna Kea, is the highest mountain in the world. Its total height, including the underwater portion, is 10,000 m (33,000 ft). The above-water part is about half as tall as Mt. Everest.

The basalt under the ocean floor is mostly covered by sediments, ranging from half a mile (.6 km) thick to over 5 miles (9 km) thick. This map shows the thickness of the sediments in various parts of the ocean. The dark blue areas have the least amount of sediments and the thickest areas are yellow, orange and red. It is obvious from this map that the thickest sediments occur close to the continents, where the water is more shallow. This makes sense when we consider what is in the sediments.

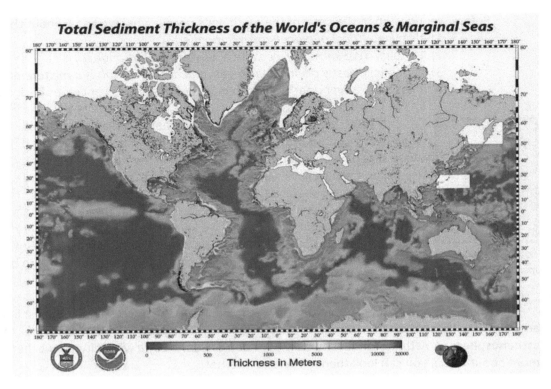

Total Sediment Thickness of the World's Oceans & Marginal Seas

Thickness in Meters

Ocean sediments are of two types: **clay** and **ooze**. We already know what clay is. Clay is made of ultra-microscopic, flat particles of SiO_2. Ocean clay is brown or red, just like the clay at the surface.

Ocean **ooze** is similar to chalk and diatomaceous earth, because it contains the shells of microscopic single-celled creatures. Chalk has coccolithophores and diatomaceous earth has diatoms. Ooze contains the shells of **foraminiferans** and **radiolarians**. These creatures, though microscopic, make some of the most fabulous shells on earth. It's hard to imagine a soft, squishy protozoan creature living in these shells, but they do. Forams are actually predators, and stick soft little "arms" out of tiny holes in the shell, in order to catch prey.

Foraminiferans (forams) build shells using calcium, like clams do.

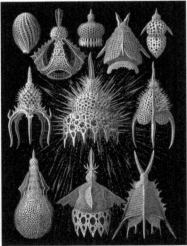

Radiolarians make shells out of SiO_2 like diatoms do.

Ocean sediments are a bit softer than the sedimentary rock on land, but they are much harder than pudding or syrup, despite the sloppy sound of the word "ooze." It's hard to imagine the immense number of microscopic creatures it would take to fill a layer of sediment half a mile deep, but oceanographers tell us this is what they find when they drill into the sea floor.

In a few places along underwater mountain ridges, there are cracks in the basalt where super hot water is leaking out. We call these sites **hydrothermal vents**. The temperature of the water can be as high as 450º C (840º F). The water is full of minerals which precipitate out and form tall "chimneys." (Sort of like stalagmites.)

We would expect magma coming up from under the basalt, but hot water? Scientists have had to come up with theories as to where this water might be coming from. This diagram demonstrates one theory. Perhaps ocean water is seeping down into the rock, going down deep where the magma is, then coming back up as steam. (Don't worry about all those chemical letters; they represent minerals in the steam.) This might look okay to a geologist, but if an engineer looks at it, he or she will tell you this can't be right. There are physical laws that govern how and where atoms and molecules will move. One of those laws is gravity, but there are many others. We can calculate the pressure of the water pressing down

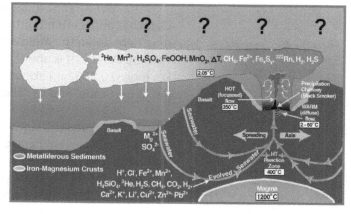

against the rock and the pressure with which the rock will resist the water. The math tells us that the rock will win and the water will not enter it. The engineer will also point out those "spreading" arrows and the magma pushing up from below. Here are two forces pushing against the direction the water is supposed to be going. An even stronger force would have to be pushing the water against these two forces. The diagram does not show any stronger forces helping the water. In the next chapter, we'll look at a theory that suggests that the water is coming from below.

Another very important aspect of the ocean floor is its **topography**: mountains, valleys, plains, plateaus, etc. The bottom of the ocean was a complete mystery until the mid 1800s. Of course, fishermen knew quite a bit about the areas of sea floor that lay close to shore; they had maps of shallow or rocky places that were dangerous to ships. But what lay at the bottom of the vast oceans?

The first clue came in the 1850s as America and Britain worked together to lay a telegraph cable all the way across the Atlantic Ocean. This cable was like a long electrical wire connecting North America to Europe. Electrical signals could be sent back and forth in a matter of minutes. Telegraph operators at each end used Morse Code to transmit messages. The original cable was made of 7 copper wires coated in rubber, then wrapped in hemp rope cov-

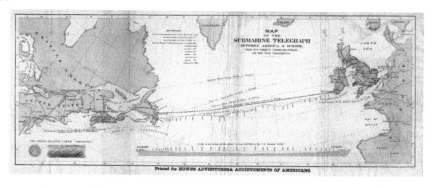

ered with tar, then wrapped with iron wires. The cables were wound onto giant spools and loaded onto ships that took them out to sea. The cables were slowly pulled off the spools and lowered into the water, and they sank to the bottom. The cables were expensive to make, so they chose the shortest distance between the continents, from Newfoundland to Ireland. Those locations were then connected to major cities such and London and Washington D.C.

The sailors who were putting the cables into the water and lowering them to the ocean floor noticed something odd at about the midpoint. Suddenly, much less cable was needed in order to hit bottom. There seemed to be a shallow place right in the middle of the ocean! They called this place the ocean's "shoal," the same word they used for shallow places close to shore. But the exact nature of this shallow place was still a mystery.

The next clue did not come until almost a century later, in the 1950s. Seismologists had discovered how to use seismic waves to detect the depth of the ocean floor. They began taking thousands of data points and putting them on maps. If enough places were measured, these measurements could be used to construct a picture that would show mountains, valleys and plains. One of the scientists who worked to put the seismic data onto maps was Marie Tharp. She is shown here with Bruce Heezen, another scientist who often worked with her and became a close friend. The maps that they made are still used today. As Marie turned the data points into a picture of the ocean floor, she discovered the long mountain range we called the Mid-Atlantic Ridge.

The Mid-Atlantic Ridge is a chain of connected mountains running right down the middle of the Atlantic Ocean. It is about 16,000 km long (10,000 mi.) about 1,500 km (1 mi.) wide, and its tallest peaks are 3 km (2 mi.) high. A few of its peaks reach the surface and create islands, including Ascension Island and the Azores.

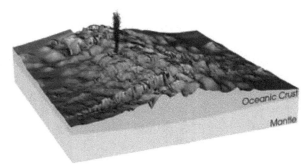

At the top of this map, you can see Iceland sitting right on top of the Mid-Atlantic Ridge. The island looks as though it slid into the ridge after the ridge was formed. The picture on the right shows a place where the Mid-Atlantic Ridge can be seen in the landscape. Many tourists visit this site.

Marie Tharp discovered many other underwater mountains. She found that the Mid-Atlantic Ridge was actually part of a world-wide feature called the Mid-ocean Ridge. This mountain chain wraps around the globe like the seam on a baseball. Look at this map and find the Mid-ocean Ridge in the Pacific Ocean. Trace it up to where it hits North America near the Baja California peninsula. Notice that this mountain chain looks as if it continues on as part of the mountain chain that goes way up into Canada and Alaska.

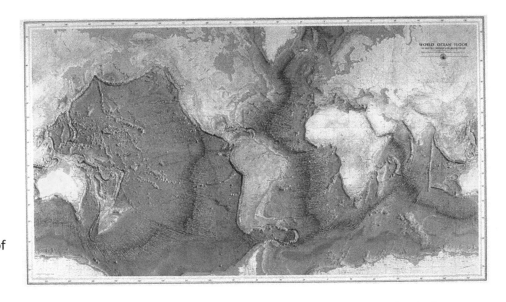

Not all underwater mountains are part of the mid-ocean chain. The western Pacific has many individual mountains that used to be volcanoes. In the next chapter, we'll look at two different ideas about these ridges and mountains.

Here's a view of the world you don't see very often. This is the floor of the Arctic Ocean (the North Pole). Trace the Mid-Atlantic Ridge up through Iceland (at the bottom, under Greenland) and see where it ends. There's another ridge parallel to it (on the left), but it has a different structure, and thus, a different origin. To the left of the ridges there is a big basin. The Arctic Ocean is mostly covered with ice. When explorers go to the North Pole they walk on top of the Arctic Ocean.

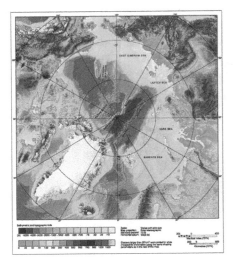

Now that we know about the layers of rocks and dirt on the ocean floor, let's see what's on the continents. First, let's look at this diagram again, to remind us of the crust's basic structure. As a general rule, the continents are made of granite and are very thick, up to 60 miles (90 km). That's a lot of rock!

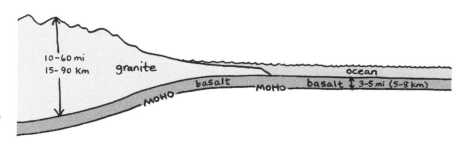

On the very top, we find all those soil horizons we learned about. If we dig down to the bedrock, we then begin to find layers of rock. It seems as though almost the entire world is covered in layers of sedimentary rock.

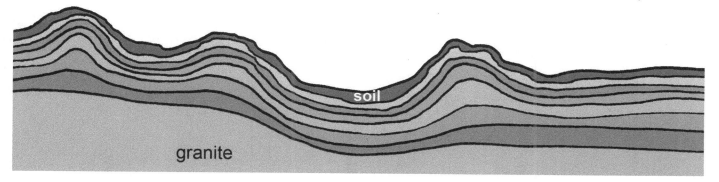

Sometimes the layers are flat and straight, but in other places they are tipped or folded. Some layers are very thin, perhaps only a millimeter or two. Thick layers might be taller than a two-story house. Sometimes the layers have very different colors and textures, but you can also find places where the layers look very much the same.

These layers are made of things that you are already familiar with: limestone, sandstone, shale, slate, coal, siltstone, gypsum, salt and quartz. Very often the layers have repeating patterns, such as: "limestone, sandstone, shale— limestone, sandstone, shale— limestone, sandstone, shale." Some layers contain fossils, others do not.

In order to study these layers, you have to be able to see them. You either have to dig a very deep hole (like the Kola hole) or else find someplace where the layers are already exposed. Both methods are used by geologists. Every time an oil company digs a borehole, they bring up core samples from the layers they go through as they drill. The man in this picture is putting core samples into a box in exactly the same order that they came out of the ground. They will be analyzed later by petrologists (people who study rocks).

Some of the information that is gathered by drilling companies ends up in government files, but information that they think is critical to their business they will keep to themselves, so that competing companies will never get access to it. Drilling a well is very expensive, so being able to find what you are looking for in as few attempts as possible can give you the advantage over your competitor. Core samples from surrounding areas can give important clues.

A much less expensive way to study rock layers is to find a place where many layers are naturally exposed, such as the side of a mountain, a cliff face, a canyon, or a tall rock formation such as a butte. In these places, natural water erosion has done the hard work for you and the layers are easily visible to anyone. Some places have so many rock layers visible that they have become classic sites to visit and attract not only geologists but tourists, too.

That picture shows a **fault**. "Fault" is the geological term for a crack. At some point in the past, a long crack developed from top to bottom in that cliff, and one side slid and sagged. This is a very unusual fault. Most faults don't sag like this. Usually the crack is just a straight-ish line.

The fault in the picture on the left is a more typical fault. The correct name for this type of fault is "normal." The picture has been marked so that you can see the ends of what should have been one continuous layer. The fault is the orange-colored diagonal line. You can see that after the crack developed, the right side slid down quite a ways. If the left side had gone up, it would be called a "thrust" fault.

Miguel Vera León - CC, https://commons.wikimedia.org/w/index.php?curid=46012511

Here are the three basic types of faults. There are variations on each, but we're going to stay simple. **Normal** faults have a section that slips downward. **Thrust** faults have a section that gets pushed up. **Strike-slip** faults are horizontal, slipping mainly side to side. Strike-slip faults are also commonly called **transform** faults.

NORMAL FAULT   THRUST FAULT   STRIKE-SLIP (TRANSFORM) FAULT

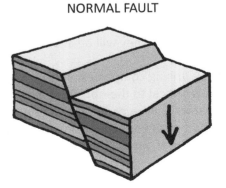
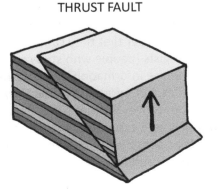
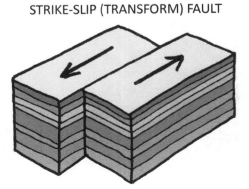

There is a famous strike-slip fault in California. The San Andreas Fault runs parallel to the coastline for 800 miles (1,200 km), and goes down at least 10 miles (16 km) deep. The red arrows show the directions that the sides have moved in the past. Movement along the fault usually coincides with an earthquake, but not always. There is a section in the middle that seems to be moving very slowly all the time.

The picture on the right shows what the southern part of the fault looks like from an airplane. Other parts of the fault are not nearly as obvious as this section. In some places, the fault is hardly visible at all.

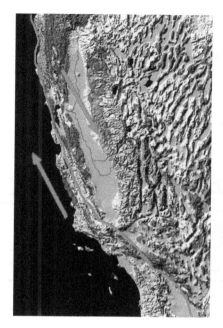

By Ikluft - Own work, GFDL, https://commons.wikimedia.org/w/index.php?curid=3106006

Geologists try to keep accurate maps of where they think fault lines occur so that we can avoid putting buildings on or near them. Sometimes roads and railroad tracks must cross these faults, but at least we know ahead of time that these places could be possible trouble spots, and we can have emergency plans in mind. Buildings and roads can sustain severe damage even if they are not on fault lines, but we can try to avoid making things even worse.

Some geologists think that earthquakes are caused by moving fault lines. Others think that the faults move because of the earthquake. What causes earthquakes to begin with? There are two main ideas that we'll look at in the next chapter. One of them puts the blame on magma sinking below that crossover depth we mentioned.

Perhaps the most well-known thrust fault is the Lewis Overthrust. It lies in both Glacier National Park, in Montana, USA, and Waterton Lakes National Park in Alberta, Canada. This aerial view shows how large the overthrust area is. The fault lies underneath this whole mountain range. This isn't the type of fault that will slip during an earthquake. The fault is horizontal, not vertical. The slippage occurred a long time ago and won't happen again.

The mountains are made of layers of sedimentary rock, mostly limestone. The exception is one thin layer of igneous rock not too far down from the top. One mountain, Chief Mountain, stands at a distance from the rest.

Chief Mountain

By Bobak Ha'Eri - Own work, CC BY 3.0, https://commons.wikimedia.org/w/index.php?curid=7034392

Here is a cross section of the overthrust area, with an arrow to indicate the direction that the overthrust occurred a long time ago. The gray part at the bottom is made of layers of sandstone and shale, but the individual layers are seldom shown. Notice that the mountains are made of layers of limestone. The colors shown here are not the colors of the real rocks, but the colors of the real rocks are different enough that you can see the layers. This overthrust is about 50 miles (80 km) long and it seems to have slid a distance of about 80 miles (128 km).

Cross section of Glacier National Park

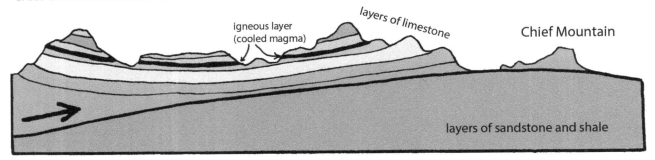

Illustrations such as this one are often used to show how these overthrust landscapes might develop:

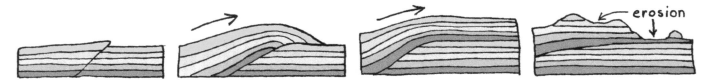

Notice that the final shape of the landscape is created by **erosion**—the gradual wearing away of the rock. In this case, the erosion was caused by glaciers that once covered the area. A **glacier** *(GLAY-sher)* is a giant sheet of ice that moves very slowly. There are still active glaciers in this area, which is why it is called Glacier National Park.

When geologists first discovered "older" rocks on top of "younger" rocks, it took them a while to figure out what had happened here. They finally concluded that a massive overthrust had occurred. But how? Can you imagine 50 miles of rock sliding? The laws of physics say that with the amount of friction (rubbing) that this would create, there must have been some kind of lubrication to make it slip more easily. The only logical possibilities are magma and water. We do see a band of cooled magma up high in the layers, but not down where it would be needed. So it must have been water. But the pore spaces in solid rock are not nearly big enough to contain that much water. Could there have been another source of water? We'll discuss this in the next chapter.

Faults and sedimentary rock layers occur in many mountains, even in the tallest mountain in the world, Mount Everest. Located in the Himalayan Mountains on the border between Tibet and Nepal, Everest's peak is about 8850 meters (29,000 feet) above sea level. It sits in a neighborhood of very tall mountains, but Everest is the peak that everyone wants to climb since it is officially the tallest mountain in the world. The rock layers at the very top of Everest hold the last thing you would ever expect to find miles above sea level.

trilobite

The peak of Mount Everest contains fossilized sea creatures such as clams, sea lilies, tiny seed shrimp, and trilobites. (Trilobites are an extinct member of the horseshoe crab family.) Obviously, these creatures never lived on a mountain top! What is now the top of the mountain must have been at or below sea level a long time ago. Somehow the mountain was pushed up to where it is now.

This diagram shows what geologists think the inside of Mount Everest looks like. The colors represent different types of rock. The colors on the top are fairly realistic: various shades of gray limestone. Then comes a yellow-ish layer that they call the Yellow Band. After that the colors are less realistic and simply represent different kinds of rock, including many metamorphic rocks such as schist and gneiss. The light red areas at the bottom represent places where magma came up and then cooled. The central part is called a **dyke** and the horizontal "arms" that go between the layers are called **sills**.

This picture also shows how geologists like to give names to groups of layers. The top limestone layers were named the Qomolangma Formation. (*Cho-mo-lung-ma* is Tibetan for "Goddess Mother of the Universe." This is the Tibetan name for the mountain. Since the mountain belongs to Tibet, not Britain, we really should use their name for it, shouldn't we?) A fault line separates this formation from the next one. The Yellow Band is at the top of the North Col Formation. Another fault line occurs, then we have the Rongbuk Formation at the bottom. The Rongbuk is probably where the continental crust begins. There is still some granite there, but a lot of the granite has been compressed and squeezed and turned into gneiss. You don't have to remember these layer names. They are just given as a good example of how layers are grouped into "formations."

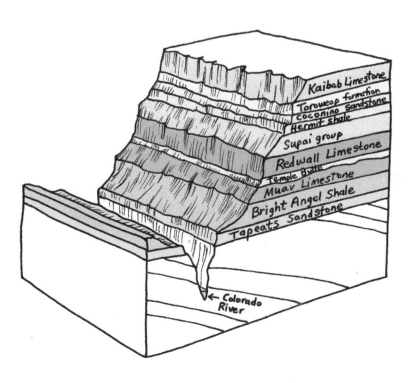

Some of the most famous named layers can be seen in the walls of the Grand Canyon. We mentioned the Redwall Limestone back in the limestone chapter. Also very famous is the top layer, the Kaibab *(kie-bab)* Limestone. It is a very hard limestone and resisted the erosion that took off several layers that used to be above it.

This pattern of repeating layers of limestone, sandstone, and shale is very typical of the sedimentary rock that covers the continents. Did these layers form one at a time, or all at once?

Notice the tiny Colorado River at the bottom. Geologists used to say that this river cut the canyon. Most now realize this is impossible. A river would have smoothed out the walls.

Another very well-known layer is the Tapeats Sandstone. This layer stretches for miles, and can be found in other parts of North America. It is called the Tapeats Sandstone no matter where it is found. Under the Tapeats we see tipped layers that look like they were cut off.

Most of these layers in the Grand Canyon contain fossils, and most of the fossils are of sea creatures. The Kaibab Limestone contains clams and other shell creatures, corals, sponges and members of the sea star family. The Tapeats Sandstone and the Bright Angel Shale contain trilobites. Many layers contain brachiopods and crinoids. Some of the layers also contain plants and insects. The Hermit Shale has fossilized fern leaves and some dragonflies.

brachiopods   crinoid (realted to sea stars)   trilobite   coral   ammonite

Animal tracks can be found in the Coconino *(ko-ko-nee-no)* Sandstone, but the animals that made them are not there. The Bright Angel Shale has lots of burrows made by some kind of sea creature, maybe a marine worm, but again, the actual animals are not there. This type of fossil, where the actual organism is missing, is called a **trace fossil,** because the animal (or plant) left a "trace" of evidence that they were once there. If you find a rock with an imprint of a shell, as if someone had pressed it in and then removed it, this is also a trace fossil.

One particularly notable fossil feature of the canyon occurs in the Redwall Limestone. There is a 1-meter-thick section that has thousands of **nautiloids**. Modern nautiloids have spiral-shaped shells, but these extinct nautiloids had straight shells. The animal that lived inside the shell was

similar to a squid. The odd thing about these fossils, besides the sheer number of them, is that quite a few of them are buried straight up and down. If they had died naturally they would have settled to the bottom and laid there in a horizontal position. It appears as though these nautiloids were victims of some kind of "sedimentary disaster" that surrounded and covered them quickly.

There are other places in the world where huge numbers of fossils occur grouped together as if in a "graveyard." Some very famous sites include the Solnhofen quarry in Germany, the Burgess Shale formation in western Canada, the Green River formation in Wyoming, USA, and the Hell Creek formation in Montana, USA. (The Hell Creek area reportedly got its name from the horrible "smell of death" that comes out of the fossils when you first crack them open.) These famous fossil beds have hundreds of thousands of plants, mammals, reptiles (including some dinosaurs), amphibians, birds, and all kinds of sea creatures.

The Solnhofen limestone quarry in Germany is the site where several archaeopteryx fossils were found. These fossils have become very famous and appear in many textbooks. (The archaeopteryx shown here is actually the author's replica, not the original.) The Solnhofen quarry is also famous for its exceptionally smooth limestone. This quarry is one of the best places in the world to get smooth, flat stones that artists can use for lithograph printing.

Another way that geologists like to classify sedimentary layers is by groups that contain similar fossils. In the early 1830s, two British geologists were studying sedimentary layers in Wales. Adam Sedgwick noticed a layer of fossils that seemed to cover a wider distance than just the spot where he was digging. He noticed that this layer contained only certain types of sea creatures. He guessed that there might be some significance to this and decided to be the first to name the layer. He chose the name **Cambrian,** using the Latin name for Wales: Cambria. Meanwhile, his friend, Roderick Murchison, was also discovering a layer. His layer, located at a higher level than Sedgwick's, also contained a very specific set of fossils. Murchison was inspired by his friend's decision to name the layer after Wales, and chose the name **Silurian**, honoring a tribe of Celtic people who had lived in Wales many centuries earlier: the Silures.

Unfortunately, there was a lot of overlap between these two layers. The fossil groupings were not quite as clear as they had originally thought. The two men argued over other locations, each one thinking that those fossils belonged to their layer. The arguments became so heated that their friendship eventually ended.

Soon after this, another geologist came into the picture. Charles Lapworth had also been studying these layers of fossils. In his opinion, there was a middle layer, between the two layers that Sedgwick and Murchison had named. Going with the theme of Welsh names, Lapworth decided to name his layer **Ordovician** *(or-do-VISH-ee-an)*, after the Ordovices, another ancient Celtic tribe that had lived in Wales.

Above the Ordovician, another layer, marked by a great number of fish fossils, was defined. It was first discovered in Devon, England, so it was named the **Devonian** layer. In other parts of the world, layers of coal appear above layers of fish, so the coal layers were given a name with the word "carbon" in it: Carboniferous. (There's that word root, "-fer," again. Remember, it means "bears" or "carries.") This carbon layer is sometimes divided into two layers that are named after U.S. states: the Pennsylvanian and the Mississippian layers. This does not mean that these layers are found only in these states. They occur in many other places, even in the Grand Canyon.

Going up from the Carboniferous layers we run into the **Permian**, named after the city of Perm, in Russia. The Permian layer contains a wide variety of sea creatures, plus plants, reptiles and mammals. Then we have the famous "dinosaur layers": **Triassic, Jurassic and Cretaceous** *(kree-TAY-shuss)*. All together, these three layers are known as the **Mesozoic**. ("Meso" means "middle.") The three "dinosaur layers" contain not only dinosaurs but also sea creatures, plants, birds, and mammals. For example, the "dinosaur layers" in some places have been found to contain bones of ducks, parrots, owls, penguins, flamingos, loons, albatrosses, and sand pipers. Dinosaur skeletons have been found all over the world, so these "dinosaur layers" are also found worldwide.

The topmost layers have gone through name changes in recent years. They used to be called **Cenozoic**, with subdivisions of Tertiary *(TER-she-air-ee)* and Quaternary. Now these subdivisions are known as the Paleogene and the Neogene.

The very bottom layer, underneath the Cambrian, is called the Precambrian, and usually contains either no fossils at all, or just single-celled organisms. It's hard to imagine fossilized bacteria and algae, but there are paleontologists who specialize in studying fossils that can only be seen under a microscope.

The order of these layer names, from the top down:

Cenozoic
  Neogene
  Paleogene
Mesozoic
  Cretaceous
  Jurassic
  Triassic
Paleozoic
  Carboniferous
    Pennsylvanian
    Mississippian
  Devonian
  Silurian
  Ordovician
  Cambrian
Precambrian

Basement rocks
  Granite
  Basalt

This is our trilobite impression.   More info about trilo's coming up!

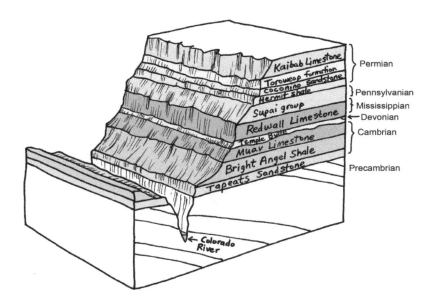

There are very few places on earth where all these layers can be seen. In the Grand Canyon, for instance, the entire Mesozoic and Cenozoic layers are missing from the top, and the Silurian and Ordovician layers are missing in the middle. The top of Mount Everest might represent the Cambrian and Ordovician layers, but nothing else.

There are reportedly about a dozen places in the world where all the layers can be seen, but it is very hard to find any actual diagrams of these places.

Unfortunately, these layer names often don't line up with the formation names, which makes all the names very confusing.

Some very useful liquids and gases can be found in and between the sedimentary layers of the crust, including water, oil and natural gas (methane).

Water that is found under the ground is called **groundwater**. As rain soaks into the ground, it goes deeper and deeper until it reaches an area where there is rock such as shale, granite or basalt, that can't absorb water. The water collects above these non-absorbing rocks and forms what we call the **water table**. In this diagram, the water table is shown as though it's as deep as a swimming pool. Most water tables don't have any open water at all. They are more like a sponge than a pool.

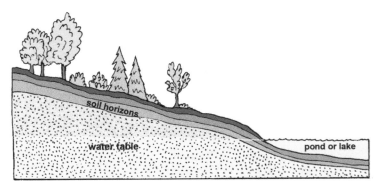

In this diagram we see five wells that were drilled in hopes of finding a source of water that could be pumped up into these houses. The top layer of rock is made of sandstone, which absorbs water very quickly. The rain water drains down through the sandstone until it hits a layer of cracked granite. Some of the water trickles down into the cracks, but a lot of it stays on top.

Well C has hit the main water table perfectly and will be a successful well. Well A also hit water, but the drillers found not the main water table, but a smaller one, called a "perched" table. ("Perched" means "sitting on top of.") The perched water is being held up by a layer of shale. Well A will run dry much more quickly than well C. Well B went all the way down into the granite and tapped into one of the cracks that is holding water. Well D also went all the way down into the granite, but it missed hitting a crack, so this well won't work. Well E was too shallow and will be dry. The drillers might wonder, though, why A and E are both the same depth and yet A has water.

Groundwater is replenished by rainwater that trickles all the way down to the water table. For water to get down that deep, you need several days of soaking rain or melting snow. Passing thundershowers in the summer might bring relief to dry gardens, but the water evaporates quickly, or is used by plants near the surface, and never gets to down to the water table.

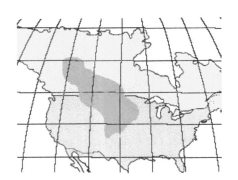

A large water table is called an **aquifer**. *(AH-kwi-fer)* ("Aqua" is Latin for "water.") Aquifers can be extremely large. In North America, for example, there is an aquifer that runs underneath several U.S. states and Canadian provinces. (There are many other aquifers in North America, but they are not shown.) This aquifer occurs in a massive layer of sandstone. Sandstone can hold a lot of water, especially if the sand grains are large and well-rounded and there aren't any tiny particles that might clog up the spaces between the grains.

Aquifers provide one fourth of the world's fresh water, and a lot of this water is used for irrigating crops. Since states and provinces have to share these aquifers, arguments easily arise over who gets to use how much water. If one state allows industrial waste to contaminate the groundwater, other states will be affected. Legal battles over groundwater can be quite fierce. Water is the most important natural resource—more important than coal, oil or minerals.

Water tables also determine where ponds and lakes can form. Temporary ponds might appear during periods of heavy rainfall, but if they are high above a water table, they'll dry up quickly. For a body of water to exist year round, it must be sitting within a water table, as shown in this diagram. The pond is replenished not just by the rain that falls directly into it, but also by the rain that falls on the nearby hills. The rain will drain through the soil horizons and then down into the sandstone.

If you've ever dug a deep hole in the sand at a beach, you may have seen water appear at the bottom of the hole. You scoop the water out, but more water begins to fill the hole. This is because you are right near a body of water. You are digging down to the top of the water table!

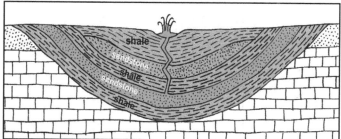

Something interesting happens if an aquifer is stuck between two layers of rock that won't let water go through them. In this picture we see an aquifer made of sandstone that is stuck between two layers of shale. There is a crack in the middle that allows water from the aquifer to rise up to the surface. This creates what we call a **natural spring**. The water in the crack will automatically rise to the level of the top of the aquifer. (This is a basic principle of how water presure works.) As long as the aquifer has a steady supply of rainwater, the natural spring will continue to bubble away at the surface. In places where there isn't a crack, a well can be drilled down into the aquifer and the water will come up without any pumping needed. This type of well is called a **artesian well**, named after the city of Artois, in France. The wells we looked at on the previous page (A through E) were not artesian wells, but were regular wells that need a pump to draw the water up out of the ground.

If groundwater happens to be near a source of heat, such as underground magma chambers, the water can become very hot and boil up onto the surface. Water that flows to the surface gently and slowly can create **hot springs.** Hot springs can be found all over the world, even in very cold places like Iceland and Greenland. If the hot water comes up quickly and suddenly, it can create a **geyser.** The most famous geyser in the world is "Old Faithful" in Yellowstone National Park in Wyoming, USA. It erupts about once every hour. After a geyser erupts, the water trickles back down and gets reheated.

Like water, petroleum is also found between layers of rock. The word petroleum means "rock oil." ("Petra" is Greek for rock, and "oleum" is Greek for "oil.") Petroleum is also called **crude oil**. ("Crude" is the natural state, before it is refined and made into useful products.) This photograph shows an oil **seep** where petroleum is oozing

up out of the ground by the same principle of physics that makes water come up out of artesian wells. The oil is under enough pressure under the ground that it naturally rises.

Seeps like this one were the way that oil was first discovered, thousands of years ago. Of course, back then there was no need to refine the oil and make it into gasoline because there were no vehicles to drive. Ancient peoples did discover that the oil would burn, though, and they were able to use it in lamps. They also noticed that oil was slippery and they found ways to use it as a lubricant. (Perhaps their cart wheels got squeaky?) It was not until the 1800s that people discovered oil deeper under ground and were able to pump it up.

Petroleum can seep up through the ocean floor, too. There are many natural seeps like this one off the coast of California and in the Gulf of Mexico. The seeps near California put about 8 million gallons (30 million liters) of oil into the ocean every year! How does this compare to an oil spill disaster? The famous Exxon Valdez oil spill accident in 1989 put about 10 million gallons of oil into the Gulf of Alaska.

Since oil is less dense than water, the petroleum rises to the surface and creates large oily patches called **slicks**. These oil slicks can be as long as 20 miles (32 km). Eventually the oil is broken down by bacteria that like to eat oil. Scientists are studying how these bacteria might be used to help clean up oil spills created by humans.

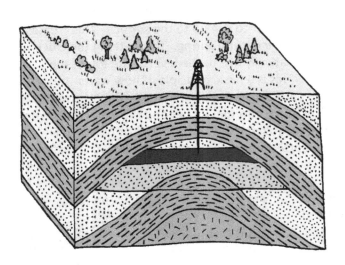

This diagram shows a simplified version of a common arrangement of rock layers that hold water, oil and gas. The dome shape that these layers form is called an **anticline**. Anticlines that occur in layers of shale are good places to look for oil and gas. Since gas is less dense than oil and water, it will rise until it gets to the layer of shale overhead. (The gas has been colored yellow.) Oil is more dense than gas but less dense that water, so it sits on top of the water. (The oil has been colored black.) Notice that all three of these—gas, oil and water —are in the sandstone. The grains of sand in the stone are large and round, which makes for plenty of spaces between them. The sandstone is acting like a very hard sponge. The oil well is pumping the petroleum to the surface.

Petroleum is put into barrels and shipped to factories called refineries where it is turned into useful products such as gasoline, diesel fuel, paraffin wax, and asphalt. Some of the oil is also turned into plastics.

Oil companies hire geologists who specialize in studying rock layers that are likely to contain oil. Seismic studies are done ahead of time to help determine the best places to drill. However, even with seismic clues, most boreholes still don't hit oil deposits. Often, gas is found along with the oil, so the gas is collected in addition to the oil. The gas is very light and is trying to escape upwards, so it does not have to be pumped like the oil does.

Natural gas is also found in areas that have coal and petroleum. The gas was probably formed by the microbes that decomposed the plants that turned into coal and oil. Unlike coal and oil, however, natural gas (methane) can be produced by other processes. Gas is produced by healthy, living algae and bacteria, making it a renewable resource that won't run out. As far as we know, coal and petroleum are not forming today, so when they're gone, they're gone.

Activity 7.1 Comprehension self-check

<u>Questions about the inside of the earth:</u>

1) The Kola Superdeep Borehole is the deepest hole on earth. How deep is it?

   a) about 1 mile (2 km)   b) about 7 miles (12 km)   c) about 70 miles (110 km)   d) about 700 miles (1120 km)

2) Which of these things did they NOT find at the bottom of the Kola Superdeep Borehole?

   a) gases like hydrogen and CO_2 b) magma   c) hot water   d) fossilized microorganisms   e) granite

3) Why did they have to stop drilling the Kola Superdeep hole?

   a) they got tired   b) they ran out of money   c) the rock got too hot and soft   d) the facility exploded

4) True or false? Geologists have drilled down into the mantle. ____

5) Which of these can a seismograph NOT do?   a) record earthquakes   b) predict earthquakes

   c) determine parts of the inner earth that are liquid or solid   d) tell the difference between granite and basalt

6) True or false? All seismographs have needles that draw lines on a roll of paper. ____

7) True or false? P and S waves are classified as "body waves." ____

8) What words can the "P" in "P waves" represent? _____ and _____

   (NOTE: Some teachers like to point out that P might also stand for "push/pull.")

9) What words can the "S" in "S waves" represent? _____ and _____

   (NOTE: In addition to these, you might also think of "shake," "shift" and "side to side." Lots of choices!)

10) True or false? Love waves are much less destructive than P waves. ____

11) True or false? When scientists use Rayleigh waves to test rocks, they end up destroying buildings. ____

12) Which of these "R" words would be helpful to associate with Rayleigh waves?

   a) red   b) really   c) resting   d) reacting   e) rolling

13) An area where seismic waves don't appear on the other side of the earth is called a:

   a) bright zone   b) dark zone   c) shadow zone   d) seismic zone

14) True or false? P waves can travel through liquids. ____

15) True or false? S waves can travel through liquids. ____

16) Which is solid? a) inner core   b) outer core

17) What elements are probably in the earth's core?   a) oxygen and nitrogen   b) carbon and silicon

   c) iron and nickel   d) copper and zinc   e) sodium and magnesium

18) True or false? We know what is in earth's core because we've drilled into it and taken out samples. ____

19) Most of the volume of the earth is the: a) crust   b) mantle   c) outer core   d) inner core

20) Which one of these is probably NOT in the mantle?   a) calcite   b) olivine   c) feldspar

21) What tool do you need to study the properties of hot magma? a) seismograph   b) borehole drilling rig

   c) diamond anvil   d) high-carbon steel forge

22) When magma drains down into the center of the earth, it reaches a "point of no return" called the

   _____ _____. Below this point magma is so dense that it can't possibly rise again.

23) At about what depth does this happen? (from number 22)

   a) about 1 mile (2 km)   b) about 20 miles (32 km)   c) about 200 miles (320 km)

24) About how thick is the mantle? (approximately)

   a) about 200 miles (320 km)   b) about 2,000 miles (3200 km)   c) about 20,000 miles (32,000 km)

25) Which of these is NOT true about the mantle?

   a) It is the largest part of the earth.   b) It is probably made of mafic rocks like olivine, pyroxenes and basalt.

   c) Seismic waves tells us it is solid.   d) It is very cold.

Questions about the crust:

26) What is the Moho?   a) the dividing line between the crust and the mantle   b) the oceanic crust

c) the dividing line between the mantle and the outer core   d) a great restaurant in New York City

27) Which is thicker, continental crust or oceanic crust? _____

28) What type of rock is the oceanic crust made of? _____

29) The tallest mountain in the world, from the base to the tip, is: a) Mt. Everest   b) Mauna Kea

30) Ocean sediments are made of either clay or ooze. What is ooze?   a) sloppy mud   b) silt

c) dead microorganisms   d) slime produced by living microorganisms

31) Because of what it is made of, ocean ooze is most like: a) chalk   b) calcite   c) quartz   d) shale

32) The shape of a landscape, with its mountains, valleys, plains and canyons, is called its:

a) topography   b) stratigraphy   c) geology   d) archaeology

33) The first clue about the shape of the ocean floor came in the 1800s during a task completely unrelated to geology. What was this event? _____

34) What structure was found during the event (in number 33)? _____

35) What island sits right on this structure? _____

36) What mountain range is an extension of this structure ? _____

37) What is the maximum thickness of the continental crust?

a) about 2 miles (3 km)   b) about 10 miles (16 km)   c) about 60 miles (90 km)   d) about 100 miles (160 km)

38) Can you name the three types of faults? _____, _____, _____/_____

39) Name a famous fault in California: _____

40) Will the Lewis Overthrust fault ever slip during an earthquake? _____

41) What caused the erosion on top of the Lewis Overthrust?   a) rain   b) snow   c) wind   d) glaciers

42) What kind of rock is on top of Mt. Everest?   a) igneous   b) metamorphic   c) sedimentary

43) What surprising thing was found on top of Mt. Everest?   a) trash   b) plants   c) fossils   d) insects   e) bones

44) Which one of these is NOT a formation found in the Grand Canyon?

a) Kaibab Limestone   b) Redwall Limestone   c) Tapeats Sandstone   d) Yellow Band

45) What river is found at the bottom of the Grand Canyon? _____

46) Fossilized tracks of animals are classified as: a) trace fossils   b) trail fossils   c) not fossils at all

47) The first geological layers discovered were named after this tiny country: _____

48) True or false? Birds can be found in the layers that contain dinosaurs. _____

49) True or false? The Grand Canyon has all the geological layers. _____

50) What kind of rock is often below a water table? (Hint: It does not let water pass through it.)

a) shale   b) sandstone   c) limestone

51) True or false? Aquifers are always small, only a few acres at most. _____

52) Where is the easiest place to dig down to the water table?   a) mountain   b) field   c) desert   d) beach

53) For a natural spring or artesian well to work, the top of the well must be:

a) at the same level as the top of the aquifer   b) higher than the level of the aquifer   c) open   d) closed

54) Boiling groundwater can burst out of the ground as a: a) hot spring   b) geyser   c) volcano   d) tornado

55) True or false? Human accidents are the only way that oil gets into the ocean. _____

56) True or false? Water, oil and gas can be found very close together in some rock layers. _____

57) What organism helps clean up oil spills by eating the oil? _____

58) Places where oil and gas leak up out of the ground are called: a) slips   b) seeps   c) creeps   d) wells

59) Natural gas is often found at the top of a: a) syncline   b) anticline   c) geocline   d) incline

60) True or false? The earth has endless amounts of coal and oil. _____

Activity 7.2 A "FAULTY" PUZZLE

Geologists are curious about the past (as we all are) and like to try to figure out what happened in and around fault lines. In this activity, you will be a geologist trying to reconstruct the order of events that led to certain formations. Write the letters in the correct order, from first event to last. The puzzles get harder as you go along.

NOTE: In all the pictures, the letter C represent an "igneous intrusion." This is an area where magma came up through the sedimentary rock and then cooled into hard rock. (The pattern can be granite or any igneous rock.)

EXAMPLE:

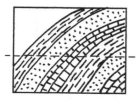
We start with sedimentary layers that have tipped up.

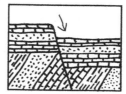
An overthrust occurs and cuts off the angled rocks.

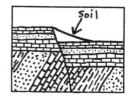
A fault develops and one side sinks.

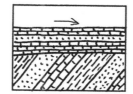
Soil fills in the gap left by the fault.

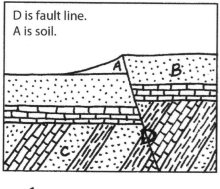

1: ___, ___, ___, ___

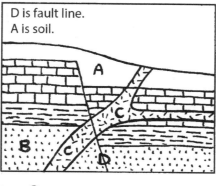

2: ___, ___, ___, ___

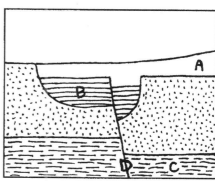

3: ___, ___, ___, ___

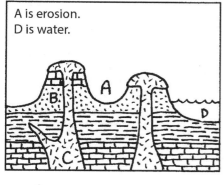

4: ___, ___, ___, ___

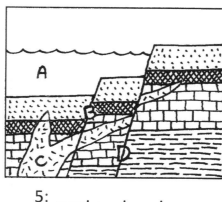

5: ___, ___, ___, ___

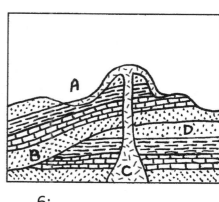

6: ___, ___, ___, ___

7: ___, ___, ___, ___

8: ___, ___, ___, ___, ___

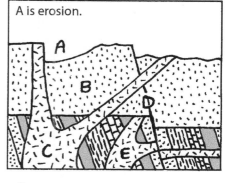

9: ___, ___, ___, ___, ___

Activity 7.3   LEARN ABOUT TRILOBITES

Apart from dinosaurs, trilobites are probably the best-known extinct creature. Their name means "three lobes." The three lobes are the cephalon (head), thorax (middle) and pygidium (*pig-ID-ee-um*) (small end part). They range in size from just a few millimeters (1/8 inch) to about 20 cm (12 inches). The modern animal they are most similar to is the horseshoe crab, but they might also remind you of woodlice (pill bugs/sow bugs/roly polies).

Trilobite fossils are found all over the world in the lower geological layers: the Cambrian, Ordovician, Silurian, Devonian, and Permian. Sedgwick and Murchison would have seen many trilobites in those layers they discovered in Wales. Other famous trilobite sites with large numbers of spectacular specimens include Canada (the Burgess Shale mentioned on page 102), Morocco, Germany, and New York and Utah in the U.S. (The native peoples of Utah wore trilobite fossils on necklaces.) Over 17,000 different species have been recorded, although we can't be absolutely sure they were all truly different species because we don't have any of their DNA to examine.

A few fossils have the underside preserved and from this we gain a little bit of information about how their insides may have worked. They breathed the same way crabs and lobsters do, using fuzzy-looking gills near the top of their legs. Judging by their body structures and their fossilized surroundings, it appears as though some of them ate bits and pieces of dead things that fell to the sea floor, while others were either predators or filter feeders.

Trilobite from Russia (By Vassil - Alias Collections., CC BY-SA 3.0, https://commons.wikimedia.org/w/index.php?curid=3200496)

Trilobites had a wide variety of body shapes. Some were very simple, but many had a complex arrangement of oddly shaped spines. The purpose of these spines is unknown. Trilobites also had an incredibly wide range of eye shapes and sizes. We have amazingly well-preserved fossils that show the eyes in great detail. Their eyes were similar to those of insects, having many tiny lenses. Their vision was probably excellent, much like a dragonfly's. Some of the eyes were shaped like tall cylinders, and a few look like they have sun shades at the top.

Some trilobites had antennae, others did not. The antennae shown below certainly are odd-looking, Is it one or three?

Trilobite eye (By Moussa Direct Ltd. - Moussa Direct Ltd. image archive, CC BY-SA 3.0, https://commons.wikimedia.org/w/index.php?curid=4437498)

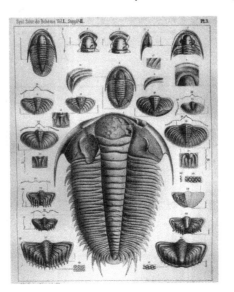

Trilobite from Morocco (By Kevin Walsh - originally posted to Flickr as Trilobite 3, CC BY 2.0, https://commons.wikimedia.org/w/index.php?curid=12244016)

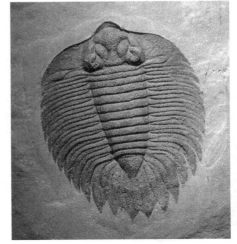

Activity 7.4 VIDEOS

As usual, there are some really helpful videos over at the YouTube channel. If you don't have time to watch all of them, at least watch the ones on the P and S waves (if you haven't already).

Activity 7.5 CANYON COLORING PAGE

This activity is designed to be used with a video that is posted on the YouTube playlist (also listed at www. ellenjmchenry.com/downloads/GrandCanyonDrawing.mp4). The video will tell you exactly what to do. However, if you don't have Internet access, here are some instructions to help you color and label it yourself. If you would like to see how the finished drawing is supposed to look, check the answer key. If the drawing looks too difficult, you can just do the parts that interest you. You could also use colors instead of those patterns.

The location of the diagram is the South Rim of the canyon, which is the most popular place for tourists to visit. It is important to remember that the canyon is hundreds of miles long. The layers change their depth, color, and mineral content in various places in the canyon. If you read information that is slightly different from what you read here, both pieces of information are probably correct, but for different locations. The canyon is very complicated!

You may notice that some of the layers have very sloped edges, while other have relatively straight (vertical) edges. The sloped layers are the ones that erode more easily. Harder layers are resistant to erosion are more like straight cliffs. The reddish color of some layers is mostly due to the presence of iron oxide (rust!) in the minerals.

1) Start by labeling the layers of the Cenozoic and Mesozoic. (Write these words between the dotted lines.) These top two layers are actually missing! These were removed before the canyon formed. (Geologists call this the Great Denudation. One very convincing idea is that a huge nearby lake spilled over and washed these layers off. Slow erosion can't explain the flat surface.) The top layer is the Cenozoic. The second layer is the Mesozoic. (The Mesozoic contains those "dinosaur layers.")

2) The next section is the Permian. There are 4 rock layers in the Permian.

3) The section below the Permian is the Pennsylvanian. (This does not mean the layer is in Pennsylvania!) There are 4 rock layers in the Pennsylvanian layer. (In other places, the Pennsylvanian contains coal, but here it does not.) The layers of the Pennsylvanian are also called the Supai Group. Often, the individual names (listed in number 8), are not used because they are hard to remember.

4) The layer below the Pennsylvanian is the Mississippian. In the canyon, the Redwall Limestone is Mississippian.

5) The Devonian layer is that little wedge-shaped piece.

6) The Silurian and Ordovician layers are missing. (Remember, we said that there are just a few places on earth where you can find all the layers.)

7) Below the wedge is the Cambrian layer. Below that (the side of that V-shaped notch at the bottom) is the Precambrian.

8) The V-shaped notch is the Inner Gorge. The tiny triangle at the bottom is the Colorado River.

9) Now we'll write the names of all the rock layers. That very tall column on the right side is for names and extra notes you might want to jot down. (Write small!) The very top box will just say that these layers are missing. The first actual layer of canyon rock is the famous Kaibab Limestone. Going on down from there the boxes should say: Toroweap Formation (you can abbreviate "Formation" as "Fm"), Coconino Sandstone (you can abbreviate "Sandstone" as "Ss"), Hermit Shale, Esplanade Sandstone, Wescogame Fm., Manakacha Fm., Watahomigi Fm., (don't worry about these hard names—you don't have to memorize them!), Redwall Limestone, Muav Limestone, Bright Angel Shale, and Tapeats Sandstone. That should bring you to that thick wavy line. (If you want to label that Devonian wedge on the left side of the Muav Limestone, it is called the Temple Butte Limestone.) All these names might sound strange and difficult to you, but to a geologist they are very familiar because these layers are studied so much.

10) The thick wavy line is called the Great Unconformity. This is a dividing line between the Cambrian and Precambrian areas. What went on here? It looks like the Precambrian layers got tipped and then cut off. Whenever there is a huge difference between two layers—something that looks abnormal or unexpected—it is called an "unconformity." There are smaller unconformities at the top and bottom of the Redwall Limestone.

11) If you want to know the names of the tipped layers, they are (from top to bottom): Sixty Mile Formation, Kwagunt Formation, Galeras Formation, Nankoweap Formation, Cardenas Lava, Shinumo Quartzite, Hakatai Shale, and the Bass Limestone. (If you'd rather not label these, that's fine. Or, you can just label the whole group as the "Grand Canyon Supergroup.")

12) Now we are ready to fill in the patterns and/or colors. (Leave that sectioned off strip blank. We will use that strip to write a few tiny notes, such as types of fossils.) If the patterns are too difficult, you can just decide to make each one a different color (with mudstone as gray). Just color over the patterns and use the colors as your key. Some of the layers are self explanatory, and have the words limestone, sandstone, and shale in their names. The ones you need more information about are listed below. (NOTE: The layers are actually much more complicated than this, and include thin stripes of mudstone, siltstone, shale, and conglomerate in almost all of the layers. The purity of the limestone varies greatly, also.)

a) The Toroweap Formation is often called the Toroweap Limestone, but it also contains dolomite, sandstone, mudstone, gypsum and shale. (Any model of the layers of the canyon is going to be overly simplified and not represent reality because the real layers are very complicated.) Since we have limited space, try this from top to bottom: one layer of limestone, some sandstone dots, a layer of shale, a blank space for gypsum, dolomite on the bottom. To indicate dolomite, just make your "bricks" look slanted.

b) The Coconino Sandstone is a "cross-bedded" sandstone. This means that there are layers going different ways, as shown in the pattern key. You can add little dots of sand, too, if you'd like.

c) In the Esplanade Sandstone there are a few layers of mudstone. Just add a few gray lines with a pencil. The Wescogame and Watahomigi Formations also have a few layers of mudstone.

d) The Redwall Limestone actually isn't solid limestone. It has a tiny bit of sandstone at the top, then a layer of mudstone, then limestone, then conglomerate, then a thin layer of limestone on the bottom. Can you fit all those into that space? Think small!

e) That rectangle in the Bright Angel Shale is a block of quartzite that is about the size of a bed or table. It's pretty small, so you might have to just color or label it, instead of trying to draw the pattern inside of it. (Unless you can fit one wavy line in the middle and some dots around it.) How did it get up there from that quartzite layer below (or somewhere else)? Perhaps this shows us that these layers were not solid rock when that block migrated upwards. You can see this block for yourself if you do an image search for "quartzite block in Grand Canyon." You can see how the layers of shale are folded around it.

f) Those tipped Precambrian layers are, from top to bottom: sandstone, shale, limestone, sandstone, basalt lava, sandstone, quartzite, shale, limestone.

g) The big areas at the bottom, around and under the tipped block, are schist. Schist is a metamorphic rock that used to be something else, in this case perhaps granite. Schist's layered look is because the original rock got squeezed, causing the atoms of SiO_2 to rearrange themselves into more compact layers. The things coming up from bottom, looking like igneous magma intrusions from below, are called the Zoroaster Granite. The schist's name is the Vishnu Schist. These names are not Native American, but are from Persia and India. (NOTE: If you look at other diagrams on the Internet, you will see that these granite intrusions are drawn in different ways. They might look very different on someone else's diagram. Remember, these are just diagrams, not photographs!)

h) Don't forget that tiny bit of Bright Angel Shale on the opposite side of the canyon. Our picture got cut off at that point so you can't see the Muav Limestone above it.

13) Now we can add fossils. You can draw simple shapes to represent the fossils, but you can also write notes in that strip on the right. You can draw the fossil symbols right on top of the patterns. (If you would like to know more about each type of fossil, use the Internet to search for information or pictures.)

a) The Kaibab Limestone has marine fossils: crinoids, clams, sponges, corals, bryozoans.

b) The Toroweap has fewer fossils, perhaps just a few brachiopods at the top.

c) The Coconino sandstone has only trace fossils, such as tracks of animals that look like they may have been reptiles or amphibians. (No bodies, just tracks.) The Coconino also has many ripple marks, made either by wind or water. (Add ripple marks.)

d) The Hermit Shale, and all 4 Supai layers (Esplanade, Wescogame, Manakacha, Watahomigi) have mostly just plant fossils, such as ferns, horsetails (which look like bamboo) and conifers (pines). There are some animal tracks in the Watahomigi.

e) The Redwall has lots of marine fossils: crinoids, clams, brachiopods (similar to clams), nautiloids (which look like squids that got stuck in long ice cream cones), bryozoans, corals.

f) The Muav Limestone also has marine fossils, but trilobites also start to appear.

g) The Bright Angel Shale has marine life such as trilobites, brachiopods, and clams, but also has trace fossils of worm burrows.

h) The Tapeats Sandstone has trilobites, brachiopods, and trace fossils such as ripple marks and trilobite tracks.

i) The tipped layers of the Precambrian that are <u>above</u> the lava layer, contain **stromatolites**, which are fossilized remains of large mats of blue-green algae (cyanobacteria). You can make some little rings to represent them, (as shown in photograph).

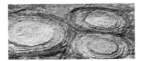

stromatolites

If you want to add some color to your drawing without covering up the patterns, you can just add a little color to the left edge, which is the surface you actually see at the canyon. Color suggestions:

Kaibab: light gray or tan; Toroweap: darker yellowish-gray; Coconino: white or cream; Hermit: rusty red; Supai group: red limestone, tan or yellow sandstone (lots of thin red strips in these 4 layers); Redwall: light red; Muav: grayish yellow-green; Bright Angel: rusty red; Tapeats: light tan; 60-mile: tan; Kwagunt: reddish purple; Galeros: greenish gray; Nankoweap: tan; Cardenas lava: dark gray; Dox: orange-tan; Shinumo quartzite: light reddish-purple (or use pink); Hakatai: red; Bass: gray.

ROCK LAYERS OF THE GRAND CANYON

This is the "South Rim" area. The top of the rim is about 7,000 feet (about 2,000 meters) above the bottom of the gorge.

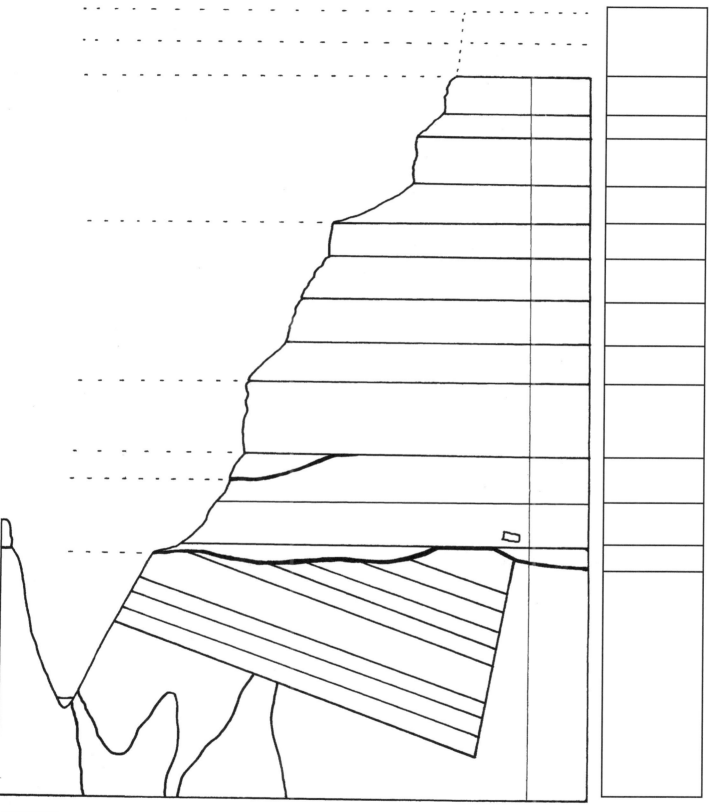

limestone   shale   granite   mudstone   schist

sandstone   cross-bedded sandstone   basalt (cooled lava)   quartzite   conglomerate

Activity 7.6   OLD FRIENDS

These guys should seem like old friends by now. How much can you remember about our friends?

SANDSTONE
I am classified as a _____ rock. My sand grains have this chemical recipe: _____. See all those tiny holes in me? One of the most important things to know about me is that I am "porous." This means that _____ drains through me very quickly. If you drill a well, the a_____ that you tap into will most likely be made of sandstone. If I get pressed and heated, I can turn into a metamorphic rock called q_____.

GRANITE
I am classified as an _____ rock even though no one has ever actually observed me forming from lava. I am made of a combination of minerals, often including clear or white _____, pink _____, black b_____, and black h_____. When I melt, I make f_____ lava because I contain so much silicon. I am an important part of the crust of the earth because I am under all the _____.

SHALE
I am classified as a _____ rock. Before I was shale, I was wet c_____, which is made of microscopic flat pieces of SiO_2 (quartz). I can be a bit flaky, and come apart in layers. If I am squeezed and heated, I can turn into s_____. Unlike sandstone, I do not let water pass through me. I cooperate with sandstone to create _____ tables. When you find me deep underground, you are also likely to find oil and natural _____.

CHALK
I am one of the softest _____ rocks. I am so soft, in fact, that I leave chalk marks all over anything I am scraped against. If you look at me under a microscope, you will see many broken shells of _____. These tiny animals took _____ out of the water and used it to make their shells.

LIMESTONE
I am also classified as a _____ rock. My basic chemical formula is _____, although I often contain small amounts of other substances. When I contain a lot of magnesium, my name changes to _____. I am by far the earth's largest source of the element _____. Large areas of limestone are called _____ landscapes and often have hidden holes called c_____.

GEODE
I am considered more of a mineral than a rock. Like my cousins the agates, my chemical formula is _____. When you find me, I look like a rock ball, and you have to crack me open. If I am hollow I will contain lots of crystals. A common crystal you might find in me is amethyst, which is a type of q_____.

GYPSUM
I don't always look like this. Sometimes I just look like a white lump. When I can develop good crystals, though, I look very attractive, don't you think? I am different from all the other rocks on this page because I don't contain any carbon or silicon. My formula is _____. Formations made of me are found in caves, suggesting that I may have helped form the cave by providing _____ acid.

CHAPTER 8: IDEAS ABOUT THE PAST

As visitors look out over the Grand Canyon, they can't help but wonder, "What happened here?!" It looks like a giant crack opened up in the crust of the earth. They might also marvel at the many layers of rock that create such colorful stripes, or the interesting fossils they contain, such as trilobites and nautiloids. Many features of our planet catch our attention and make us wonder how they formed.

Perhaps the most frustrating thing about geology is that so much of it occurred in the past, and we are stuck in the present. We can test rocks for their mineral content or heat lava in a diamond anvil, but these experiments only give us information about the chemistry of the sample we are testing. Chemistry experiments can't tell us exactly how or when the sample formed.

Things we can actually observe are classified as "facts." It's a fact that quartz is made of SiO_2. We can do chemistry experiments on samples of quartz to find out what it is made of. All the properties of minerals are facts that came from observation. We can make factual diagrams of where fossils are located in the Grand Canyon. We can measure soil horizons and determine how much clay is in each one. We can measure erosion on Surtsey Island. All these observations are facts because we can observe them directly.

We often use facts to make an **inference**. (We saw this word back on page 91.) An inference is a very educated guess made by using facts. An excellent example of this is how we use facts about P and S waves to make inferences about what is inside the earth. Our *facts* are the measurements of how long it takes for P and S waves to reach various seismometers around the globe. From these measurements, we make *inferences* about the nature of the inner earth. It's a fact that S waves don't travel through the very center of the earth. Our inference is that the

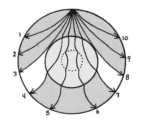

outer core is a liquid. We can use facts that we know about the elements on the Periodic Table to infer that this liquid core is probably made of very heavy metals such as iron and nickel. But no matter how certain we are that the core is made of these elements, this will always be an inference, not a fact, because we can't actually observe the core. Inferences are often correct, as this one is likely to be. However, more data about the core might become available in the future, and this inference could need to be adjusted. Some scientists are already saying they think that the core has many more elements in it.

This distinction between fact and inference is very important when scientists start making guesses about what happened in the distant past. It can be a lot of fun to try to solve the mysteries of how things got to be the way they are, but these guesses can never become facts, because facts are things that you can actually observe. Now, if we had a time machine and could go back into the past, we would be able to actually observe what happened, so these observations could be classified as facts. And what amazing facts these might be! But until someone invents a time machine, we are stuck in the present, making our educated guesses.

Ideas about the past get really complicated because life is complicated. Scientific study of the natural world is just one part of life on planet earth. There are so many aspects to life—art, music, literature, drama, sports, law, politics, philosophy, religion, cultural heritage, charity, and so on. So when we look at the world, we look at it through a complex set of "lenses" that include many factors. Two people can look at the same bird or rock or tree and have very different ideas about them. This book isn't going to tell you which factors should be most important to you. That's for you to decide. We'll just look at a few ways that geological features have been interpreted and see how these ideas came about. (And let's hope the guys with the time machine behave themselves.)

People in the past had a very limited number of facts about the crust of the earth. As we learned in the last chapter, no one knew about the mid-ocean ridge until the 1800s. Seismology in the 1900s brought in a large amount of new data, and we found out about large cracks (fault lines) in the continental crust. These fault lines were traced and mapped, and you've probably seen a diagram of them, similar to the one shown here. The pieces between the cracks are known as **plates**. In this picture, the plates have been given different colors so that you can see them easily. (Seeing them on a globe would be even better, though, because the world isn't flat like this map!)

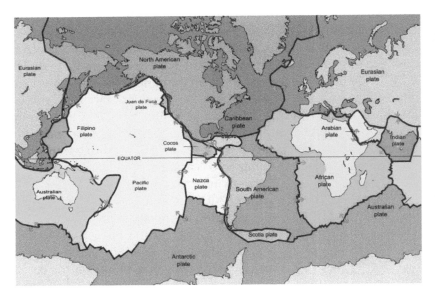

The invention of GPS (Global Positioning System) gave us a new type of data. GPS can be used to sense the motion of these plates. The world map below shows the directions of movement over the course of several years. The arrows are long and large just to make them easier to see. The continents did not move this much! They only move a few centimeters per year. Notice that more arrows point to the Pacific Ocean than anywhere else.

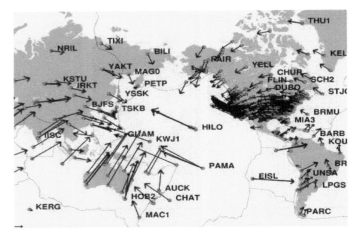

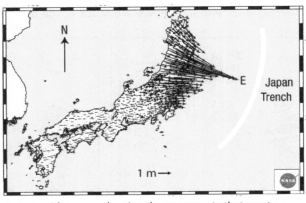

Japan, with arrows showing the movements that went on during the March 2011 earthquake.

The map of Japan (on the bottom of page 116) shows movements detected by GPS during the earthquake that occurred on March 11, 2011. Each arrow indicates a shift of one particular point that the GPS was monitoring. The letter E represent the epicenter of the earthquake, where the quake started. In this case, the epicenter was not at the surface, but 20 miles (32 km) below the surface, right along a fault line. This fault is a particular type, called a **Benioff zone**. The crack is wide and flat, and goes down into the mantle, right under Japan. Imagine taking a cake knife and sticking it down into that picture right at the top of that curved white line; then lean the knife handle to the right so that the point is to the left, under Japan. Then make the knife follow that curved white line, slicing down and under Japan. That is the shape of this Benioff fault.

Epicenters can be very deep.

Earthquakes often occur at the fault lines between the plates, so many geologists say that the movement of the plates is what causes earthquakes. However, quite a few earthquakes occur in the middle of those plates, too, so moving plates can't be the only cause of earthquakes. And what causes the plates to move in the first place?

Until the 1900s, most people thought that the continents did not move at all, which is understandable since the movement is only about the speed at which your fingernails grow. When you consider how large the continents are, that much motion is incredibly small. The edges of the continents are probably eroding away faster than that. People thought that either the earth has always looked the way it does now, or else the earth has been gradually shrinking, causing wrinkles (mountains and valleys) in the crust. Some thought that the continents might be moving up and down, rising and sinking, but no one thought they were moving side to side.

The earliest record of a geologist suggesting that land masses moved was a man named Abraham Ortelius in the year 1596. However, no one at that time took any notice, and the idea quickly faded away. The next person to suggest that continents could move was Roberto Mantovani in 1889. He had learned quite a bit about the geology of both Africa and South America. He noticed that the rocks on the edges of these continents were very similar and he wondered if that could mean that the continents were once much closer together, if not actually joined. Two decades later, in 1908, a geologist named Frank Taylor proposed that the continents were "creeping" toward the equator due to earth's recent capture of the moon during the dinosaur era. Taylor also thought that the Himalayan mountains were the result of a "collision" between India and Asia. A few other scientists were saying similar things, but rarely are any of them ever mentioned in textbooks.

One man has gone down in history as the person who proposed continental movement: Alfred Wegener (in 1912). Wegener read the papers written by Mantovani, Taylor, and others, and put all the ideas together into one theory he called "continental drift." He believed that the continents had started out as one large land mass that he called **Pangea**. ("Pan" means "all," and "gea" means "earth.") He said that Pangea had broken apart and the pieces had started drifting in different directions at a speed of about 250 centimeters (100 inches) a year. He had some very good evidence that the continents had been joined together at one time, such as similar fossils on both sides of the Atlantic Ocean. However, Wegener's proposal did not include a mechanism for how the continents broke apart in the first place and what force had caused their movement. Because of these two issues (and perhaps the fact

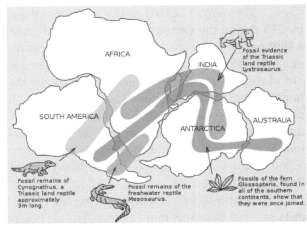

Similarity of fossils across the continents

that Wegener did not have a degree in geology), his theory was not accepted by the major geological societies of his day. In fact, his theory was very unpopular for almost 40 years. In the 1940s, geology professors at universities in England were still ignoring Wegener's theory, and sometimes they even made fun of it. One professor called it "nothing by moonshine." This is an important point to remember. New theories are often set aside or rejected at first.

The general acceptance of Wegener's theory only came about after yet another theory was proposed in the 1960s. One of the geologists who helped to develop this new theory was named Robert Dietz, and he will come back into the story again later in this chapter. Dietz was one of a group of scientists (along with Bruce Heezen who is shown next to Marie Tharp) who developed a theory that would explain not only what was moving the continents, but also why there was a long mountain range in the middle of the ocean. Ever since the mid-ocean ridges had been discovered in the late 1880s, scientists had been puzzling over how they might have formed. Why was there a mountain range running right down the middle of the Atlantic Ocean? Why not just scattered mountains here and there? Why a line? According to this new theory, which they called "plate tectonics," the mid-ocean ridges are places where magma is coming up from the mantle. It reaches the surface at the center of these ridges and spreads out and to the sides. When the magma cools, it turns into new oceanic crust. Hydrothermal vents are found along these spreading centers.

If new ocean floor is constantly being formed at these mid-ocean ridges, then what happens to all the old ocean floor? The theory goes on to propose that edges of some continental plates are diving down and disappearing into the mantle. They call these edges **convergent zones**, because plates "converge," which means "come together." The mid-ocean ridges were named **divergent zones**, because they "diverge," or "go apart." Go back and look at the colorful world map on page 116. The red arrows are there to indicate whether each line is a convergent or divergent boundary.

Now look really carefully at that map. Find three places where the plates come together to make a T intersection. Are there any red arrows at these places? Would putting arrows here be a problem? What would be happening right at the T?

Plate tectonic theory also suggested a reason that the plates were moving. Wegener had proposed a tidal force coming from the moon, but the physics of this did not add up. The moon could never provide enough force to move an entire continent. But it was an interesting idea, for sure. Plate tectonics said that

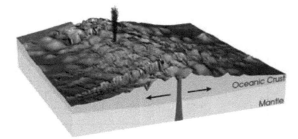

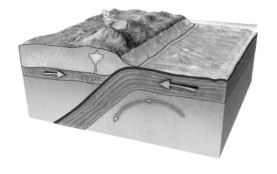

perhaps the mantle underneath the plates was moving and the plates were just going along for the ride. The theory said that **convection** is happening in the mantle. **Convection** is when heat from one area spreads to another area. If you have a heater running in a room, the heat doesn't stay right near the heater—it rises and fills the room because of convection. When a moving air molecule hits a neighboring molecule, it starts its neighbor in motion. The motion keeps getting transferred from molecule to molecule and soon they are all moving. The fast-moving air molecules spread out and therefore become less dense, so they rise. Boiling water in a pan also experiences convection.

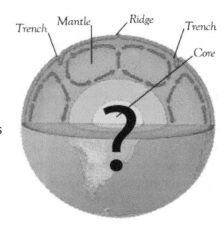

Convection can create flow patterns, and sometimes currents will form. This is what they proposed for the mantle. Hot magma, they said, was rising up through the mantle to reach the surface. Now, this theorizing was going on back in the 1960s. The diamond anvil experiments with magma had not been done yet. These scientists in the 1960s did not have information on the cross-over depth of magma. They did not know that magma can't rise if it is below about 300 km (220 miles). They also did not have as much seismic data as we have today. Seismic data is confirming that the mantle is solid.

After a few decades, people stopped questioning the theory of plate tectonics, and this is the theory that now gets published and printed in just about every textbook. But does anyone have any other ideas? If so, are other scientists likely to listen to them? (What happened to Wegener?)

One of the scientists who provided seismic data to people who were creating maps of the ocean floor was N. Christian Smoot. Mr. Smoot's job was to study the ocean floor and provide information to the U.S. Navy so they could make maps for their submarines. Submarines need maps, too. They don't want to get lost in a maze of undersea trenches and mountains. From 1966 to 1998, he went on 67 scientific ocean cruises and collected not only seismic data but magnetism and gravity measurements, too. He probably knows more about the ocean floor than anyone else in the world. In the 1990s he began looking over all the data he'd collected over the years and realized that plate tectonics can't explain many features of the ocean floor. In fact, some features almost seemed to prove that plate tectonics is wrong. For starters, look back at that colorful map on page 116. Do you see any red arrows around Antarctica? Mr. Smoot says, "The realization that the Antarctic plate has no convergent margins whatsoever should have raised a red flag."

Smoot has a whole list of problems with plate tectonics. He points out that the Eurasian plate has the same problem as the Antarctic plate: it has no convergent margins. This means that old sea floor should be disappearing under these plates and he found zero evidence that this is happening. Another huge problem can be discovered if you add up the number of kilometers where the new seafloor is supposedly coming out and spreading, and you compare it to the number of kilometers where the old seafloor should be disappearing. These numbers should be about equal. There are about 148,000 kilometers of mid-ocean ridges (where spreading should occur) and only 30,500 kilometers of trenches, where seafloor should be disappearing. That's not even close. Another glaring error, he says, is the term "mid-ocean." Look at where the ridge (or "rise") is in the Pacific Ocean. It's off to one side, not even close to the middle. Is the floor of the entire Pacific Ocean being pushed westward? He finds this very hard to believe. And speaking of the mid-ocean rise in the Pacific, there isn't really any ridge there at all—that's why they call it a "rise." It doesn't look anything like the Atlantic Ridge, and Smoot has found no evidence that it is spreading. He also is very critical of the theory for not explaining the curved shapes around the outside of the Pacific Ocean.

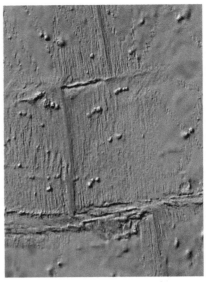

East Pacific Rise and two of the fracture zones that have offset it

And what about the mechanism by which these plates are supposedly traveling—the currents of magma that are slowly convecting in circles and dragging the plates along with them? Smoot points out that this breaks a law of physics. A simple way to explain his argument is to say that the magma has nothing to push against. Not only that, but Smoot's own seismic data strongly suggests that the mantle is not convecting. Places that plate tectonics says should be hot are actually cold, and places that should be cold are actually hot!

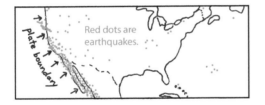

Red dots are earthquakes.

plate boundary

Smoot also studied earthquakes and noticed something we've already mentioned, that many earthquakes do not occur along those fault lines between the plates. Plate tectonics says that earthquakes are the result of plates rubbing and slipping past each other. Smoot gives many examples of places where this is obviously not true.

Smoot has proposed an alternative to plate tectonics that he calls "surge tectonics." He says that his data suggests that magma might follow channels that are deep under the crust. These channels would be under every single geological feature on the planet. The driving force for magma would be the rotation of the earth. As water in a bowl goes in the same 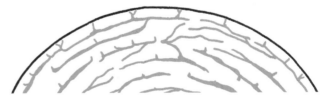 direction as the bowl if you spin the bowl, so the magma inside the earth goes in the same direction as the earth. He thinks that occasionally a lot of magma comes to the surface, emptying the channels. This allows the channel areas to cool, and then this causes it to crack, break, and sink down, creating ocean trenches. If the magma gets stuck in a channel, it starts to build up and thus creates upward bulges. (Since this is a new theory, there aren't a lot of specifics to discuss. Perhaps in a few years there will be websites and articles to recommend.)

Now we return to the office of Robert Dietz, but we fast forward to the 1980s. Sitting in his office is a scientist named Walter Brown. Dr. Brown is asking Dr. Dietz many of the same questions that Mr. Smoot asked about plate tectonics, though at this point in time Mr. Smoot has not yet written his books, so Dr. Brown doesn't know that someone else is asking the same questions. Dr. Brown is disappointed that the reply to most of his questions is always the same: "Walt, I haven't got a clue." Dr. Brown ponders these problems with plate tectonics and thinks that there must another option out there, perhaps something that no one has thought of yet. He begins studying maps, reading research journals, and traveling around the country doing his own observations of many geological features.

In his search for clues, he also does something that most geologists would find quite shocking. He reads ancient texts to see if people who lived long ago left any clues about what the world looked like thousands of years ago. What he finds is that almost every culture on the planet has a story about a great flood. Not surprisingly, these stories have some differences because they are so old that they were handed down through many generations before writing was invented. Stories often change over time as they are told over and over again. However, these stories also have much in common, and one theme that runs through most of them is that the flood waters covered the earth. What did these ancient people see? Are they trying to tell us something? We might laugh at the idea of taking ancient legends seriously, but if something amazing really did happen back then and they wanted to tell future generations about it, what were they supposed to do—make a video? They could not even write a book because the alphabet had not been invented yet!

A flood painting from Persia, painted thousands of years after the time when the story began.

The version of the flood story that seems to be the most factual is the one recorded in the Hebrew Torah. It has the greatest number of measurements and other technical details and the fewest improbable details such as talking animals. It states that on a certain day, the "fountains of the great deep burst forth" and this is what started the flood. What does this mean? Could this be an important clue? Could erupting water explain some of earth's strange features? Dr. Brown decided to find out.

He started by assuming that this phrase was literal and that water had come up from below the earth somewhere. He calculated how much water must have been down below in order to have covered the earth. But wait—how did he know that the water had not greatly changed the crust? Could mountains or ocean trenches have formed as a result? What about continental drift? Could this water explain why they had separated?

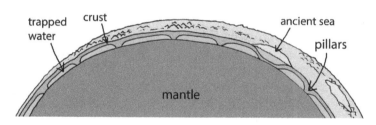

trapped water — crust — ancient sea — pillars — mantle

Dr. Brown suggested that not only were the continents once connected, but that the granite crust of which they were a part was actually a complete sphere, forming the earth's outer shell. This shell might have been anywhere from 10 to 60 miles thick (15 to 90 km). Underneath this would have been the water, and then under that would have been what we now call the mantle, made out of olivine and basalt. The outer crust of granite was not floating on water, of course. Water was trapped underneath it. He guessed that there would have been places where the crust bowed down and touched the mantle, creating wide pillars.

So then what happened? At this point he widened his vision a bit and thought about the moon. Today, the moon pulls the tides back and forth. What would the moon have done to water trapped under the crust? Same thing—it would have pulled the water back and forth. The water would already have been under pressure, and this extra pulling would have made things worse. Laboratory experiments have shown that when water is under this kind of pressure it gets far hotter than anything we've experienced in the kitchen, even with a pressure cooker (canner). Water goes "super critical." If this sound dangerous, you're right, it is. If you could see super critical water, it would look a bit like steam, but the droplets would be much smaller, perhaps consisting of only a few dozen molecules. Tiny water droplets like this have two important properties: 1) they can dissolve things like you wouldn't believe, and 2) they can cool almost instantly. That super critical water would have been eating away at the bottom of the granite crust, and the top of the basalt mantle. Dissolved minerals would be floating in this water and also probably collecting in piles at the bottom as they precipitated out of the solution and gravity pulled them down. However, everything was fine as long as the granite crust remained intact. But, what if... a crack developed?

If the earth's crust had cracked open, shouldn't we be able to detect the scar? Dr. Brown guessed that perhaps this is what the Mid-ocean Ridge might be. However, it would not be the actual crack, but the place right under where the crack was. The crack would have allowed that super critical water to escape from below, and it would have come out with such tremendous force that the crack would have gotten wider very fast, as erosion took place. As the crack widened, the mantle below would have started to bulge upward because it was no longer under pressure from above. This process has been observed at quarries. After a large amount of rock has been taken off a wide enough area, the "floor" of the quarry will bulge up a bit.

As the mantle bulged upward, two things would have happened simultaneously. The first was that the rock directly underneath that bulge would have been pulled up by that bulge's upward motion. Then the rock underneath that rock would have risen along with it. And the rock under that would have been pulled up, too. At these depths, air would not be able to get down to fill in any gaps, so the upward pull on the rocks would have gone deeper and deeper. When this pulling got to the opposite side of the earth, it would have created the opposite of a bulge. The force would have pulled the crust in, toward the center of the earth. Thus, on the opposite side of the Mid-Atlantic Ridge we would expect to see a sunken place. Indeed, we do find that the deepest trenches of the Pacific Ocean are directly opposite the Mid-Atlantic Ridge. If this pulling force had cracked a large portion of the granite shell on the Pacific side of the world, it would have done so according to known laws of physics, making those C-shapes (cusps) that outline the Pacific Ocean, as well as the Benioff fault lines (shear lines) that run under the edges of the Pacific.

The second thing that would have happened is that the bulge would have tipped the edges of the continents up a bit, causing them to start sliding. This continental drift would have happened quite quickly, not slowly like the other tectonic models. These continental plates were riding on a layer of water. (From this, the theory gets its name: the "hydroplate theory." "Hydro" means "water.") Eventually the water would have completely escaped and the continental plates would have started sliding against the mantle rock below. The plates would have come to a screeching halt and crumpled up in places, especially at the leading edge. This would explain why the Andes and Rocky Mountains are parallel to the Mid-Atlantic Ridge.

These diagrams show the crust much thicker than it really is. Also, the Mid-Atlantic Ridge is too large. This has been done to make the ideas presented in the diagrams easier to see.

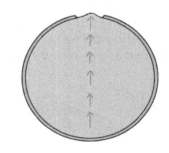

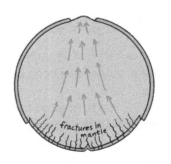

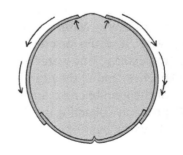

When the crack first opened up, the water would have been coming out with tremendous force, shooting up into the atmosphere. Some of it would have come back down again as rain (the 40 days of rain in some of the flood stories?), but some of it might have kept going, traveling out into the solar system. Out in space, the water droplets would have frozen and been pulled together by gravity to form comets. (The water content of comets has always mystified astronomers.) Additionally, the water could easily have taken many chunks of eroded rock up with it, explaining the origin of meteors and asteroids and their chemical similarity to earth's crust. Asteroids are basically flying rock piles made of rounded rocks that look as if they have been smoothed by water in a giant rock tumbler. Asteroids such as *Itokowa*, shown here, puzzle astronomers because of their peanut shape. They look like they are made of large rocks that merged together at a funny angle and stayed that way. The laws of physics say that objects in space can only do this if they are traveling in the same direction at the same speed. The hydroplate theory provides an explanation of how this could have happened.

Itokowa is a peanut-shaped asteroid.

The water that came up out of the crack would also have carried with it the mineral sediments that had been dissolved by the hot, super critical water. After the initial burst, the pressure became much less, and the water would have just gushed up and out onto the surface of the crust. Unimaginable amounts of mineral-rich water would have poured over the land. It would no longer have been super hot, because as it expanded, it cooled quickly. As it cooled, huge amounts of $CaCO_3$ would have precipitated out of the solution to become limestone. When magnesium was present, dolomite ($MgCO_3$) would have formed. This solves the problem of how so much limestone could have formed without putting toxic amounts of CO_2 into the air, as discussed in chapter 5.

Water that contained dissolved quartz would have seeped into dead animals and plants, fossilizing and petrifying them. Lab experiments have shown that to petrify wood, you need water that has at least 140 ppm (**p**arts **p**er **m**illion) SiO_2. This means that if you collected one million molecules of this solution, 140 of those molecules would be SiO_2 and the rest of the million would be water molecules. Rain dripping on granite over many years can only produce a solution that is about 6 ppm. Boiling water is better, but can still only dissolve granite to about 60-80 ppm. How can you get to 140 ppm? You have to apply *pressure* to the boiling water. Pressure canners work on this principle; they have locking lids that keep the steam from escaping. As the pressure builds, the temperature can go well over the boiling point. There isn't any mechanism on the surface of the earth that can apply enough pressure to hot water to create a silica solution of 140 ppm. Boiling water from magma or hot springs simply evaporates into the atmosphere. However, super critical water under the crust of the earth would definitely have generated solutions of at least 140 ppm, if not higher.

When the water was at its highest level, huge waves would have swept over and around the globe many times. The wave action would have helped the sediments to sort themselves out according to their density and particle size, a process known as **liquefaction**. The sorting action of liquefaction, working at the same time as wave action, might help to explain some of the striped layers of sedimentary rock we find in places where there is very little erosion between layers. Most importantly, these layers could have been in place before the continents slid to a halt. The sudden jolt would have caused these soft layers to fold and bend, explaining the many folded sedimentary layers we see in so many places. The folded layers would have become hard in the years following.

Well, if it's a time machine, it shouldn't be gone too long, from my perspective.

Ah, yes...

As with all rainy weather, puddles would have been left behind, and in this case HUGE puddles the size of the Great Lakes of North America. (Or maybe the Great Lakes ARE puddles?) Natives of Tibet have a legend that a huge lake used to exist on what is now a dry plateau in the Himalayas. Geologists never took this story seriously... until they found fossilized fish on that plateau. There is evidence of two lakes having once existed to the east of the Grand Canyon. Part of the hydroplate theory goes on to suggest that these lakes could have spilled over and carved

the Canyon, explaining many of the mysterious features of the canyon, such as side canyons that go the wrong way. Speaking of the Canyon, remember that quartzite block that was stuck in layers of shale in the side of a canyon wall? (from the coloring page) This is easily explained if those layers were still wet when this happened. The block could have been lifted by a sludgy mix of wet sand and mud. (Geologists have seen this happen in other places, so we know it is possible.) The surrounding layers of shale hardened after the block was in place.

The hydroplate theory also offers an explanation for what causes earthquakes. Perhaps earthquakes are not caused by the edges of continental plates rubbing together, but rather, the edges rub together because of the earthquakes. Earthquakes might be caused by fault lines slipping deep in the mantle as the earth tries to recover from the damage it has sustained. The mantle might have many cracks and be unstable. When the faults in the mantle slip, the rocks rub together and magma is produced. If the magma is below the cross-over depth, its volume will shrink and it will begin to flow down the crack, toward the outer core. (When seismologists detect a channel of magma connecting the crust to the outer core, they infer that it is a plume of magma coming up. However, seismometers can't sense whether the magma is moving up or down. They could be seeing one of these cracks with magma draining down.) Because the magma is shrinking, the surrounding rocks move closer together. When they move, the effects are felt all the way up to the crust, and the continental plates suddenly lurch, slipping along visible fault lines. Seismologists are baffled by very deep earthquakes because they can't possibly be caused by plates slipping at the surface.

If the mantle really has experienced the kind of cracking and shearing that the hydroplate theory suggests (on page 122), then this slipping of deep fault lines isn't just random. The laws of physics say that spinning spheres, even large ones like planets, want their mass evenly distributed, with no thick places and no empty holes, so they will try to correct any imbalances that occur. The deep ocean trenches along the western Pacific act like empty holes in the crust. Therefore, the mantle will keep making little shifts from time to time, as if it is trying to fill these holes. This could explain why so many of those motion arrows in the map at the bottom of page 166 point to the western Pacific. Those trenches have "missing mass," so to speak. This missing mass causes them to have less gravity. Less gravity means less weight on a scale. If you weigh yourself while sailing over an ocean trench, the scale will say that you weigh less!

These theories (plate tectonics, surge tectonics, and hydroplate theory) are certainly quite different. Which one is true? Since they involve ideas that can't be directly observed, only inferred, we will never know for sure. Our confidence can be higher or lower in one theory or another, but they will always remain theories.

Geology would be a lot simpler if there weren't any fossils. The fact that we have petrified plants and animals means that biology is tied into geology somehow. What a complication! A time machine would let us solve the mystery of how those dinosaurs got into the ground, but until someone invents one, we can only take educated guesses based on things we can observe. The problem is that our observations give us conflicting data. Some observations seem to suggest that fossils are very old, while other observations suggest that they are not so old. Since fossils are found in and around other types of rock, we might also wonder how old rocks are, in general. It's not much of a jump, then, to wonder how old the entire planet is, and how it came into being in the first place. The study of how stars and planets (and even the universe) might have formed is called **cosmology**. This is not a book about cosmology, so those last questions are outside the scope of our discussion. We will just look at some reasons why people might think rocks (or fossils) are very old, or not so old.

Why would someone think that rocks are very old, perhaps even millions or billions of years old? Probably the main reason is because of **radiometric dating**. To understand how radiometric dating works, we need to go back to chapter one. That basic information about atoms at the beginning of the book turns out to be very important here at the end of the book.

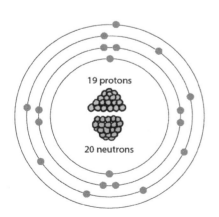

Potassium atoms ALWAYS have 19 protons. They usually have 20 neutrons, but occasionally one might have 19 or 21.

The nucleus of an atom contains positively charged protons and neutral (no charge) neutrons. The number of protons in the nucleus is what determines which element that atom is. The atom shown here is K, potassium. It has 19 protons and 20 neutrons. It is number 19 on the Periodic Table. The table also lists another number for potassium, 39.098. This number is the **atomic mass** ("weight") and is calculated by adding up the total number of protons and neutrons. But how can you add 19 and 20 and get 39.098? You can't, of course. The decimal number is actually an average, not the mass of any one particular atom. They've calculated the mass of thousands of atoms, then taken the average. This means that every once in a while you will find a potassium atom that has 21 neutrons, instead of 20. This brings the average up from 39 to 39.098. (A few rare potassium atoms even have 18 or 22 neutrons.) Each of these variations of potassium is called an **isotope**. Every element has a variety of isotopes, but they are all still that element because the number of protons does not change.

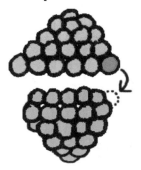

19 protons

21 neutrons

Potassium 40 has 19 protons and 21 neutrons. This is not as stable as potassium 39. As a result, a proton can turn into a neutron. To do this, the proton captures an electron.

One particular isotope of potassium is helpful in dating rocks. This isotope is called "potassium 40" and is written like this: ^{40}K. This isotope has 19 protons (required in order to be potassium!) and 21 neutrons. It turns out that this isotope is not stable. Stable isotopes never change. Unstable isotopes can change in a number of ways. In this case, one of ^{40}K's protons will turn into a neutron. Yes, you heard that right—a proton will turn into a neutron! And yes, that will indeed change the identity of the atom. It will no longer be potassium, but will become an atom of argon, with 18 protons. (The mass will be the same, so it will be "argon 40" or ^{40}Ar.) The element argon is a non-reactive gas that does not combine with any other atoms, and is not produced in any other way, as far as we know. This is what makes it so useful for radiometric dating.

Imagine a lump of basalt that has some potassium in it. (Basalt contains a lot of silicon, iron and magnesium but it also has small amounts of calcium, sodium, aluminum and potassium.) A small amount of that potassium is ^{40}K. Some of this ^{40}K may have already turned into ^{40}Ar. Now the basalt gets super-heated and melts into magma. Any atoms of argon gas that are in the magma will be able to bubble to the surface and go into the air. When the magma starts to cool, it should have almost no argon gas molecules in it. After it hardens, any argon gas atoms that form after that time will be trapped inside the rock.

Scientists have calculated the rate at which ^{40}K turns into ^{40}Ar. Therefore, by measuring the amount of ^{40}Ar they find in a rock sample, they can estimate how long that argon has been accumulating. The more ^{40}Ar they find, the older the rock sample. Using this "potassium-argon" method, rock samples have been dated at anywhere from 2 million years to 2 billion years. This dating method is only a rough estimate, and with a margin of error of several million years. For example, samples of the same rock might date 325 million, 280 million and 385 million. The geologists then find other ways to choose which of these dates they think is most accurate. Sometimes other elements can be used to check the K-Ar dates. For example, one of rubidium's isotopes will turn into strontium. (In this case, a neutron turns into a proton.)

This is the machine they use to measure ^{40}Ar.
By Radiogenic - Own work, CC0, https://commons.wikimedia.org/w/index.php?curid=28544036

Precambrian rocks tested by this potassium-argon method return dates of around 600 million to 2 billion. Rocks that give dates of 485-540 million are assigned to the Cambrian layer. The famous Jurassic period covers dates of 20 to 145 million. The potassium-argon method can't be used to find dates that are less than 2 million.

This landscape is in the Grand Canyon area. Though it looks much different from the view seen from the South Rim (as in the coloring page) many of the layers are the same. The basalt of the Cardenas lava layer has red Dox sandstone on either side of it at this location.
By High Contrast - Own work, CC BY-SA 3.0, https://commons.wikimedia.org/w/index.php?curid=2851264

Notice that this dating method can only be used to date igneous rocks since it is based on accumulation of gases inside solidified magma. Sedimentary rocks did not start out as magma, so this dating method can't be used on them. Geologists try to find stripes of basalt that run through sedimentary areas. They date the basalt and then assume that everything above that stripe is younger and everything below is older. (The photo here shows the lava layer that runs through the bottom layers of the Grand Canyon.) They also look for familiar layers such as the Cambrian, Silurian, Devonian, Pennsylvanian, Jurassic, etc., and do comparisons of similar layers in different places, in conjunction with the datable layers of basalt. Dating sedimentary rock is more complicated than igneous.

Sedimentary layers are also dated using estimates of how long it takes for layers of dirt and minerals to accumulate. This method is based on two assumptions: 1) **uniformitarianism** and 2) **superposition**. These words might look difficult, but their meanings are easy to understand.

"Uniform" means "the same." "Uniform-i-tar-ian-ism" means that we assume that geological processes have always been basically the way they are now. Erosion, for example, has always happened at the same rate that it is happening today, so we can measure the current rate of erosion and then calculate backwards to figure out what happened in the past. Also, sediments accumulate at a certain rate. If we assume that the sediments have always been accumulating at this rate, we can measure the current yearly accumulation, then calculate how many years it took for an inch, or a meter, or half a mile of sediments to accumulate. Both these processes, erosion and sedimentation, happen very slowly today, so the age estimates of deep layers of dirt and rocks often go into the millions.

"Super" means "over, above, or on top." Superposition is the assumption that things on top are more recent than things underneath. This is easy to understand from examples in everyday life. When clutter accumulates on your desk, the things on top of the stacks are the most recent additions to the piles. Superposition alone can't tell you how old anything is; it only compares layers to say which one is older or younger than another one. This assumption seems pretty foolproof, but surprisingly, there are a few places where this doesn't hold true. Where rivers dump sediments into bays, you can get layers forming all at the same time, as the sediment particles settle according to their density and size. Also, during flood conditions, layers have been observed to form according to density. However, in many cases, superposition may indeed hold true.

There is a certain type of sedimentary layer that geologists are particularly interested in counting: a very thin layer called a **varve**. A varve can be as thin as a sheet of paper or as thick as a pencil. One place varves are forming today is at the bottom of lakes. With the turning of the seasons, different types of particles are brought into the lake, or fall on top of the lake, then gradually settle to the bottom. During the summer, organic particles such as pollen and diatoms, will be present. During the winter, these organics will disappear. Because the particles that come into the lake vary according to the seasons, the varves (layers) at the bottom of the lake will be different colors or textures. By counting the varves, geologists estimate the number of years they have been forming.

Lake Suigetsu, in Japan, is the best example of a modern lake that has many varves. It also has layers of volcanic ash every once in a while, which is not surprising because of the many volcanoes around Japan. Scientists have been studying these varves and watching them form for over 30 years. They are very sure that one varve equals one year. They've taken core samples below the lake and counted over 50,000 varves.

Another famous varve site is the Green River Formation in Wyoming. Here, the lake in question is fossilized, so the varves have turned to stone. Geologists have counted several million varves. If a varve equals one year, this suggests an age of several million.

Green River varves

Here I am waiting again.. This is crazy!

What have we got here? I'm guessing they're trilobites.

Why would someone think that rocks are not that old, perhaps only thousands of years old? The answer most frequently given is another type of radiometric dating, this time using the element carbon. To understand how carbon can be used for dating, we again go back to chapter one, to the Periodic Table. Carbon has atomic number 6, which means it has 6 protons. Its mass is listed as 12.011. The mass is the number of protons added to the number of neutrons. $6 + 6 = 12$, so carbon should have 6 neutrons. Where does that .011 come from? Occasionally you find a carbon atom that has more than 6 neutrons, so when thousands of carbon atoms are measured and the results are averaged, those few heavier carbons cause the average to go a bit above 12.

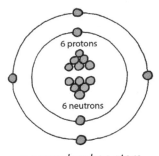

a normal carbon atom

The isotope of carbon that is useful for radiometric dating has a mass of 14 instead of 12. This means it has 8 neutrons. Is it still carbon? Yes, because it still has 6 protons and it's the number of protons that determines which element the atom is; having a few extra neutrons doesn't change its identity. Only one in a trillion carbon atoms has a mass of 14, but carbon atoms are so abundant that ^{14}C ("carbon fourteen") still adds up and can be detected by high-tech atom counting machines.

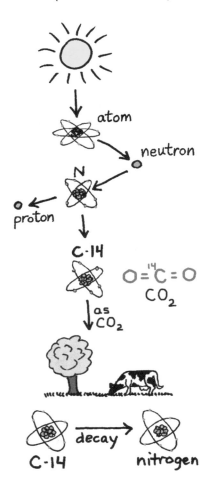

As far as we know, the source of ^{14}C is mainly the upper atmosphere. High-energy cosmic rays from the sun strike the air molecules in the atmosphere. These dangerous rays can damage atoms. Sometimes they hit a nucleus and cause a neutron to go flying off. If the loose neutron happens to hit the nucleus of a nitrogen atom, it can dislodge a proton, causing the proton to go flying off. Nitrogen atoms normally have 7 protons and 7 neutrons, giving them an atomic mass of 14. When a loose neutron comes in and replaces a proton, the mass stays the same but the atom's identity changes because the number of protons drops from 7 to 6. Having only 6 protons, the atom is now carbon. It still has a mass of 14, so we call it carbon 14.

Atoms of ^{14}C usually find a pair of oxygen atoms floating past and join with them to make carbon dioxide, CO_2. This "heavy" CO_2 acts the same as normal CO_2, so it is eventually taken in by plants and used for photosynthesis. When animals eat the plants, the ^{14}C is then transferred into their bodies. Plants and animals take in carbon, including ^{14}C as long as they are breathing and eating.

Carbon 14 is an unstable atom. After all, it used to be nitrogen. To regain its stability, one of the neutrons will change back into a proton. We've already learned that protons can turn into neutrons, so this change, from neutron to proton, shouldn't surprise us. As the neutron "decays" into a proton, the atomic number goes back up again, from 6 to 7. The atom is now nitrogen again. This decay process is happening right now in your body, because your body contains a lot of carbon. Carbon atoms are the basis for most organic molecules: proteins, sugars, fats, and DNA. What would happen if one of the carbon atoms in your DNA suddenly turned into nitrogen? It's a terrible thought, but this is happens about 2,000 times every second in your body! Some scientists think this might be one of the things that causes aging.

When a plant or animal dies, it stops taking in carbon. As time goes on, more and more of the carbon atoms in the dead organism will decay back into nitrogen. After about 5,700 years, half of the ^{14}C atoms will have turned into nitrogen. After another 5,700 years, half of those remaining ^{14}C atoms will be gone. After every 5,700 years, half of the remaining ^{14}C will decay. The number 5,700 is called the "half-life" of ^{14}C. After about 50,000 years, almost all of the ^{14}C will be gone. This means that radiocarbon dating can only give dates that are less than 50,000.

This is an actual photograph of a nuclear blast that occurred on Bikini Atoll (in the Pacific Ocean) on July 25, 1946. It was a planned test, not an act of war. The US navy just wanted to see how many ships it could destroy. The results were disappointing for the navy and devastating for the environment. This area had to be evacuated for the test, and remains uninhabited to this day because of high levels of radioactive waste.

Radiocarbon dating assumes that the ratio of normal carbon to carbon 14 has always been about the same as we see it now. If this is not true, the age estimates will not be accurate. Recently, scientists have discovered that the levels of ^{14}C in the atmosphere can change quite a bit. For example, in the 1940s and 1950s, a number of atomic explosions occurred, mostly as a result of the USA's atomic weapons program. These explosions created a lot more ^{14}C atoms in the atmosphere. People who were alive during that time have more ^{14}C in their bodies than people who were born after that time. Also, the level of ^{14}C rose all during the 20th century, as more and more coal and oil were burned as fuel.

With all this talk of plants and animals, how can ^{14}C be used to date rocks? Radiocarbon dating is almost useless for igneous rocks as they are made of mostly silicon, oxygen, iron, magnesium, potassium, calcium, sodium and aluminum. They rarely contain any carbon. Sedimentary (and some metamorphic) rocks are where you find carbon in many different forms. Carbon 14 has now been discovered in limestone, coal, oil, natural gas, diamonds, and fossils. A fossilized dinosaur (mososaur) skeleton, which had been given a date of 80 million years by other dating methods, was found to contain ^{14}C which means it has to be less than 50,000 years. Which is correct?

Even stranger yet, remains of biological tissues such as blood cells and cartilage have been discovered in many fossilized dinosaurs, including T-rex, triceratops and hadrosaurus. The remains are not in perfect condition, but still, finding any remnants at all was very surprising. At first, most scientists refused to believe it and were angry at those who made the discoveries.

Experiments with DNA have shown that it is a fairly unstable molecule. In our bodies, our DNA is constantly being repaired by little cell parts that know how to fix it. The estimate for how long a DNA molecule can survive outside a cell is only in the thousands of years, not millions. DNA should not be in any fossils at all. However, DNA was extracted from an insect in fossilized in amber that was said to be over 130 million years old.

soft tissue from a hadrosaur

Polystrate fossils also seem to suggest that sedimentary layers might not have taken millions of years to form. ("Poly" means "many," and "strate" means "layer.") A polystrate fossil goes through many sedimentary layers. This picture shows a tree trunk going through layers of rock that many geologists say were laid down over millions of years. Common sense tells us that something is wrong here. Could the tree have stayed there, in place, for millions of years while sediments built up around it? Plant scientists say no, this is not possible. Trees die if their trunks are buried. If the top of the tree was already fossilized when the layers started accumulating, we should see significant erosion at the top and much less erosion at the bottom. Other polystrate fossils include fish, leaves, and groups of animals, including a school of jellyfish. How could a school of jellyfish go through several layers that each span millions of years? And just the fact that the jellyfish were fossilized at all suggests that they were buried and petrified very rapidly. Jellyfish rot quickly.

The magnetic field of the earth (shown as blue lines in this diagram) also suggests that the inner earth might not be millions of years old. The strength of earth's magnetism can be measured very precisely, and it has been measured for several decades. What they've found is not what anyone would have predicted. Far from being steady and consistent, the magnetic field is getting weaker at an astounding rate, about 10 percent in the last 150 years. If we calculate backwards only one million years, the magnetic field would have been strong enough to generate intense heat, causing the oceans to evaporate. The magnetic fields of other planets are also getting weaker very quickly, though their magnetism is not nearly as strong as the earth's.

Coal samples from uranium mines in Utah have been found to contain spherical "halos" caused by the decay of radioactive uranium and polonium. The halos have been pressed into ovals, as have the logs of coal themselves. The radioactive atoms are assumed to have entered the logs along with the water that was necessary to turn them to coal. The logs were obviously flattened before the necessary heat and pressure were applied, or else they would have shattered instead of being nicely flattened. (Soggy wood flattens nicely when pressed, but coal crushes into pieces.) All these bits of evidence together suggest that this coal seam was formed quickly, not over millions of years.

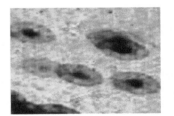
Flattened radioactive "halos" in coal, as seen under microscope

If we think that science is going to provide all the answers for us, it looks like we are going to be disappointed. Do we believe potassium-argon dates or carbon 14? Should we prioritize varve evidence over magnetic fields? How can we reconcile polystrate fossils with the general pattern of the geological layers? And what about dinosaur soft tissue? Science itself seems to be giving us conflicting evidence. Because of this, people begin to broaden their view beyond the bounds of observational science, to see what else might help them to interpret these conflicting bits of data. Philosophical and religious ideas are then brought into the discussion, and cosmology (the study of how everything began) often becomes the deciding factor in how geological features are interpreted. Two people can look at the same rock or geological feature and come up with two very different theories to explain it, based on their ideas about where the universe came from. For each piece of information given in this chapter, you will be able to find researchers who disagree on its interpretation. Fortunately, ideas about the past don't affect anyone's ability to study and appreciate rocks and dirt. All of us can use the minerals found in rocks, study mineral chemistry, find ways to conserve our precious soils, and enjoy the natural beauty of our rock collections.

Activity 8.1   Comprehension self-check

1) An "epicenter" is the center of: a) a fault line   b) an earthquake   c) a volcano   d) a continental plate

2) A Benioff zone is: a) a fault line   b) an earthquake   c) a spreading region   d) a subduction region

3) Before continental drift was proposed, people thought the continents might be moving:

   a) up and down   b) side to side   c) north to south   d) east to west

4) True or false? Earthquakes always occur at the fault lines between continental plates. ____

5) The first person to ever suggest that the continents were moving was:

   a) Ortelius in 1596   b) Mantovani in 1889   c) Wegener in 1912

6) True or false? Wegener was the first person to notice that there are similar rocks and fossils on the east coast of South America and the west coast of Africa. ____

7) True or false? Scientists are always very quick to pick up on new theories. ____

8) How fast did Wegener think the continents were moving?

   a) 2 centimeters per year   b) 250 centimeters per year   c) 2 kilometers per year   d) 200 kilometers per year

9) What was the main reason that Wegener's continental drift theory was not accepted at first?

   a) It was too complicated.   b) It could not explain the ocean tides.

   c) The rate of drift was just too high.   d) It did not have a mechanism to explain how the continents moved.

10) What does "diverge" mean? a) come together   b) go up   c) go apart   d) go down

11) What does "converge" mean?   a) come together   b) go under   c) go apart   d) come apart

12) The plate tectonic theory says that the Mid-Atlantic Ridge is a _____ zone.   a) divergent   b) convergent

13) According to plate tectonics, what mechanism is moving the continents?

   a) ocean waves   b) the pull of the moon   c) magnetism of the inner core   d) circulation of magma in the mantle

14) Which of these is NOT a problem for the plate tectonic theory?

   a) The mantle appears to be solid.   b) The crust of the earth appears to be cracked into large pieces.

   c) The Pacific Rise does not appear to be spreading.   d) Antarctica has no convergent zones around it.

15) Which of these was NOT known in the 1950s, when the plate tectonic theory was being proposed?

   a) the thickness of the mantle   b) the behavior of P waves through the mantle

   c) the locations of the cracks between the continental plates   d) the cross-over depth of magma

16) Mr. Smoot's theory is called _____ _____.

17) In Mr. Smoot's theory, the driving force behind the movement of magma is the _____ of the earth.

18) What is the primary cause of tides? a) the sun   b) the moon   c) the circulating magma in the mantle

19) The hydroplate theory suggests that as the crust split and eroded away, the mantle below came bulging upward

   because it was no longer being pressed down by the crust. Where can this principle been seen on a small scale?

   a) mountains   b) valleys   c) mines   d) quarries

20) True or false? All theories about continental plates agree that they have moved from an original starting point. ____

21) If you want to heat water over the boiling point, what do you need to add?

   a) more heat   b) minerals   c) pressure   d) gravity

22) True or false? Carbon 14 levels in the atmosphere have never changed. ____

23) It is the number of _____ in the nucleus that determines which element an atom is.

24) The atomic mass is basically the number of _____ added to the number _____.

25) Atoms of the same element that have different numbers of neutrons are called:

   a) isotopes   b) isotherms   c) ions   d) radioactive

26) True or false? When one of an atom's protons turns into a neutron, it changes into a different element. ____

27) Potassium 40 will eventually change into what element? _____

28) A rock that has a lot of argon in it will give a date that is: a) very old   b) very young

29) Which layer of rock returns the oldest age estimates using the potassium-argon method?

   a) Jurassic   b) Triassic   c) Cretaceous   d) Cambrian   e) Precambrian

30) True or false? Radiometric dating can only be used on igneous rocks. ___

31) The assumption that the mechanisms of geology are the same today as they were a long time ago is called:

   a) superposition   b) unequivocalism   c) uniformitarianism   d) uniformation

32) A thin layer of rock or dirt is called a v_____.

33) How many protons does an atom of carbon 14 have?   a) 14   b) 12   c) 8   d) 6

34) When carbon 14 decays, what does it turn into?   a) argon   b) oxygen   c) carbon 12   d) nitrogen

35) How long would it take for almost all carbon 14 to disappear in a rock or fossil?

   a) 1,000 years   b) 5,7000 years   c) 50,000 years   d) 5 million years

36) Is the magnetic field of the earth getting stronger or weaker? _____

Activity 8.2   More info: Flooding on Mars

Mars has some very mystifying surface features, including erosion channels that look very much like they were created by water. Since Mars has no liquid water on its surface, scientists wonder how this could have happened.

Recently, the Mars Reconnaissance Orbiter has recorded information about one particular area near the equator called Marte Vallis. This area is part of a wide plain that has huge erosion channels. The satellite data shows fault lines and old lava flows under the surface. Geologists have interpreted this data to mean that there might have been large reservoirs of water under the surface, and that cracks in the surface allowed the water to come up and gush out, causing a huge flood. The flooding then caused the visible erosion channels on the surface. The crust shifted around during this time, melting rocks and turning them to lava.

Does this theory sound a bit familiar? Is water beneath the surface of Mars easier or harder to believe than water below the Earth's crust?

The Mars Reconnaissance Orbiter is a satellite that stays in orbit and send out signals that gather information.

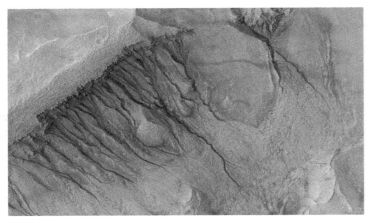

This is a satellite picture of some of Mars' erosional channels.

Activity 8.3 Even more info: A little more about Benioff zones

This illustration shows an actual image of a Benioff zone. Thousands of seismometers around the world gathered P wave data from earthquakes. The data was analyzed by a computer, then used to make this image. The blue areas indicate places where P waves traveled 6% faster. The image doesn't tell you what the blue area is, only that P waves traveled 6% faster. The faster P wave travel time is inferred to be because this area is either colder or denser.

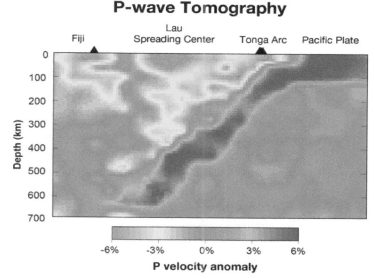

One theory is that the blue area is a continental plate diving down into the mantle. Because the plate is colder, it is sinking down into the hotter mantle. This idea lines up with plate tectonics. Another theory is that the blue area is denser, and is the "scar" left by a fault that slipped as the continental plate cracked and sunk. The slippage would have caused great heat, melting the rock. When the rocked cooled again, it would have been more dense. Densities of cooled magma are known to be more dense than the parent rock from which the magma came.

This is one of those cases where a scientist's beliefs about the origins of the earth and the universe will determine how he or she interprets the data. The data doesn't tell you what happened—you must decide what the data means and draw an inference. Your interpretation will depend on your beliefs.

However, if the blue area is a diving continental plate, how did it begin to dive in the first place? Why didn't the moving plate just crumple up against the edge of the other plate and create a mountain? In 2004, Robert Stern, a plate tectonics expert, wrote this: "In spite of its importance, it is unclear how subduction starts. This remains an unsolved mystery" (Earth and Planetary Science Letters, Vol. 226). Also, if the blue area is a cold ocean plate, its temperature should be about the same as other parts of the ocean floor, ranging from 2° to 30° C (36° to 100° F). Applying a math formula known to be true for P waves at cold temperatures, the blue area should be a region where P waves travel about 400% faster than normal.   If fact, the measured speed of the P waves was only 6% faster, which is exactly what you would expect for an area that is denser, not colder.

Activity 8.4 A challenging question

The last sentence of this chapter pointed out that our ability to use and enjoy rocks, minerals, and soil doesn't depend on our ideas about the past. Come to think of it, does our view of the past limit us in any way from participating in science and technology (STEM)? Think about each invention or discovery and put a check mark on the line if it depended on a particular view of the <u>distant</u> past.

___ airplane	___ telephone	___ radar	___ submarine
___ nuclear power	___ cell phone	___ radio	___ Periodic Table
___ rockets	___ combustion engine	___ light bulb	___ GPS
___ electric motor	___ satellites	___ television	___ computers
___ antibiotic medicines	___ medical imaging	___ anesthesia	___ brain surgery
___ Internet	___ no-till farming	___ limestone chemistry	___ microscopes
___ calculus	___ discovery of DNA	___ soil chemistry	___ refrigeration
___ indoor plumbing	___ smoke detectors	___ recycling	___ mining

Activity 8.5   Fact or inference?

Sometimes it isn't easy to tell the difference between facts and inferences. We use inferences all the time, and, more often than not, they appear to be true. In chapter 7 we saw how inferences can be drawn about the inside of the earth, using facts about how P and S waves travel through the interior of the earth. Scientists can be so confident about their inferences that when they publish them in books they present them as facts. This can be misleading to the reader, and dangerous to the scientist's reputation, if future data contradicts the inference.

In this activity, you <u>don't</u> need to determine whether the statements are true or false. All you need to do is figure out whether it is a fact (a piece of data from actual observation of some kind) or an inference (a conclusion drawn from the data). Again, this is NOT about whether any statement is true or false. If something is an inference, this does not mean it is false, only that it cannot be directly observed.

Write "F" on the line if you think it is a fact (piece of actual data) and "I" if you think it is an inference. The answer key will have a short discussion about each one.

1) Fungi in soils decompose leaves into humus. ____
2) Earthworms like living in soil. ____
3) S waves travel through the mantle of the earth. ____
4) The mantle of the earth is solid. ____
5) The mantle of the earth convects. ____
6) The inner core of the earth is made of iron and nickel. ____
7) Carbon-14 levels in the atmosphere can change. ____
8) Caves were carved out by rain water that contained a lot of carbonic acid. ____
9) Mars had water underneath its surface at one time. ____
10) The earth had water underneath its surface at one time. ____
11) Neutrons can turn into protons. ____
12) Some dinosaur species could run faster than a horse. ____
13) Argon levels in the atmosphere are the same now as when rocks were forming. ____
14) Earthquakes occur in the middle of continental plates, not just on their edges. ____
15) Earthquakes are caused by magma shrinking and causing deep faults to shift suddenly. ____
16) Granite is made when lava cools very slowly. ____
17) It only takes a few decades after the eruption for a volcanic island to become habitable. ____
    ("Habitable" means plants and animals can live there.)
18) All the continents used to be connected. ____
19) Coal is made of ancient plants. ____
20) Obsidian is lava that flowed into water and cooled quickly. ____

Activity 8.6   "Odd one out"

In each set of four words or phrases, one of them doesn't belong and is the "odd one out." The other three have something in common. Circle or underline the one that doesn't belong.

EX: proton, ion, electron, neutron   *(Ion does not belong. The others are "subatomic" particles that make atoms. An ion is a whole atom, but one that is electrically unbalanced, having too many, or too few, electrons.)*

1) diamond, amber, coal, graphite
2) solid, inorganic, naturally occurring, mafic
3) sulfur, silver, copper, gypsum
4) structure, size, porosity, texture
5) basalt, granite, gabbro, diorite

6) nematode, mite, mosquito, springtail
7) limestone, chalk, calcite, dolomite
8) amethyst, corundum, olivine, mica
9) luster, density, porosity, streak
10) geyser, aquifer, hydrothermal vent, hot spring

Activity 8.7   Coloring page (review of sedimentary layers on continents)

   Here are stratigraphic columns for you to color and compare. The columns represent information gained by doing not only seismic studies, but bringing up core samples from drilling projects such as wells. You will need to fill in the patterns in the key on the right. (Page 46 might help.) The light gray areas are shale. You can fill in the pattern for shale or color code it. The salt should stay white, but you can choose colors for all the other types of rock. You can use realistic colors or bright colors, but make sure you don't cover the patterns. Notice the salt dome in Michigan Basin, and also the fact that salt is often found with gypsum. In the Michigan basin column, "W" is water, "S" is salt, "G" is gypsum. The little V's on the Vancouver column indicate some kind of volcanic rock, in this case, basalt. (You can make all the lavas one color.) What happened there in Vancouver? Wow, what a lot of volcanic activity there was at some point in time!   Also, take note of the depth scale on the left of each column. 1,600 meters equals one mile.

WILLISTON BASIN, North Dakota   MICHIGAN BASIN, Michigan   VANCOUVER ISLAND, Canada

135

Activity 8.8 Extrapolation: Making a little bit of data go a long way

Some of the information in this chapter (such as radiometric ages and speed of continental drift) came from either "extrapolation" or "interpolation." "Inter" is Latin for "between," and "polare" is Latin for "to polish, refurbish, or alter." Originally, the word interpolation was used only by mathematicians to describe the process of adding numbers to a series. If you have a group of numbers that form a series of some kind, and you can figure out the pattern that created the series, you can then insert appropriate numbers between your original numbers. "Extra" means "outside of." To extrapolate means to extend the series outside of, or beyond, your last number.

Apart from mathematics, interpolate can mean "to insert something extra," such as illustrations into a text, or extra paragraphs into a piece of writing. The general meaning of extrapolate is "to draw a conclusion about the future by looking at present trends." Trends could be mathematical or not related to math at all. In this exercise, we'll stick to data that is mathematical.

Let's extrapolate! We'll start with extrapolation of a simple number series. All you need to do is figure out what the pattern is, then add the next two numbers in the series. For example, if the number series is "2, 4, 6, 8..." you might easily guess that the pattern is counting by 2s, and the next two numbers would be 10, 12.

1) 2, 3, 5, 8, 13, 21, _____, _____   3) 2, 4, 6, 12, 14, 28, 30, 60, 62, _____, _____
2) 1, 2, 6, 24, 120, _____, _____   4) 4, 7, 13, 25, 49, _____, _____

Now let's extrapolate some data. In each case, you will again look for a pattern, then extend that pattern.

A 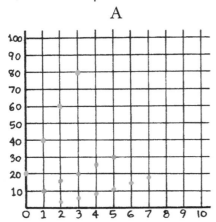   B 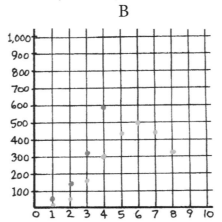   C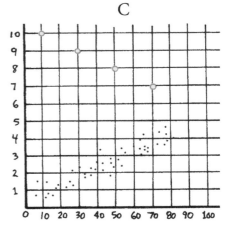

In graph A, we have three sets of colored dots. Let's imagine that the colors represent chemicals. The numbers 10 through 100 will represent temperature and the numbers 1 through 10 will represent grams of the chemical used. (This is an imaginary experiment so we don't need to know anything more about it.) Let's look at the red dots first. When one gram of red chemical was mixed into the experiment, the temperature in the test tube rose to 10 degrees. When 2 grams of the red chemical were used, the tempterature rose to 15 degrees. When 3 grams were used, the temperature was 20 degrees. The experiment stopped after 5 grams, so we don't have any more data points. However, we can draw an imaginary line (or a real line with a pencil, if you want to) that connects the red dots to find out what probably would have happened had the experiment continued. The experiment was done again, but this time with the green chemical. Then it was done again with the blue chemical. In each case, you can draw a line to extend the trend and make a good guess as to what would have happened if the experiments had continued.

5) What will the temperature be if we use 8 grams of red chemical? _____ 10 grams? _____ 0 grams? _____
6) What will the temperature be if we use 4 grams of green chemical? _____
7) Is there any amount of green chemical that will give us a temperature of less than 20 degrees? _____
8) How much blue chemical will achieve a temperature of 0? _____ (How accurate do you feel your answer is?)
9) What will the tempterture be if you use 10 grams of blue chemical? _____ 12 grams? _____
10) Now imagine that the graph goes on well past 10 and well above 100. (You can draw extra lines if you want to.) How many grams of green chemical are needed to achieve a temperature of 200 degrees? _____ Could you have figured this out without drawing any lines?

In graph B, we have only two sets of dots, purple and orange. Go ahead and try to connect the dots. You'll end up with two curves. The purple one will go up and off the graph. The orange one goes up and then back down again. Assume that this pattern will continue and the curve will look the same on both sides. Let's not specify what the experiment is about. We can answer questions by just using the numbers.

11) If the number 5 is used for the purple experiment, what will be the corresponding hundreds number? _____ What about the number 6? (Just give your best guess.) ____

12) In the orange experiment, what will the hundreds number be at the 10 mark? _____

13) In the orange experiment, what is the corresponding bottom number when the hundreds value is about 450? Is this question a problem? Why? _____

14) In the orange experiment, what will happen as the bottom number goes into the teens? Can you think of at least two options? _____

15) In the purple experiment, is there a limit on how far out we could extend the bottom numbers? If we wanted to use a bottom value of 99, and we had paper large enough, could we plot this data point? _____

In graph C, we have two new scenarios. It will be easy to draw a line through the yellow dots, but what happens with those tiny black dots? Many experiments produce graphs that look like a whole bunch of scattered tiny dots. In this case, you just do your best to create a line that goes right through the middle of them. This type of line is called a "best fit" line because you just do the best you can knowing that you are going to miss most of the dots. Lay a ruler down so that the edge goes through the middle of those dots and then draw a line, extending it up to the edge of the graph.

16) In the yellow dot experiment, what will the bottom number be when the side number is 5? ____

17) How confidently would you be able to predict more values for the yellow experiment? _____

18) In the black dot experiment, use the "best fit" line to predict a side value when the bottom number is 100. ____

19) If you actually ran this black dot experiment with a bottom value of 100, does it look likely (judging by the other data points) that you would hit your predicted number exactly? _____

Now, let's do some interpolation. This just means that you figure out a data point between two other data points, instead of extending the data points.

20) In the yellow dot experiment, what would the number on the left be, if the bottom number was 40? ____
And in the red dot experiment, estimate the temperature for 2.5 grams. ____

****
USING EQUATIONS TO HELP US

In some situations, you simply can't gather all the data you need. For example, if you want to know the temperature in the mantle, you can't drill a hole and stick a thermometer down. We read about the deepest hole in the world, and it did not go all the way down to the mantle. To estimate temperatures at depths deeper than the Superdeep Borehole, estrapolation was used. While drilling the borehole, they took temperatures at various depths. A graph was then created using temperature as one number line and depth as the other. Real data points could be plotted down to the borehole depth. Then, a line was drawn on the graph, extending down below the last data point. They used this extended line to predict what the temperatures would do below the borehole.

Scientists are also able to create mathematical formulas by looking at a graph. The formulas allow them to plug in any number and get an answer without even looking at the graph. For example, in the green dot experiment, we could write a formula for the line: **y = 20x + 20** The letter y represents any number on the left, and the letter x represents any number on the bottom. (The letter x is right next to the 20. This means they are multiplied.)

Let's choose one set of matching numbers for a green data point and see if these equation is true. When the number 1 on the bottom is used, the corresponding number will be 40. Do these numbers work in the formula? 40= 20 times 1, plus 20. (40= 40.) Yes! How about the numbers 2 and 60? 60 = 20 times 2, plus 20. (60= 40+ 20) Yes, again! Now, use the formula to figure out what y will be when x is 10: _____ = 20 times 10, plus 20. What about when x is 100 (which is WAY off the chart)? _____ = 20 times 100, plus 20 When y is 18, what would x be? 180 = 20 times _____, plus 20

Activity 8.9   More information about radiometric dating: Argon-argon

This extra information is for readers who might be curious to know the very latest in radiometric dating techniques. If this is more than you want to know, you are not required to read it, as the basic ideas behind radiometric dating have already been explained.   However, science keeps moving on and refining its techniques and some people like to be aware of the very latest developments.

Potassium-argon dating has some problems. First, if the rock is heated (but not melted) at some point in its history, the argon gas can actually escape. This would mess up the age estimate because the rock would contain less argon than it should. Second, argon gas actually can become trapped in a rock by processes other than decay of potassium. Third, the levels of potassium and argon are extremely hard to measure. This third reason is the one that drove scientists to find a better way to measure argon levels.

Potassium is a solid and argon is a gas. Different machines are needed to measure them, so you end up comparing the measurement taken on one machine with a measurement taken on a different machine. This makes our final calculation of age estimates less accurate. It would be like comparing your weight on your bathroom scale to someone else's weight taken on their bathroom scale. You would not be able to compare your weights down to ounces or grams, because you would not know if both scales were calibrated properly. (Bathroom scales were notoriously inaccurate before they went digital. Even with digital scales, though, there can be slight variations.) To gain better accuracy, you would need to weigh both yourself and your friend on the same scale. We need a way to measure potassium and argon with the same machine.

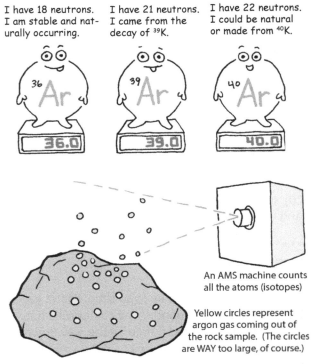

I have 18 neutrons. I am stable and naturally occurring.

I have 21 neutrons. I came from the decay of ^{39}K.

I have 22 neutrons. I could be natural or made from ^{40}K.

Nuclear scientists came up with a very clever solution, and figured out a way to turn potassium into argon. In this method, however, potassium-39 is the target isotope, not potassium-40. (^{39}K is the most common form of potassium. Only about 1/10,000 potassium atoms is ^{40}K.) They put the rock sample into a nuclear reactor and bombarded the rock with free neutrons. Some of the neutrons hit the nuclei of ^{39}K atoms and replaced one of the protons. This turned the ^{39}K into ^{39}Ar. An AMS machine can measure the levels of all three isotopes of argon: ^{36}Ar, ^{39}Ar and ^{40}Ar. Each isotope is there for a different reason, so the levels of each one, along with their levels in the atmosphere, are used to determine the age of the rock.

To increase accuracy, when they measure the argon they do so very slowly, so they can tell if a whole bunch comes out at the same time. The testing method involves heating the rock to release the argon gas, so they simply turn up the heat a little bit at a time. If they get a sudden burst of argon, they assume that this argon came from a process other than natural aging. The bursts are then subtracted out of the equation.

An AMS machine counts all the atoms (isotopes)

Yellow circles represent argon gas coming out of the rock sample. (The circles are WAY too large, of course.)

The age estimates using argon-argon dating are not all that much different from the potassium-argon dates, always in the millions of years and sometimes even into the billions. That lava layer at the bottom of the Grand Canyon was given a date of 1104.3 millions years old using the argon-argon method.

Of course, this method must make certain assumptions in order to work, such as the stability of the levels of argon isotopes in the atmosphere. To figure out how much argon was in the rock to start with, the levels of argon isotopes in the rock are compared to the levels of argon isotopes in the atmosphere right now. If the levels in the atmosphere have changed significantly since the rock was formed, this would change the interpretation of the data. The age estimates might go way up or way down.

This is what an AMS machine actually looks like. AMS stands for Accelerator Mass Spectometry.

INDEX

Periodic Table of Elements

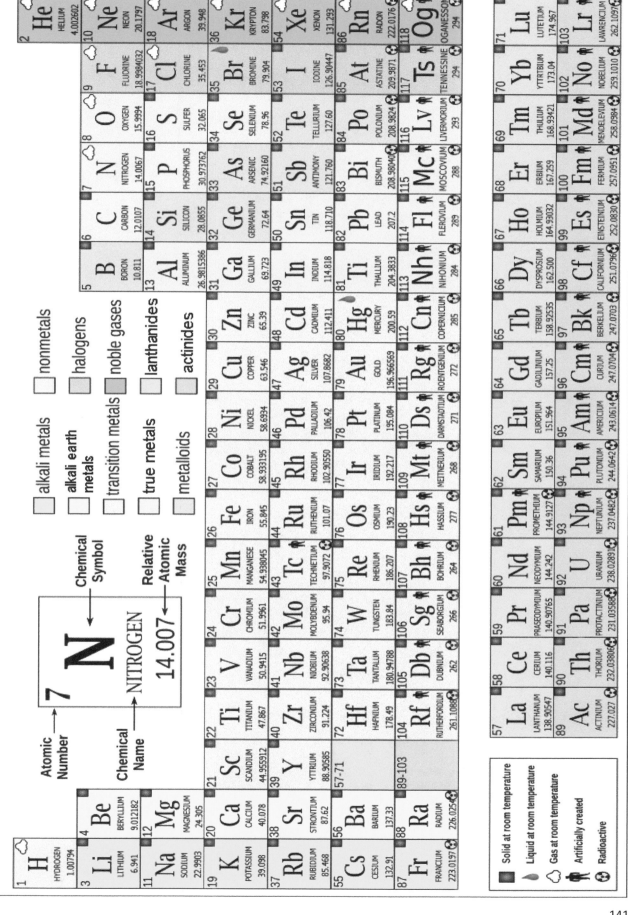

ANSWER KEY

ANSWER KEY

Chapter 1
1.1
1) False   2) b   3) Fire, water, air, earth   4) no   5) turn things into gold, cure any disease   6) bismuth
7) Elements make minerals, minerals make rocks.   8) phosphorus   9) "demon" (Because the nickel ore would not produce copper. Not understanding why, the miners thought the ore was cursed.)   10) Periodic Table   11) John Dalton
12) number of protons   13) 15   14) True   15) True

1.3
1) Hydrogen   7) Phosphorus   13) Copper
2) Carbon   8) Sulfur   14) Chlorine
3) Oxygen   9) Fluorine   15) Silicon
4) Sodium   10) Calcium   16) Potassium
5) Magnesium   11) Iron   17) Gold
6) Aluminum   12) Silver   18) Zinc

Chapter 2
2.1
1) F   2) T   3) coal, amber   4) Bronze is man-made, not naturally occurring.   5) F   6) carbon   7) red, blue   8) salt
9) pyrite   10) barite   11) hexagonal   12) rhombohedral   13) chalk or talc (or another if you know of one)
14) Lets light through but you can't see through it.   15) diamond, quartz, calcite, others you know of/ any opaque mineral
16) Rub it on a streak plate and it will leave a black mark.   17) talc, diamond   18) no   19) corundum   20) luster
21) T   22) biotite   23) water   24) gold   25) specific gravity

2.3
1) native elements, turquoise does not belong   2) silicates, apatite does not belong
3) carbonates, gypsum does not belong   4) sulfides, bismuth does not belong
5) sulfates, calcite does not belong   6) oxides, feldspar does not belong   7) halides, barite does not belong

2.4
ACROSS:
2) gypsum   3) magnetite   8) malachite   9) feldspar   10) topaz   11) talc   15) pitchblende   17) fluorite   18) bauxite
19) apatite   20) pyrite   22) ulexite   23) sphalerite
DOWN:
1) diamond   2) galena   4) graphite   5) barite   6) beryl   7) hematite   12) cinnabar   13) calcite   14) corundum
16) halite   18) biotite   21) quartz

Chapter 3
3.1
1) silicon and oxygen   2) 4   3) 8   4) covalent   5) ionic   6) hydrogen   7) 6, 2   8) iron (Fe) and magnesium (Mg)
9) olivine   10) pyroxenes   11) asbestos   12) hornblende   13) mica   14) dark: biotite   light: muscovite
15) yes   16) false   17) aluminum (Al), potassium (K), and sodium (Na)   18) No. Possibly because all the olivine, pyroxenes, amphilboles, and mica used up all the Fe and Mg when they were forming.   19) framework   20) true
21) b) gypsum   22) It produces a small electrical current. (piezoelectric effect)   23) aluminum   24) topaz
25) beryl, emerald   26) They are used as abrasives.   27) tourmaline   28) corundum   29) chalcedony
30) hidden   31) d) geode   32) jasper   33) opal   34) carnelian   35) heliotrope

3.3
1) J   2) M   3) E   4) K or L   5) K or L   6) F   7) R   8) I   9) O   10) N   11) Q   12) G   13) C
14) A or D or P   15) A or D or P   16) A or D or P   17) H   18) B

3.6
1) 125   2) 500   3) 50   4) 1175   5) 840

Chapter 4

4.1

IGNEOUS

1) extrusive   2) lava   3) plutonic, Pluto   4) a   5) iron, Fe, and magnesium, Mg   6) olivine   7) feldspar, silicon
8) T   9) b   10) andesite, Andes   11) c   12) b   13) b   14) d   15) a   16) b   17) c   18) b   19) d   20) a
21) a   22) felsic, mafic   23) c   24) b   25) d

SEDIMENTARY

26) cobble   27) clay   28) T   29) b   30) F   31) T   32) more   33) F   34) shale   35) siltstone   36) sandstone
37) sandstone   38) c   39) smooth, angular   40) T   41) T   42) b   43) a   44) F   45) halite   46) d   47) c
48) F   49) 2   50) a   51) c   52) F   53) F   54) T   55) F   56) c   57) It makes sparks (good for lighting fires).
58) egg   59) T   60) a

METAMORPHIC

61) change, shape   62) gneiss   63) schist   64) heat   65) heat and pressure   66) granite, slate
67) a   68) split   69) d   70) T

4.3

IGNEOUS PAIRS don't have to be the order listed. They can be reversed, except for number 10.
1) gabbro, granite   2) pumice, obsidian   3) scoria, pumice   4) basalt, rhyolite   5) granite, obsidian
6) diorite, andesite   7) basalt, tuff   8) gabbro, basalt   9) tuff, obsidian   10) Continents are granite, ocean is basalt.

4.4

1) G   2) E   3) F   4) B   5) C   6) J   7) D   8) I   9) A   10) H

4.5

1) coal   2) chert   3) slate   4) cobbles   5) gypsum   6) flint   7) halite (salt)   8) marble   9) limestone
10) granite   11) quartzite   12) scoria   13) sandstone   14) pumice
What is not so useful? I vote for siltstone, mica schist, travertine and conglomerate.
A lot of the lesser quality ingenous rocks, such as tuff or rhyolite, can be used as building stone if nothing better is in the area.
They are not ideal but they have been used in the past.

4.9

Students may use colors as well as patterns. Sometimes geologists use both at the same time, putting a light layer of color on top of the patterns.

Notice that magma is called lava after it comes out of the volcano.

The order of the layers of sedimentary rock is variable.

This does NOT represent an exact location that a geologist actually observed. It is only intended as a way to practice lithologic patterns and to apply general concepts learned in the chapter.

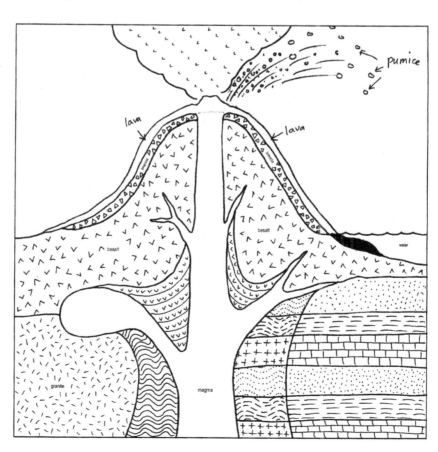

Chapter 5

5.1

1) T   2) yes   3) The molecules arrange themselves into different crystal shapes.   4) b   5) b
6) a   7) d   8) b   9) d   10) b   11) T   12) see picture on right.
13) covalent   14) ionic   15) alive   16) F   17) a   18) b   19) a   20) T
21) making pickles, sugar production, adding calcium to fruit juices, making tortillas, sewage treatment
22) T   23) T   24) 2   25) d   26) c   27) T   28) c   29) b   30) F
31) F   32) caves   33) F   34) T   35) blue holes

12)
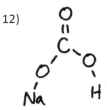

5.3

1) Kentucky   2) Georgia   3) eyes   4) bat poop   5) fungus   6) Big Room   7) salamander
8) sound   9) centipede   10) spider   11) white   12) fleas   13) glowworms   14) crayfish   15) copepod   16) gypsum

Chapter 6

6.1

1) b   2) b   3) A   4) B   5) C   6) B   7) E   8) A   9) C   10) d   11) F   12) T   13) silt loam   14) clay loam   15) clay
16) T   17) T   18) F   19) F   20) e   21) a   22) F   23) a   24) c   25) leaching   26) c   27) a   28) T   29) c   30) f
31) F   32) b   33) hydrogen ions (protons)   34) a   35) rainforests   36) d   37) b
38) T   39) c   40) c   41) b   42) F   43) F   44) F   45) T   46) b   47) a   48) b   49) d   50) F   51) b   52) F   53) F
54) 5%   55) 45%   56) c   57) mites, spiders   58) woodlice, centipedes, millipedes   59) c   60) answers will vary
61) c   62) e   63) a   64) c   65) b   66) a   67) wet, good for growing   68) e   69) California
70) planted trees, rotated crops, planted cover crops, better plowing methods, drilled wells, used fertilizers

6.3

1) dirt   2) limestone   3) profile   4) loam   5) porosity
6) ped   7) cover   8) heave   9) leaching   10) Iceland
11) lichen   12) potassium   13) hydrogen   14) adsorption
15) enzymes   16) water   17) hummus   18) macromolecule
19) carbon   20) nitrogen   21) germinate   22) nematode
23) slug   24) ammonium   25) California

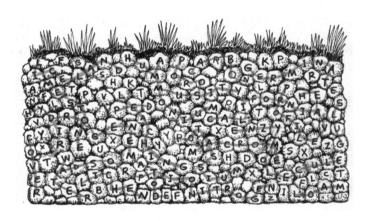

6.5

Names of mushrooms:
Top row: shiitake, enoki, morel, yellow oyster
Bottom row: fly amanita, truffle, white button
Clues:
1) mycelium   2) carbon   3) hyphae   4) Oregon   5) spores   6) erosion   7) communicate   8) fuel   9) microscopic
10) phosphorus   11) K   12) decomposers   13) woodlice   14) texture   15) no-till farming

6.7

PART 1:
1) J   2) E   3) K   4) H   5) P   6) A   7) B   8) F   9) C   10) O   11) L   12) N   13) D   14) G   15) M   16) I
PART 2:
Three-letter words: 1) ore   2) fix   3) ped   4) ion   5) air   6) red   7) tin   8) low   9) gas   10) egg
Four-letter words:
1) cube   2) salt   3) talc   4) gold   5) iron   6) lead   7) opal   8) tuff   9) jade   10) clay   11) sand   12) silt
13) coal   14) meta   15) cave   16) soil   17) dark   18) rock   19) loam   20) mica   21) cell   22) lime

Chapter 7

7.1
1) b 2) b 3) c 4) F 5) b 6) F 7) T 8) primary, pressure 9) secondary, shear 10) F 11) F 12) e
13) C 14) T 15) F 16) a 17) c 18) F 19) b 20) a 21) c 22) crossover depth 23) c 24) b 25) d
26) a 27) continental 28) basalt 29) b 30) c 31) a 32) a 33) Trans-Atlantic telegraph cable
34) the Mid-Atlantic Ridge 35) Iceland 36) Rocky Mountains 37) c 38) normal, thrust, strike-slipo (or transform)
39) San Andreas 40) no 41) d 42) c 43) c 44) d 45) Colorado River 46) a 47) Wales 48) T 49) F 50)a
51) F 52) d 53) a 54) b 55) F 56) T 57) bacteria 58) b 59) b 60) F

7.2
1) CBDA 2) BDAC 3) CBDA 4) BCAD 5) BCDA 6) BDAC (overthrust) 7) DACB 8) BCEDA 9) EBDCA

7.6
SANDSTONE: sedimentary, water, aquifer, quartzite   GRANITE: ingeous, quartz, feldspar, biotite, hornblende, felsic, continents
SHALE: sedimentary, clay, slate, water, gas   CHALK: sedimentary, coccolithophores, calcium/$CaCO_3$
LIMESTONE: sedimentary, $CaCO_3$, dolomite, carbon, karst, caves   GEODE: SiO_2, quartz   GYPSUM: $CaSO_4$, sulfuric acid

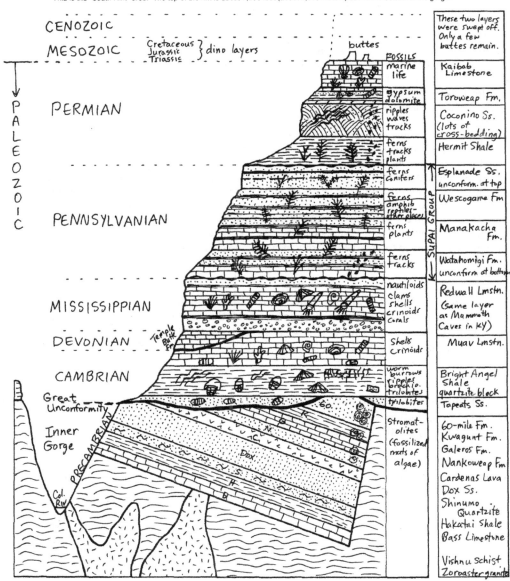

Chapter 8

8.1

1) b   2) a   3) a   4) F   5) a   6) F   7) F   8) b   9) d   10) c   11) a   12) a   13) d   14) b   15) d
16) surge tectonics   17) rotation   18) b   19) d   20) T   21) c   22) F   23) protons   24) protons and neutrons
25) a   26) T   27) argon   28) a   29) e   30) T   31) c   32) varve   33) d   34) d   35) c   36) weaker

8.4

None of these require any particular viewpoint about the past.

8.5

1) F (We can observe this under a microscope and with chemistry experiments.)   2) I (There is no way to know what an worm is thinking!)
3) F (As we have learned in the text, seismic waves are just about the only direct observation we can do of the Earth's interior.)
4) I (We can compare the behavior of P and S waves in rocks that we know are solid to unknown places inside the earth such as the mantle. When P and S waves act in a certain way, we infer that this is because they are the same density and texture. In this case, we infer that the mantle is solid.)   5) I (No one has ever directly observed the mantle.)
6) I (No one has ever directly observed the earth's interior. We only have P and S wave data, and facts about the various elements behave.)
7) F (Carbon-14 levels were observed to change in the mid 1900s, after the nuclear explosions.)
8) I (We did not observe caves forming. We see water dripping inside of caves now and we can measure carbonic acid levels in rain, but we can't go back into the past and see if conditions were the same back then as they are now.)
9) I (The actual data showed fault lines and various densities of rock under the surface. Everything else was inference.)
10) I (Again, with no time machine available, we can't observe the past.)
11) F or I (If you said "fact" that's fine. This type of fact is technically an inference, since we can't actually "see" a proton or neutron. However, the high-tech methods of gathering data are so good that it is pretty obvious what is going on, and that a neutron has turned into a proton. Interestingly enough, the structure of an atom is still actually theory, not fact. The basic structure was worked out by Ernest Rutherford, using logic to explain why particles behaved the way they did when they were shot through a thin piece of gold foil.)
12) I (We were not there to see dinosaurs running. All we have is fossilized bones. We can compare dinosaur skeletons to the skeletons of other animals and make guesses about how they would have walked or run, but we will never know for sure.)
13) I (No scientists were alive when the rocks were forming so we have no direct data. Inferences are made from levels of various gases in rocks and polar ice cores, but always using the principle of uniformitarianism, which assumes that Earth has not been through anything as drastic as a worldwide flooding event.)
14) F (Earthquake epicenters have been detected in the interior of plates.)   15) I (We can't observe the mantle directly.)
16) I (This was a tricky one, and not really something you could have been expected to know for sure. However, it's a good opportunity to sneak in a little extra fact about granite. Granite has NEVER been observed forming from melted rock. This means that technically, it is not really igneous ("from fire"), although we put it into this category. Once granite melts, that beautiful speckled pattern is gone for good. We can't do an experiment where we let granite cool slowly for thousands of years, let alone millions of years, so we'll never know what would happen in this case. However, knowing that things tend to settle out according to density, we might predict that granite's minerals would settle out into striped layers, with the most dense minerals at the bottom.)   17) F (We can observe Surtsey Island off the coast of Iceland.)
18) I (We were not there to see the continents connected.)   19) F (Some coal still looks like logs! We can clearly see plant structure in coal.)
20) F (There are many places around volcanoes where this has been observed by volcano scientists.)

8.6

1) amber (others are made of only carbon)   2) mafic (others are characteristics of minerals)   3) gypsum (others are pure elements)   4) size (others are characteristics of soils)   5) basalt (is fine grained, others are coarse grained)   6) mosquito (does not live in soil, others do)   7) dolomite (others are all $CaCO_3$, whereas dolomite is $MgCO_3$)   8) corundum (does not contain silicon, others are silicates)   9) porosity (others are used for mineral ID)   10) aquifer (water stays in ground, other come up)

8.8

1) 34, 55 (You add the two previous numbers. This pattern is common in math and science and is called the Fibonacci series.)
2) 720, 5040 (You multiplty by 2, then by 3, then by 4, then by 5, etc.)
3) 124, 126 (You multiply by 2, then add two, multiply by 2, add 2, etc.)
4) 97, 193 (You double the number, then subtract 1.)
5) 45, 55, 5   6) 100   7) no   8) 1   9) 25 to 30, 30 to 35   10) 9   11) about 1000; around 2,000?
12) about 100   13) You get two answers: 5 and 7   14) It might stop, or it could go back up again and repeat the pattern.
15) yes   16) no   17) very   18) about 6   19) not likely, since the data is so scattered   20) 8.5; around 17 or 18
USING EQUATIONS TO HELP US: 220, 2020, 8

BIBLIOGRAPHY

Books:

Cambridge Guide to Minerals, Rocks and Fossils by A.C. Bishop, A.R. Woolley and W. R. Hamiltion. Published by the Cambridge University Press, UK, 1999. ISBN 0-521-77881-6

Eyewitness Handbook of Rocks and Minerals by Chris Pellant. Published by DK Publishing, NY, 1992. ISBN 1-56458-033-4

In the Beginning by Walter Brown. Published by CSC, Phoenix, AZ, 2008 (8th edition). 978-1-878026-09-5

Flood Legends; Global Clues of a Common Event by Charles Martin. Published by Master Books, Green Forest AZ, 2009. ISBN 978-0-89051-553-2

Manual of Mineralogy (21st edition), by Cornelis Klein and Cornelius Hurlbut, Jr. Published by John Wiley & Sons, New York, 1993. ISBN: 9780471574521

Modern Earth Science by Robert Sager, William Ramsey, Clifford Phillips and Frank Watenpaugh. Published by Holt, Rinehart and Winston, 1998. ISBN 0=03-050609-3

Physical Geology; Exploring the Earth by James S. Monroe and Reed Wicander. Published by West Publishing Company, St. Paul, Minnesota, 1995. ISBN: 9780314042736

Sedimentary Petrology by Harvey Blatt, Univ. of Oklahoma. Published by W. H. Freeman and Company, New York, NY, 1992.

Tectonic Globaloney by N. Christian Smoot. Copyright 2004 by N. Christian Smoot. ISBN: 1413437281

The Dating Game by Cherry Lewis. Published by Cambridge University Press, UK, 2000. ISBN: 521790514

Periodicals:

"Sulfuric acid speleogenesis of Carlsbad Cavern and its relationship to hydrocarbons, Delaware Basin, New Mexico and Texas." by C. A. Hill, 1990. American Association of Petroleum Geologists Bulletin.

Websites:

Alchemy
https://en.wikipedia.org/wiki/Alchemy
https://en.wikipedia.org/wiki/Mary_the_Jewess
https://en.wikipedia.org/wiki/Jabir_ibn_Hayyan
https://en.wikipedia.org/wiki/Georgius_Agricola

Definition of mineral
http://geology.com/minerals/what-is-a-mineral.shtml
https://en.wikipedia.org/wiki/Mineral
http://www.rocksandminerals4u.com/what_is_a_mineral.html

Hardness
https://en.wikipedia.org/wiki/Mohs_scale_of_mineral_hardness

Mineral colors
http://www.treasuremountainmining.com/index.php?route=pavblog/blog&id=38

Formation of minerals on Surtsey
http://www.surtsey.is/pp_ens/geo_4.htm
https://en.wikipedia.org/wiki/Palagonite
https://en.wikipedia.org/wiki/Opal

Silicates
https://www.esci.umn.edu/courses/1001/minerals/amphibole.shtml
https://en.wikipedia.org/wiki/Amphibole
http://geology.com/minerals/muscovite.shtml
http://www.minerals.net/mineral/chalcedony.aspx
http://geology.com/minerals/garnet.shtml
https://en.wikipedia.org/wiki/Talc
http://www.gly.uga.edu/railsback/Fundamentals/BowensRxnSeriesSketch07P.pdf
http://news.psu.edu/story/140648/2001/09/01/research/how-do-agates-form

Igneous rocks
https://www.youtube.com/watch?v=cjyF-te4lQI
https://www.youtube.com/watch?v=Zbz4e-9pjY4
http://www.rense.com/general20/graniteal.htm
http://www.geologyclass.org/Igneous%20Concepts.htm
https://en.wikipedia.org/wiki/Igneous_rock
http://www.skidmore.edu/~jthomas/fairlysimpleexercises/brs.html

Sedimentary rocks
http://www.thisoldearth.net/Sedimentary-Rocks-sorting.cfm
https://www.youtube.com/watch?v=uozyWZ6XQzM
https://www.youtube.com/watch?v=Etu9BWbuDlY
https://www.youtube.com/watch?v=G0ru-kLpedo
https://www.youtube.com/watch?v=1JLa392qA-k
http://geology.com/rocks/sedimentary-rocks.shtml
http://www.geomore.com/sedimentary-rocks/
https://en.wikipedia.org/wiki/Fossils_of_the_Burgess_Shale
https://www.youtube.com/watch?v=XjxJUZOJXN4

Metamorphic rocks
https://www.youtube.com/watch?v=ckypDmkbhb8&t=3s&index=27&list=PLRaq2fWj05NMh6byzC2hYZGtSYMxqwApc
https://en.wikipedia.org/wiki/Metamorphic_rock

Limestone
http://geology.com/usgs/limestone/
https://www.gsa.gov/portal/content/111930
http://www.sandatlas.org/limestone/
http://pecconsultinggroup.com/the-effects-of-impurities-on-lime-quality-featured-industry-pec-consulting-group
https://www.youtube.com/watch?v=9LDG9cnGlDo
https://en.wikipedia.org/wiki/Calcium_oxide
https://www.thenakedscientists.com/articles/questions/what-happens-when-acid-reacts-limestone
https://phys.org/news/2015-03-seashells-mineral-differently-seawater.html
http://www.discoveringfossils.co.uk/chalk_formation_fossils.htm
http://www.discoveringfossils.co.uk/dover_kent_fossils.htm
https://news.agu.org/press-release/giant-algal-bloom-sheds-light-on-formation-of-white-cliffs-of-dover/
http://www.stoneagetools.co.uk/what-is-flint.htm
http://www.sciencenutshell.com/rapid-growth-of-plankton-due-to-rising-carbon-dioxide-levels/
https://www.nps.gov/cave/learn/nature/lechuguilla_cave.htm

Global carbon estimates

1) U. Siegenthaler and J. L. Sarmiento, "Atmospheric Carbon Dioxide and the Ocean," *Nature*, Vol. 365, 9 September 1993, pp. 119–125
2) Bert Bolin, "The Carbon Cycle," *Scientific American*, Vol. 223, March 1970, pp. 125–132
3) P. Falkowski et al., "The Global Carbon Cycle: A Test of Our Knowledge of Earth as a System," *Science*, Vol. 290, 13 October 2000, p. 293.

Karst

http://www.fillmoreswcd.org/documents/Sinkhole3.pdf
http://www.differencebetween.net/miscellaneous/difference-between-cave-and-cavern/

History of geology

http://slideplayer.com/slide/4315242/ (quick overview using ppt)

Layers in Grand Canyon

https://pubs.usgs.gov/bul/1395j/report.pdf (Supai formation)
https://3dparks.wr.usgs.gov/coloradoplateau/lexicon/vishnu.htm
https://www.nps.gov/grca/learn/nature/fossils.htm
https://www.hitthetrail.com/kaibab-limestone-to-supai-formation/
http://asset.emsofl.com/ONLINE%20CLASS/EarthSci/_labdata/unit7/pages/strata/supaigrp.html
https://3dparks.wr.usgs.gov/coloradoplateau/lexicon/supai_group.htm
http://grandcanyonpark.weebly.com/toroweap-formation.html

Dendritic formations in rocks

http://www.sandatlas.org/dendritic-growth-in-crystals/

Cross over depth of magma

http://www.spring8.or.jp/pdf/en/res_fro/06/113-114.pdf
Okayama University, Japan School of Natural Science and Technology

Cold seeps and hydrothermal vents

https://en.wikipedia.org/wiki/Cold_seep

Soil

https://www.youtube.com/watch?v=mg7XSjcnZQM
https://www.youtube.com/watch?v=uP-nXy34KFA&t=1s
https://www.youtube.com/watch?v=p166fVxwyuY
http://eldoradochemical.com/fertiliz1.htm
texture and triangle https://www.nrcs.usda.gov/wps/portal/nrcs/detail/soils/edu/?cid=nrcs142p2_054311
structure: http://passel.unl.edu/pages/informationmodule.php?idinformationmodule=1130447039&topicorder=4&maxto=10
http://www.deeproot.com/blog/blog-entries/what-is-soil-structure-and-why-is-it-important
https://opentextbc.ca/geology/chapter/5-4-weathering-and-the-formation-of-soil/
http://www.soilminerals.com/Cation_Exchange_Simplified.htm
http://www.eebweb.arizona.edu/courses/182-spring2009-Bonine/182-KEB-PlantNutrition-lect6-CH38-40-sp2009postx6.pdf
http://www.nutrientstewardship.com/implementation/article/soil-ph-and-availability-plant-nutrients
http://www.wildmadagascar.org/overview/rainforests2.html\
Humification controversy: http://www.nature.com/nature/journal/vaop/ncurrent/pdf/nature16069.pdf
https://en.wikipedia.org/wiki/Soil_microbiology
https://www.researchgate.net/figure/232365028_fig1_Fig-2-Geometrically-optimized-three-dimensional-structure-of-the-SOM-model-of-soil
http://passel.unl.edu/pages/informationmodule.php?idinformationmodule=1130447042&topicorder=2&maxto=8
https://www.sciencedaily.com/releases/2008/10/081015110240.htm

Surtsey

http://www.surtsey.is/pp_ens/geo_5.htm
https://en.wikipedia.org/wiki/Surtsey
http://www.icelandontheweb.com/articles-on-iceland/nature/volcanoes/surtsey
http://humusandcarbon.blogspot.com/
http://extension.psu.edu/plants/crops/soil-management/soil-quality/earthworms

https://en.wikipedia.org/wiki/Dust_Bowl
https://en.wikipedia.org/wiki/Mycelium
https://en.wikipedia.org/wiki/Mycelium
https://en.wikipedia.org/wiki/Diatomaceous_earth
http://entnemdept.ufl.edu/creatures/nematode/soil_nematode.htm

Seismography
https://www.e-education.psu.edu/earth520/content/l4_p3.html
http://eqseis.geosc.psu.edu/~cammon/HTML/Classes/IntroQuakes/Notes/waves_and_interior.html
https://en.wikipedia.org/wiki/Seismometer
https://en.wikipedia.org/wiki/Rayleigh_wave
http://geophile.net/Lessons/PlateTectonics/PlateTectonics2_02.html

Kola Superdeep Borehole
https://en.wikipedia.org/wiki/Kola_Superdeep_Borehole
http://www.zmescience.com/science/geology/drilling-to-the-mantle-6-unexpected-discoveries-from-the-worlds-deepest-well/
http://www.dailykos.com/story/2015/3/11/1367130/-The-Kola-Borehole-The-Deepest-Spot-on-Earth

Inside earth
http://www.smithsonianmag.com/science-nature/decades-long-quest-drill-earths-mantle-may-soon-hit-pay-dirt-180957908/
http://geology.com/articles/mohorovicic-discontinuity.shtml

Ocean sediments
https://en.wikipedia.org/wiki/Pelagic_sediment

Ocean topography
https://www.gislounge.com/marie-tharp-and-mapping-ocean-floor/
https://en.wikipedia.org/wiki/Transatlantic_telegraph_cable
http://theory.uwinnipeg.ca/mod_tech/node79.html

Sediment layers and faults
https://en.wikipedia.org/wiki/Core_sample
https://www.thoughtco.com/geology-of-mount-everest-755308
http://www.noahcode.org/the-geologic-column---falsified.html
https://en.wikipedia.org/wiki/Lewis_Overthrust
http://www.talkorigins.org/faqs/lewis-overthrust.html

Groundwater
http://academic.emporia.edu/schulmem/hydro/TERM%20PROJECTS/2008/Pick/dakotaaquifer.html
https://water.usgs.gov/edu/gwartesian.html
Lineaments   http://www.amyhremleyfoundation.org/php/education/features/Aquifers/Geology.php

Fossils
https://www.trilobites.info/trilointernal.htm
https://en.wikipedia.org/wiki/Trilobite
https://www.nps.gov/grca/learn/nature/fossils.htm#CP_JUMP_2241953
http://grandcanyonnaturalhistory.com/pages_nature/fossils/cover_fossils.html
http://kgov.com/files/docs/Austin-et-al-nautiloids-GSA-abstract.jpg
https://en.wikipedia.org/wiki/Silurian
https://en.wikipedia.org/wiki/Cambrian
https://www.adn.com/arctic/2016/07/30/new-knowledge-about-ancient-arctic-creatures-is-reshaping-understanding-of-dinosaurs/

Petrification
http://www.grisda.org/origins/05113.htm (quotes Anne Sigleo research on ppm silicon)
http://www.sciencedirect.com/science/journal/00167037/42/9 (for paid access to Anne's research paper)

Petroleum

http://response.restoration.noaa.gov/oil-and-chemical-spills/oil-spills/resources/what-are-natural-oil-seeps.html
http://www.whoi.edu/oil/natural-oil-seeps
http://www.livescience.com/5422-natural-oil-spills-surprising-amount-seeps-sea.html
http://geologylearn.blogspot.com/2015/08/is-there-oil-beneath-my-property-first.html

Continents

https://en.wikipedia.org/wiki/Continental_drift
https://en.wikipedia.org/wiki/Plate_tectonics

Dating

https://en.wikipedia.org/wiki/K%E2%80%93Ar_dating
https://en.wikipedia.org/wiki/Beta_decay
https://en.wikipedia.org/wiki/Proton_decay
http://www.indiana.edu/~ensiweb/lessons/varve.ev.pdf
http://www.oldearth.org/varves.htm
http://geology.com/articles/green-river-fossils/
https://thenaturalhistorian.com/2012/11/12/varves-chronology-suigetsu-c14-radiocarbon-callibration-creationism/
http://science.howstuffworks.com/environmental/earth/geology/carbon-142.htm
http://www.c14dating.com/int.html
https://en.wikipedia.org/wiki/Radiocarbon_dating
http://www.smithsonianmag.com/smart-news/nuclear-bombs-made-it-possible-to-carbon-date-human-tissue-20074710/
https://str.llnl.gov/str/Knezovich.html
https://ageofrocks.org/2015/04/13/argon-argon-dating-how-does-it-work-is-it-reliable/

Flooding on Mars

http://science.time.com/2013/03/11/the-great-and-recent-martian-flood/
http://www.space.com/20111-mars-megaflood-underground-radar.html

TEACHER'S SECTION

GAMES
CRAFTS
LABS
ACTIVITIES

NOTE: You don't have to do all of these activites.
Choose the ones that are best suited to your
student(s) and your situation.

CHAPTER 1

1) PERIODIC "SCAVENGER HUNT"

Copy the following page and have the students use a Periodic Table to solve the clues. (The Periodic Table at the top of the page will not be sufficient. It does not contain enough information.) There are lots of very nice Periodic Tables online. Just Google "Periodic Table."

Answer key:
1) 79   2) Cf, californium   3) uranium, 92   4) tellurium   5) 3
6) K, Na, Fe, Ag, Au, Sn, Sb, W, Hg, Pb   7) 43 technetium   8) Br, bromine, and Hg, mercury
9) Ge, germanium, Sn, tin, and Pb, lead   10) Rn, radon   11) I, iodine, and At, astatine
12) U, uranium, Np, neptunium, Pu, plutonium   13) Po, polonium and Fr, francium, Cm, curium   14) Mg, magnesium, and Ca, calcium   15) Er, erbium, Tb, terbium, Y, yttrium and Yb, ytterbium   16) Z: 2   A: 7   P: 9   J: 0
17) N, nitrogen, O, oxygen, P, phosphorus, S, sulfur   18) 99   19) Pb, lead   20) Sn, tin

2) "POINT TO THE ELEMENT" laser pointer challenge (group game)

You will need:
• a Periodic Table poster
• two laser pointers

This is a team game, although you could play it with as few as 2 players. If you can possibly find two laser points with different colors, such as red and green, this is ideal. However, you can make it work even with the same color. The kids will know which dot is theirs.

Divide the students up into two teams. The students will take turns using the laser pointer, one person per team. The adult in charge will call out clues and the first team to land the red laser dot on the correct element wins the round.

Suggested clues:
  1) You might want to start out with just finding the letters. Call out the letters of common elements such as carbon, C, or nitrogen, N, or helium, He. For harder elements, choose minerals they are likely to see in this book. Take a look at the poster on page 69 for ideas.
  2) If your students are familiar with a lot of the names and letters, call out just the name, such as "Magnesium."
  3) If your students know a little more about the elements, you might want to create suitable clues for their knowledge base. Easy clues might be, "I am thinking of an element that you can put into a balloon," or "I'm thinking of the element that the Romans used to make pipes." Tailor the clues to what you think your students might know or be able to figure out.

3) LEARN MORE about a scientist

If you need to lenghten the amount of time you spend on this chapter, you can always ask the students to choose a scientist from the chapter and learn a little more about him/her. Books and websites are fine, but you can also tap into video documentaries via YouTube (with parental supervision). (Or Hulu, or other video services if you have them.)

PERIODIC TABLE SCAVENGER HUNT

*Lanthanide series

**Actinide series

1) If you're digging for gold, you are searching for element number _____

2) The most expensive element in the world is number 98. It is a man-made element and costs about 27 million US dollars per gram to produce. What is the name of this element? _____

3) In general, elements get larger as the atomic numbers get higher. The largest element (highest number) that is natural and not man-made is _____ at number _____. (HINT: Use the symbol key in the lower left corner.)

4) The element selenium was named after the moon. The scientist who discovered the element right underneath it thought that the moon should be near the earth, so he used the Latin name for "earth" and named it _____.

5) The best batteries in the world have the element lithium in them. Lithium is a small atom. Its atomic number is ___

6) There are 10 elements with symbols that are different from the first letters of their name. What are the symbols?
__K__, _____, _____, _____, _____, _____, _____, _____, _____, _____

7) What is the smallest (lowest number) radioactive element? _____

8) Which two elements are liquids at room temperature? _____ and _____ *(according to the table in this book)*

9) The elements carbon, C, and silicon, Si, are key ingredients in many minerals. They are both in the same column, so that means that they have similar chemical properties. What other elements are in this column? _____, _____, and _____.

10) The elements in the column on the far right are called the noble gases. They are the only elements that have no interest in making molecules. All of the noble gases are used in light bulbs except this one, because it is radioactive: _____

11) The column just to the left of the noble gases is called the "halogens" (salt-makers). They combine with the very first column on the left side of the table to make salts. Only two halogens are solids. They are _____ and _____

12) Three elements are named after planets. Which ones? _____, _____, and _____

13) Marie Curie discovered these elements. She named one after her home country of Poland and the other after her adopted country, France. They are _____ and _____ (Which element is named after her? _____)

14) Iron and magnesium are very important elements in the chemistry of volcanic rocks. Calcium is the key to understanding limestone. Which two of these elements are in the same column? _____ and _____

15) Four elements are named after Ytterby, a small town in Sweden: Erbium, Terbium, _____ and _____

16) How many element *symbols* begin with the letter Z? _____ the letter A? _____ the letter P? _____ the letter J? _____

17) There is only one place on the table where you will find four elements sitting in a 2x2 square, where all four elements have only one letter in their symbol. What four elements are these? _____, _____, _____, and _____

18) What is the atomic number of the element named after Albert Einstein? _____

19) In radioactive decay, an atom's nucleus falls apart, often spitting out two protons and two neutrons. If a uranium atom decays 5 times (2 protons lost each time) what elements does it end up as? _____

20) Which element has the fewest number of letters in its name? _____

CHAPTER 2

1) "MAKE FIVE" A CARD GAME ABOUT MINERAL RECIPES

NOTE: This game also appears in the curriculum titled "The Elements" by the same author. If you have already done *The Elements* curriculum, and therefore have already played this game, you can decide to skip it here. Or you can get out your old game and play it again! (Your students will find that they have a much better understanding of those chemical recipes than they did the last time they played.)

You will need:
• copies of the pattern pages copied onto card stock (If you purchased a hard copy of this book and would like to be able to print pages directly from your computer printer or take the file to a print shop, use this web address to download digital copies of the pattern pages: www.ellenjmchenry.com/rocks-and-dirt)
  (NOTE: Another option would be to cut these pages out of this book and use them.)
• scissors
• white glue (if you are assembling the paper dice)
• If you are using wooden cubes for the dice, you'll also need one or more markers.
  (In a pinch, just take a fine point marker (red?) and write on real die. Everyone can just ignore the dots.)

NOTE: If you can get three wooden cubes, this is the best option. Most craft stores sell wooden cubes individually or in small units and fairly inexpensively. If you want this game to be sturdy enough to survive future uses, consider using wooden cubes.

Preparation:
  1) Cut out the dice patterns (copied onto heavy card stock) and make into cubes, using small dabs of white glue on the tabs. Or, write the symbols on wooden dice or even regular dice.
  2) Cut apart the 16 mineral cards.

How to play:
        Place the mineral cards on the table, face up, so they form a 4 x 4 square. Each player will have a turn rolling all three dice at once. The goal is to roll the ingredients to form a mineral. (One roll of the three dice per player per turn.) For example, if the first player rolls Cu, Fe, and S, he should notice that those are the ingredients of chalcopyrite. Therefore, that player picks up the chalcopyrite card. If the next player rolls Ca, C, and WILD, he could make the wild card into O, and be eligible to pick up calcite. (To speed up the game, you can allow players to roll twice on their turn. The second roll can be one, two or all three cubes.)
        The first player to collect five cards wins the game.

2) CUT-AND-ASSEMBLE PAPER CRYSTAL SHAPES

You will need:
• copies of the pattern pages printed onto card stock (any color)
• scissors
• white glue (or a really good glue stick, the kind made for adults)
• paper clips or clothespins to hold joints while they dry (optional, but recommended)

Directions:
  1) Copy the pattern pages onto card stock. (Tip: Card stock feeds through most computer printers, so if getting to a copy shop is hard, keep a supply of card stock on hand so you can print using your computer's printer.)
    NOTE: If you have a hard copy of this book and would like a digital file to print from, go to: www.ellenjmchenry.com/rocks-and-dirt
  2) Cut out the crystal shapes. Cut on all solid lines.
  3) You may want to "score" the dotted fold lines using a ruler and a very sharp pencil. (Actually, a compass point works the best. But you can also use a nail, a dead ball point pen, or a scissor point if you are gentle.) Run the sharp point along each fold line. Press hard enough so that the paper is slightly dented (scratched). This will make folding very easy.
  4) Pre-fold along all the fold lines. (Don't be overly concerned about folding the wrong way because any pre-folding is better than none at all.)
  5) Put a SMALL amount of white glue on one or two of the tabs. White glue is very strong and you don't need a lot of it. Press and hold those one or two joints and count to ten slowly. If you don't have the time or patience to hold the joints, clip them with paper clips or clothespins and let them dry for a few minutes.

NOTE: The most common mistake students make when assembling paper projects is to use too much glue. If glue oozes out of the cracks when you press the joint, you've used too much glue. One way to help students avoid using too much glue is to tell them not to squeeze it directly from the bottle onto the tab. Have them put a few drops onto a piece of paper and just dip the tip of a finger into it, or use a toothpick or cotton swab to apply the glue.

TIP: You may want to work on two models at a time. While one set of joints is drying, you can be working on another one.

6) When you get to the last joints, you just have to do the best you can to get the joints to stick. You can try folding the tabs just barely enough to get them in, so that after they are in they will apply a little counter pressure and push back up against the surface. Or, you could resort to using a flattened out paper clip that you can poke into the adjacent corner, giving you an extra "hand" (albeit a skinny one) inside the figure. You might not have to resort to this though. The best tool you've got when doing a project like this is PATIENCE.

Display suggestion: The complete figures make a nice mobile. (They are also less likely to be damaged while hanging up.)

3) OBSERVING TINY MINERAL CRYSTALS

You will need:
• a 40x magnifier (or at least 20x)   (TIP: Amazon sells some good magnifiers at reasonable prices.)
• salt
• epsom salt (easily purchased from the pharmacy section of any store, and very inexpensive)
• sugar
• sand (several kinds of sand from different sources is ideal)

What to do:
Look carefully at each type of crystal. The salt should look like little cubes, and the epsom salt should like little 6-sided hexagonal crystals. Sugar crystals might be a bit 6-sided, but they also might have random shapes. Sand is very interesting to look at as long as your magnification is high enough. You should be able to see tiny quartz crystals, bits of pink feldspar, and some black mica, as well as some miscellaneous rounded (and probably unidentifiable) sedimentary rocks. The mix of rocks in minerals in sand is unique to each location on earth. Sand from various beaches around the world have their own particular blend of minerals.

If you don't have a magnifier available, you can use the pictures on the following pages. Real life observation is better, of course, because you can see the crystals in 3D, which is so much better than a flat picture.

4) LAB: MINERAL IDENTIFICATION

Purchase a mineral identification kit from a science supply store. Follow the instructions that come with the kit. If you search Amazon, you can find a wide variety of sizes and prices. The kits start at about $15 USD and go up to $50 USD. If you are on a budget and don't want to spend money on this particular activity, that's okay. You don't absolutely have to do this lab. (Perhaps you can wait until you are done with chapter 4, and then go out in your neighborhood and look at some local rocks and try to identify them.) Another substitute for doing this lab is watching some mineral ID vidoes on YouTube.

5) LAB: MEASURING DENSITY— a lab for older students (See activity 6 for a lab for all ages)

This lab is a bit too tedious for students younger than 11 or 12.
Allow at least an hour. Doing it in a group slows you down A LOT. In a group setting, the entire lab might take 90 min.
Make copies of the following lab pages. (You can print from a digital copy by going to www.ellenjmchenry.com/rocks-and-dirt) If you want to shorten the time needed to do the lab, precut the food cubes ahead of time. If you want to do this lab with younger students (under 12) precutting is highly recommended in order to avoid issues with knives and razor blades. You don't have to cut a set for each student, however. Cubes can be shared around.
NOTE: This seems like a lot of bother just to explain density, but as my group worked through it, I could see several of them having "lightbulb" moments. ("Hey, look! Even if I cut the apple cube in half, the density doesn't change!")

table salt   40x

Epsom salt 20x

Himalayan salt   20x

raw cane sugar 40x

DENSITY LAB

Name _____

INTRODUCTION:

Density is the word we use to define how tightly packed the molecules are in a certain substance. Some substances, such as stone, have many atoms packed closely together, giving them a high density and making them feel heavy. Other substances, such as foam, have many fewer atoms in the same amount of space, making them feel light. The size of the atoms themselves also affects the measurement of density. For example, atoms of gold, mercury and lead are very large, much larger than aluminum or copper atoms. The larger atoms have more protons and neutrons in their nuclei, adding to the overall mass and density of the element.

Density is a clue that can help mineralogists figure out the identity of a sample. Two minerals might look exactly the same but have different densities. Therefore, all rock and mineral guide books will list the density along with the other properties of the rock or mineral. However, some books prefer to use the term "specific gravity" instead of density. Specific gravity compares the density of the mineral to the density of water. Since the density of water is exactly 1.0, the specific gravity number is always the same as the density number. It doesn't make any practical difference which term you use. Specific gravity is technically more correct, but you still see the word density, too.

Since we will actually be comparing some substances to water, we'll go ahead and switch over to using the term specific gravity. Don't let it throw you! We are still talking about density.

PART 1: Comparing the specific gravity of various substances to the specific gravity of water

You will need:
- a small graduated cylinder (10 ml is best for this first part, but 25 ml is okay. 50 ml will do if you don't have a smaller one.)
- a balance (digital scale)   (Smaller cylinders give you a more accurate measurement.)
- a sharp knife or razor blade (and adult supervision)
- a bowl of water
- vegetable oil
- a pipette (eye dropper)
- a metric ruler
- a variety of solid "waterproof" foods, such as potato, apple, pear, banana, cheese, carrot, squash, cucumber, eggplant, broccoli stem, egg white, cantaloupe (You can use non-food items, too, such as foam. Just make sure anything you choose will be able to go into a bowl of water and not disintegrate. That eliminates bread. Also, make sure you will be able to cut the substance safely. Stay away from nuts or other hard items that might make the knife or razor slip onto your finger.)
  You don't have to use this entire list. Just choose a handful of items from what you have around the house.
  TIP: Try to use apple, if you can, as it will give an interesting result.

What to do:
1) Choose 4-6 solid items that you will test. Use the metric ruler and the knife or razor to cut cubes of each food that are EXACTLY one centimeter on a side. Be patient and try to get your centimeter cubes cut as accurately as possible. The more accurate you are, the better your measured results will be.

2) Use the balance (set on grams) to record the mass of each cube. Write the name of the substance on the line and then record its mass in grams. (Be sure to keep track of which cube is what, since some of the cubes might look very similar.)

MASSES OF MY CUBES: *(you don't have to use all the lines)*

_____ = _____ grams   _____ = _____ grams   _____ = _____ grams

_____ = _____ grams   _____ = _____ grams   _____ = _____ grams

_____ = _____ grams   _____ = _____ grams   _____ = _____ grams

4) Your cylinder is marked with lines. If you are using a 25 ml or a 50 ml cylinder, those lines will probably represent 1 milliliter (ml). If you are working with another size, figure out what each line represents. Pour water into your cylinder until the top of the water is EXACTLY at 20 ml for the 25 ml size, or 40 ml for the 50 ml size. Use your pipette (eye dropper) to adjust the level of the water until it looks exactly right. (TIP: If you see a slight dip in the surface of the water, this is called the meniscus. Read the level of the water from the bottom of the meniscus (the low point) not from where the water hits the sides of the cylinder.)

Read from bottom

5) Put your cylinder onto the balance and then turn it on. The scale might read "0." This is good. If your scale reads some-thign other than 0, try hitting the TARE button if you have one. The TARE button resets the starting point for 0. In effect, the TARE button tells the balance to ignore all that weight that is on it, and just weigh what is coming next. If you don't have a TARE button, just write down the mass it is reading right now so you don't forget. You'll just add one to that number in step 7.

6) Take your pipette/dropper and add water until it looks like you have added exactly one ml to the cylinder. Remember to read from the low point of the water (the meniscus).

7) Look at the number of grams the balance is reading now. Did it go up by exactly 1 gram, or very close to that? If you could be perfectly precise, it would go up by exactly 1.

8) Now try the reverse. Don't look at the cylinder while you are dropping in water. Watch the numbers on the balance. When you've added exactly one gram, stop and look at the water level in the cylinder. Did you add 1 ml?
(Don't throw out that water. Leave it in the cylinder.)

1 ml of water has a mass of 1 gram. What is the density of water? Density is calculated like this:

$$\text{Density} = \frac{\text{grams}}{\text{milliliters}}$$

In this case, we have 1 gram/1 ml. 1 divided by 1 is 1, so **water has a density of 1**.

Water is the standard by which all other substances on earth are judged. If a mineral has a density of 2, that means it has a density twice that of water. A substance with a density of 0.5 has a density half that of water. Everything is compared to water.

Substances sort themselves out according to their densities. Higher densities go down, lower densities go up. This principle works in air, in liquids, and sometimes in solids as well. It is easy to see this principle at work in water. Substances with a density greater than 1.0 will sink, and those with a density less than 1.0 will float. Let's test our food cubes!

9) Let's test the density of another liquid. You will add exactly 1 gram of oil to the cylinder of water. Use the pipette/dropper to add the oil drop by drop, as you watch the numbers go up on the balance. Add exactly 1 gram.

10) Now look at the level of fluid in the balance. The oil will be on top of the water. Did you add exactly 1 ml oil? What does that tell you about the density of the oil? (The fact that it is floating on the water confirms what your numbers are telling you.)

11) The size of your cubes was not chosen at random. One solid cubic centimeter happens to have the same volume as one liquid milliliter. That makes calculating the density of these foods very easy! Using the density formula, Density= g/ml, we can put in "1" as the bottom number. That means we'll be dividing by 1, which means our number won't change. So those measurements we did on the cubes ARE the densities. All we need to do is look at those numbers and see which ones are greater than or less than 1.

12) Pick up a cube, check the number of grams, and then predict whether it will sink or float. Then put it into the bowl of water and see if you are right.

13) Lastly, put one of the sinking cubes into the cylinder. How much do you think it will raise the water level? (Make sure you read the cylinder level before you put the cube in.) Do your results confirm that 1 ml = 1 cc? (cc = cubic centimeter)

EXTENSION QUESTIONS:

1) If you cut a food block in half, will half a block have the same density as the whole block? You might want to test two of your blocks, one that floats and one that sinks. Does a very small piece act the same way as the whole piece did? What would happen of you put a whole apple/potato/zucchini/carrot into a large tub of water? Can you accurately predict whether the whole fruit or vegetable will float or sink?

2) Liquid mercury has a specific gravity of 5.43 g/ml. What would happen if you poured some liquid mercury into water? What would happen if you poured the mercury into a glass of vegetable oil?

2) Copper has a density of 8.9. What would happen if you put a piece of copper into liquid mercury? Titanium has a density of 34.5. What would it do in liquid mercury?

PART 2: Calculating the specific gravity of some mineral samples

You will need:
- a small graduated cylinder (25 ml or 50 ml, depending on your sample sizes)
- a balance
- water
- some small minerals or rocks (must fit into your graduated cylinder)
- a calculator

In the last section, our samples were perfect cubes—cubic centimeters that were equal to 1 ml. It was easy to find the density because the number on the bottom of the equation was 1. In this section, we will use things that are not cubic and will need to use the graduated cylinder to help us find their volume, plus a calculator (or your brain) to do the math.

What to do:
1) First, hold your mineral samples and compare how heavy they are. They are probably of different sizes, so comparing densities will be difficult. If you can, try to judge their densities based on their weight and size. Which one would you predict as being the most dense? Which one is the least dense? Take a guess before you measure them. See how good your brain is at estimating density. You might be surprised at how accurate your brain turns out to be!

2) Weigh each of your mineral samples. Keep track of which is which by numbering them or giving them names if you know what they are (calcite, granite, etc.). Write the name or number on a line and then now many grams it weighs. Include the number to the right of the decimal point, such as "19.45." (You might not need all these lines.)

_____ = _____ grams   _____ = _____ grams   _____ = _____ grams

_____ = _____ grams   _____ = _____ grams   _____ = _____ grams

2) Now choose one of the mineral samples and a graduated cylinder. You want to choose the smallest size cylinder that the sample will fit into. Smaller cylinders always give you a more accurate measure.
3) Fill the cylinder about halfway full with water, but make sure that the water line is exactly at a nice even number such as 20 or 30. Use the dropper to get the water line (the bottom of the meniscus) exactly at the mark.
4) Slide the sample gently down into the cylinder. You don't want water splashing out because you just carefully measured the level of the water. Make sure the sample is completely covered by the water. You will get an incorrect reading if it is sticking up out of the water. The sample will "displace" a volume of water equal to its volume. In other words, the amount that the water rises is the volume of the sample.   Record the volume of each sample.

_____ = _____ ml   _____ = _____ ml   _____ = _____ ml

_____ = _____ ml   _____ = _____ ml   _____ = _____ ml

Remember, 1 ml equals 1 cubic centimeter, so this ml measure is also giving us the volume of the sample in centimeters.

5) Finally, calculate the density of each one by dividing the grams by the mls. Density = g/ml

_____ = _____ density   _____ = _____ density   _____ = _____ density

_____ = _____ density   _____ = _____ density   _____ = _____ density

6) Calculate the specific gravity of each sample. This is a no-brainer. The specific gravity of a sample is its density divided by the specific gravity of water, which is 1. Any number divided by 1 stays the same. So the density and the specific gravity are the same. No calculations needed!

7) Compare your results to your original guesses. Did you get any right?

6) LAB: OBSERVING DENSITY (a qualitative lab for all ages)

You will need:
• a tall glass jar (I used an old olive jar-- perfect size and shape, and has lid)
   TIP: Tall glass jars will come in handy in future chapters, too.
• water
• vegetable oil
• various small items from around the kitchen, such as dried beans, raisins, nuts,
   dried fruit, carrot, apple, pasta noodles, etc. (TIP: Include walnuts, if possible.)
• optional: empty gel caps (oblong capsules that powdered vitamins and medicines
come in) If you have vitamin gel caps, you can pull the capsules apart and empty them.

What to do:
   1) First, fill the jar halfway with water. Then add an inch or so of oil.
   2) Screw on the lid and turn the jar upside down. Then turn it back again. Students
should notice that no matter which way you turn it, the oil always goes to the top.
This is because the oil is less dense than the water. Less dense things will always rise
(if they can) and say above more dense things. This principle holds for solids, liquid,
and gases.
   3) Begin adding small items, one at a time. You might even want them to predict
what will happen. Things that are more dense than water will sink to the bottom.
Things that are less dense than water, but more dense than oil, will float at the dividing line between the oil and the water. Things that are less dense than oil will float
on top of the oil.
   Raisins, carrots and dried beans are guaranteed to sink. Apples, fresh or dried,
should float. Nuts of various types will behave differently. Peanuts and almonds will
sink. Walnuts seem to be more oily and will float at the dividing line. Try many items and see what happens!

   NOTE: If you had an object that was exactly the same density as water, it would be able to stay right in the middle of the water. Neutral buoyancy is very hard to accomplish. If you happen to have an empty gel cap, fill it with water and see if you can
get it to float in middle of the water. Fill another one with oil and see what happens. Fill one with a little oil and a little water.
This last one will be interesting to watch. The capsule will hang right at the water/oil line with the water half down and the oil
half up. The dividing line between the oil and water inside the capsule will exactly match the line in the jar.

   ALSO NOTE: Some objects will want to stick to the oil. Almonds, for example, will hang at the water/oil dividing line for a few
minutes before sinking. The oily surface of the nuts will be attracted to the oil and might be able to overcome gravity, at least
for a while. To really test their density, shake the jar a few times and let everything settle out again.

7) LAB: CRYSTAL GROWING KIT

   If you want to further explore crystals, consider purchasing a crystal growing kit. These kits usually give you everything
you need to grow several kinds of mineral crystals. Kits are available in any well-stocked toy store, or any online science store.

barite

$$BaSO_4$$

Often found in limestone or hot spring areas. Usually white or light brown. Sometimes crystalizes into rose shapes, which are popular with collectors.

zircon

$$ZrSiO_4$$

Found in nearly all igneous rocks, although in very small amounts. Because it is so hard, it is often used as a gemstone in jewelry.

hematite

$$Fe_2O_3$$

Hematite is a major ore (source) of iron. The name "hematite" comes from its blood-red color ("hema" means blood).

cinnabar

$$HgS$$

Cinnabar has a reddish color and is very dense (heavy) because of the mercury (Hg). Pure mercury is a liquid at room temperature, but it is a solid when bound to sulfur.

cuprite

$$Cu_2O$$

Cuprite forms cubic crystals. It is sometimes called "ruby copper" because of its color. When exposed to air it changes to CuO.

fluorite

$$CaF_2$$

Fluorite is used in the production of steel. It has a glassy luster and can look similar to a quartz crystal, except for its tetragonal (4-sided) shape.

quartz

$$SiO_2$$

Quartz is used in electronics, as a gemstone, and in the manufacturing of glass (where it is the main component). Sand is made of very tiny pieces of quartz.

galena

$$PbS$$

Galena is very dense (heavy) because of the lead in it. During the era of musket rifles, galena was used as the source of lead to make musket balls.

pyrite

$$FeS_2$$

This mineral is often called "fool's gold" because of its golden color and shiny luster. It has no actual gold in it. It leaves a black streak, not gold.

Cards for "Make Five"
Copy onto card stock and cut apart.

corundum  Al_2O_3 Corundum is very hard. It is so hard that it is used in industry as an abrasive (like sand paper). Blue corundum is called a sapphire and red is a ruby.	**talc**  $Mg_3Si_4O_{10}$ Talc is extremely soft. In fact, you can scratch it with your fingernail! Talc is the main ingredient in talcum powder (used to dry off after a shower).	**calcite**  $CaCO_3$ Calcite is the main ingredient in limestone. It is one of the most common minerals in the world. Caves are made of limestone.
gypsum  $CaSO_4$ Gypsum is a soft mineral. It is one of the main ingredients in plaster and plasterboard. One type of gypsum is called alabaster and was carved by ancient peoples.	**chalcopyrite**  $CuFeS_2$ Chalcopyrite is pinkish-purple with flecks of gold. It is found wherever copper is mined. The copper can be taken out of it by using chemical processes.	**epsom salt**  $MgSO_4$ This mineral dissolves into water very easily. It is often used in medical treatment of wounds on hands and feet. It helps in the healing process.
diamond/graphite  C Strangely enough, both priceless diamonds and the stuff in your pencil are made of the same thing: pure carbon. The difference is how the atoms are bonded together.		

Cards for "Make Five"
Copy onto card stock and cut apart.

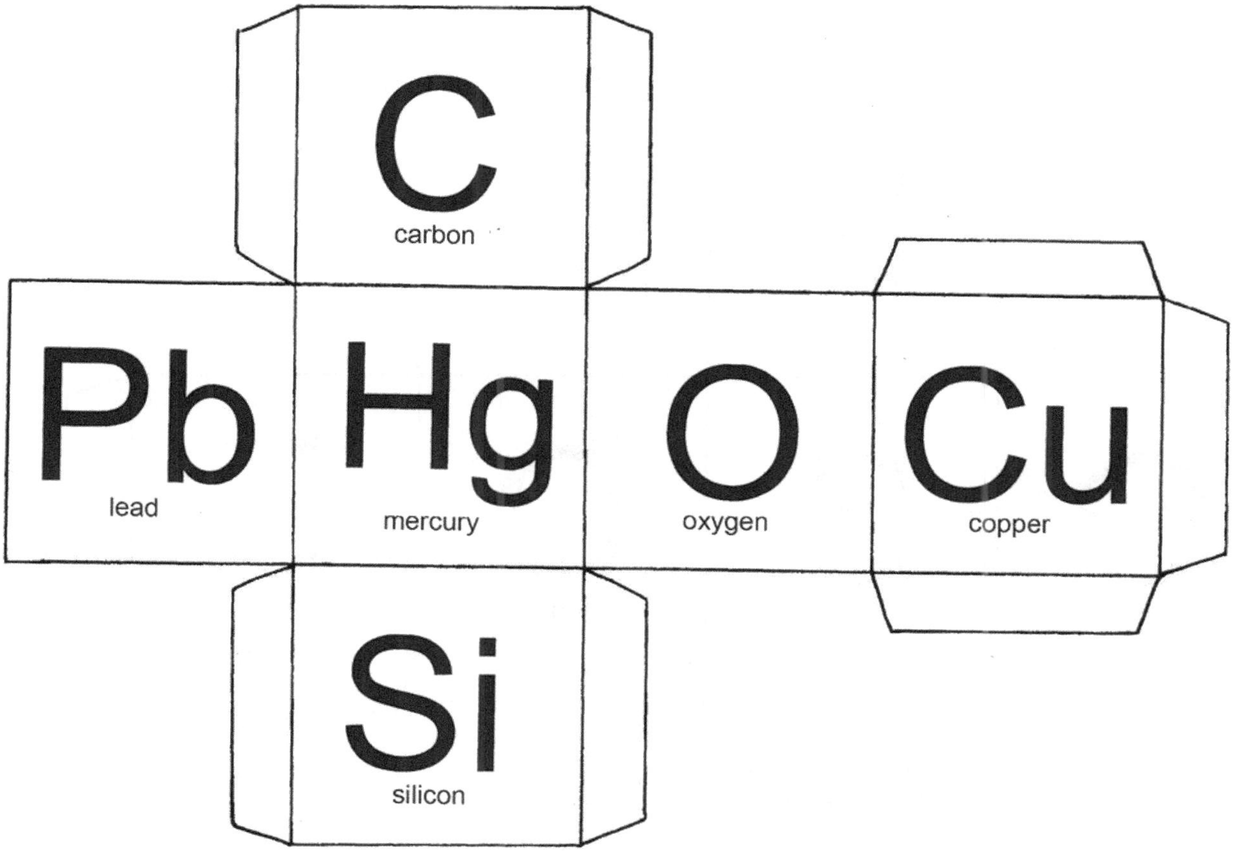

C
carbon

Pb
lead

Hg
mercury

O
oxygen

Cu
copper

Si
silicon

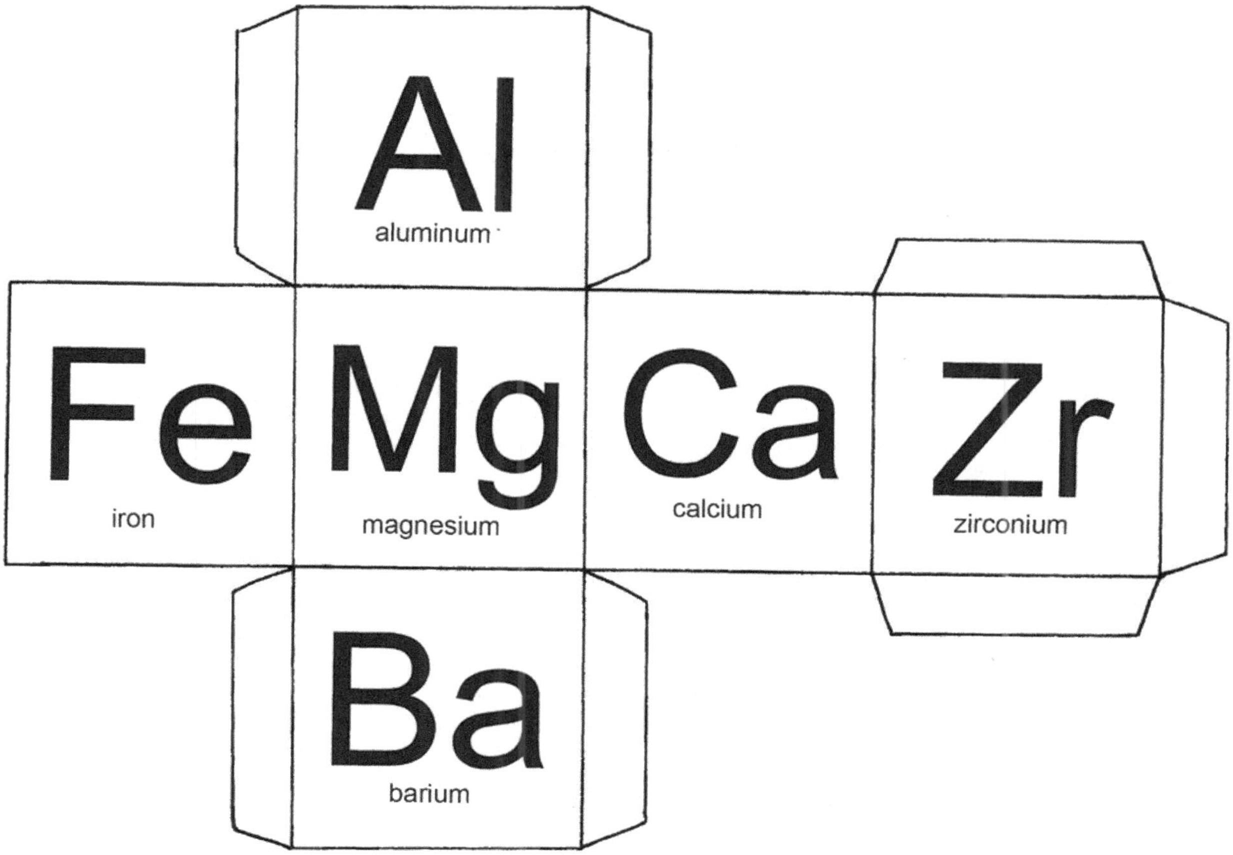

Al
aluminum

Fe
iron

Mg
magnesium

Ca
calcium

Zr
zirconium

Ba
barium

("Make Five" game)

COPY ONTO CARD STOCK

171

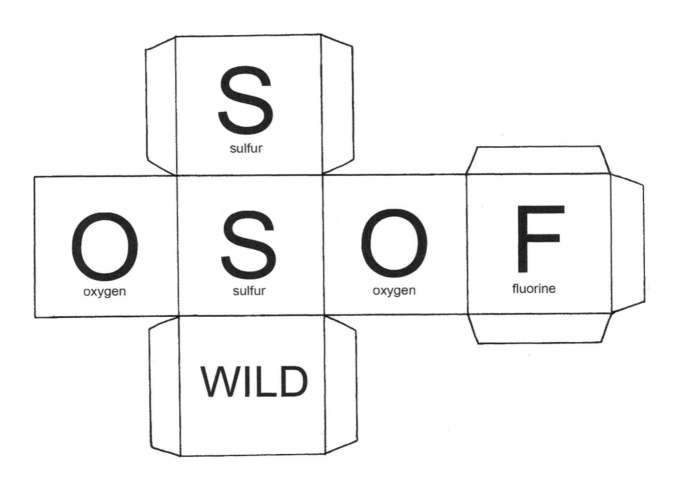

COPY ONTO CARD STOCK   ("Make Five" game)

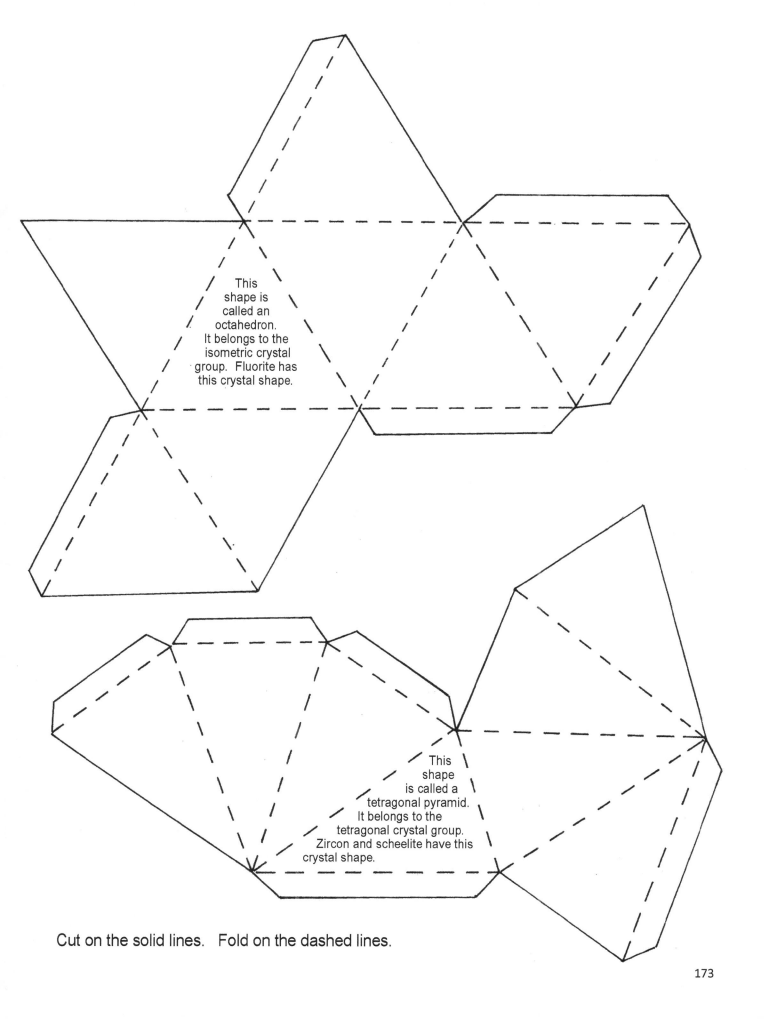

This
shape is
called an
octahedron.
It belongs to the
isometric crystal
group. Fluorite has
this crystal shape.

This
shape
is called a
tetragonal pyramid.
It belongs to the
tetragonal crystal group.
Zircon and scheelite have this
crystal shape.

Cut on the solid lines.   Fold on the dashed lines.

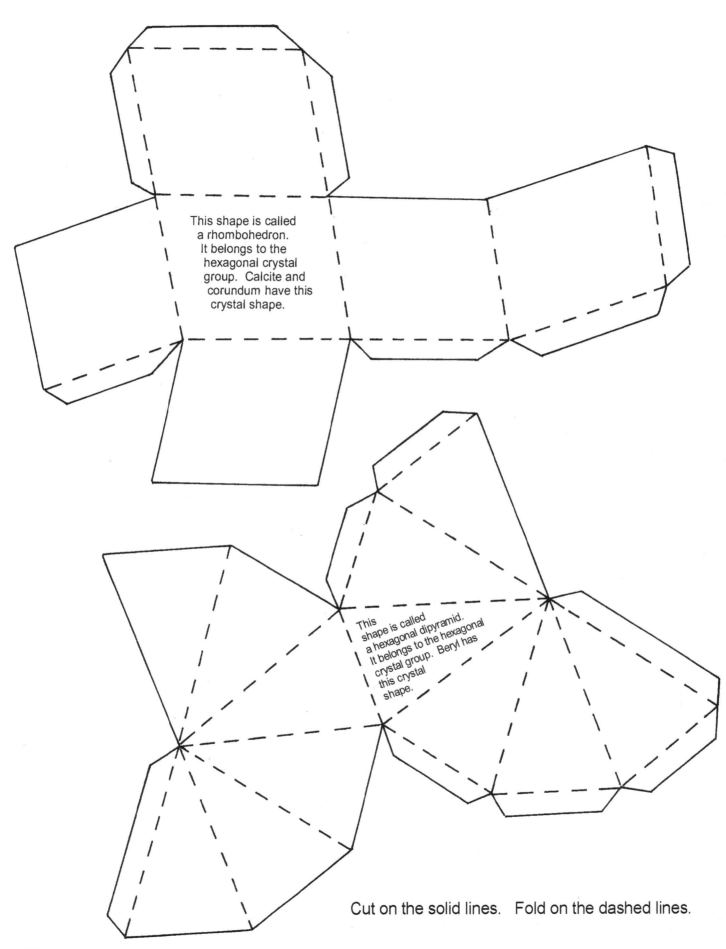

This shape is called a rhombohedron. It belongs to the hexagonal crystal group. Calcite and corundum have this crystal shape.

This shape is called a hexagonal dipyramid. It belongs to the hexagonal crystal group. Beryl has this crystal shape.

Cut on the solid lines.   Fold on the dashed lines.

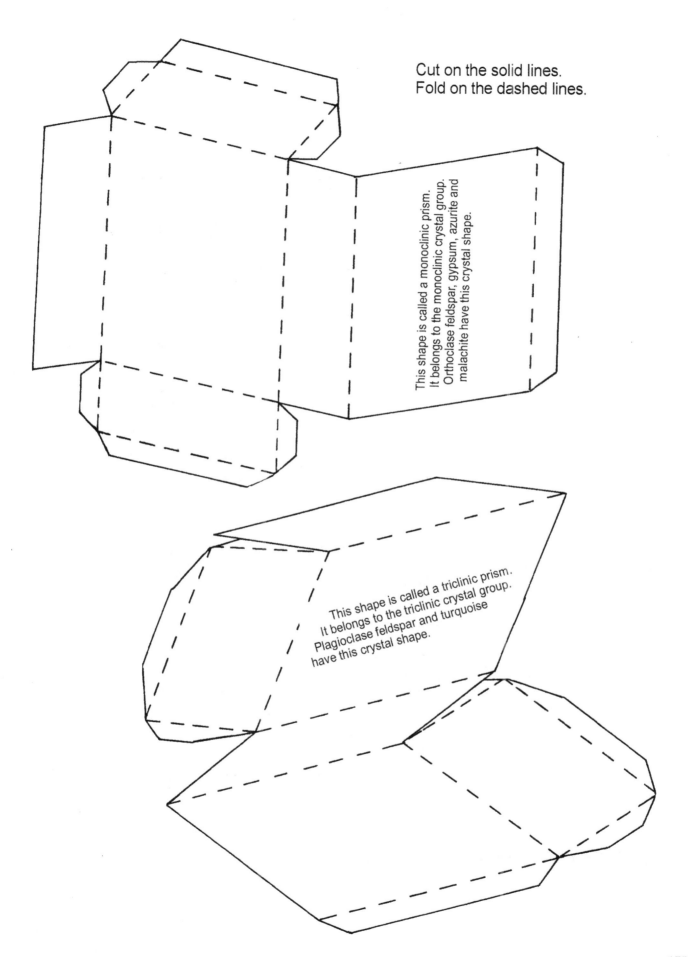

Cut on the solid lines.
Fold on the dashed lines.

This shape is called a monoclinic prism. It belongs to the monoclinic crystal group. Orthoclase feldspar, gypsum, azurite and malachite have this crystal shape.

This shape is called a triclinic prism. It belongs to the triclinic crystal group. Plagioclase feldspar and turquoise have this crystal shape.

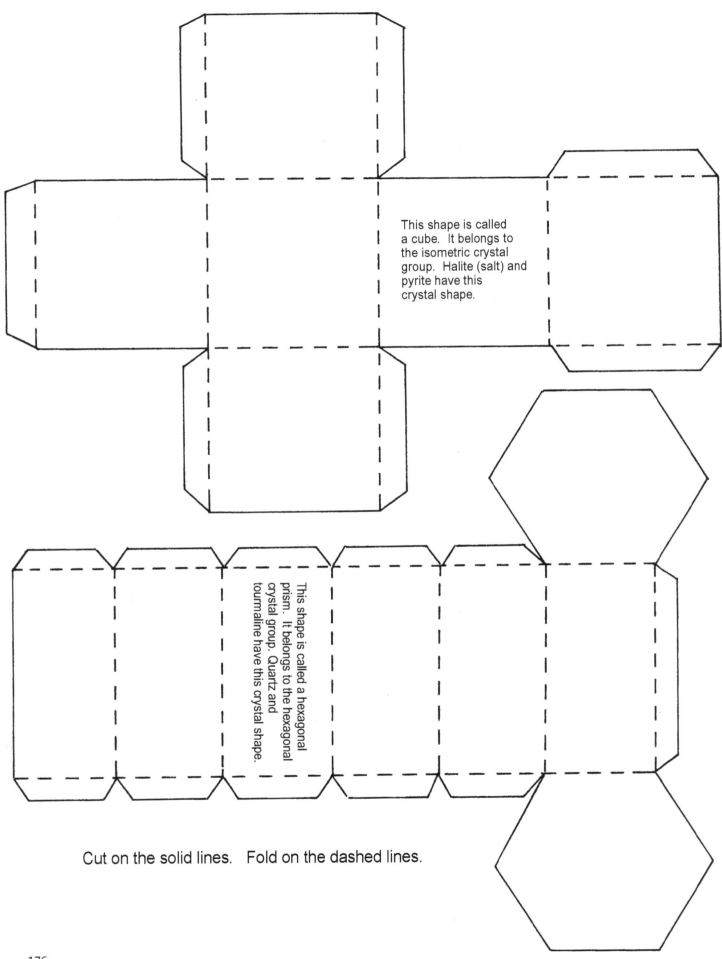

This shape is called a cube. It belongs to the isometric crystal group. Halite (salt) and pyrite have this crystal shape.

This shape is called a hexagonal prism. It belongs to the hexagonal crystal group. Quartz and tourmaline have this crystal shape.

Cut on the solid lines.   Fold on the dashed lines.

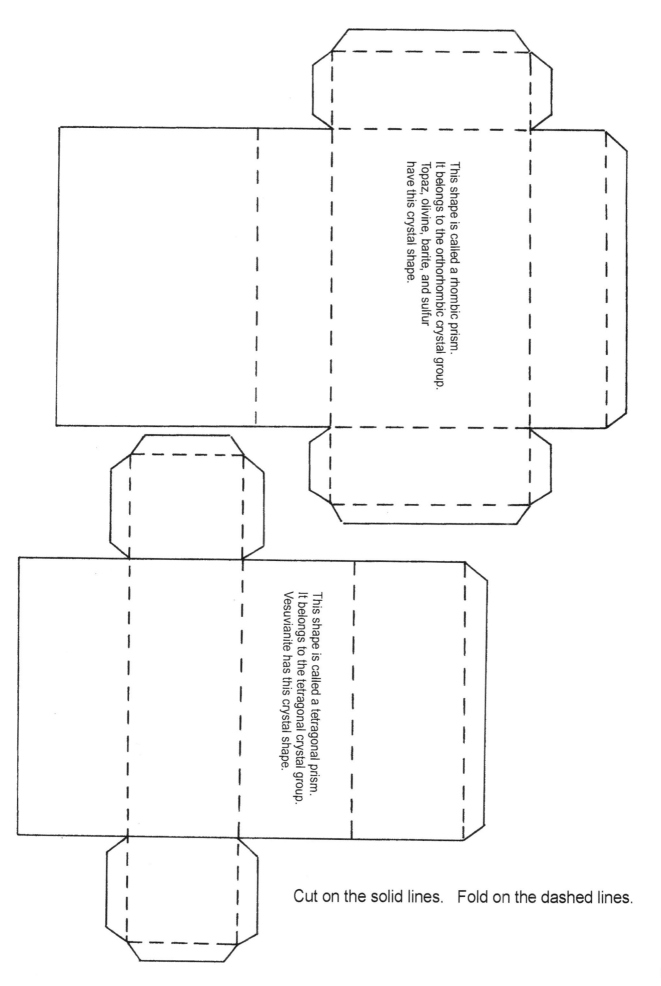

This shape is called a rhombic prism.
It belongs to the orthorhombic crystal group.
Topaz, olivine, barite, and sulfur
have this crystal shape.

This shape is called a tetragonal prism.
It belongs to the tetragonal crystal group.
Vesuvianite has this crystal shape.

Cut on the solid lines.   Fold on the dashed lines.

CHAPTER 3

1) MINERAL CARD GAMES

There are 48 mineral cards, with which you can play three different games. You can choose which games are most suitable for your students. You can also add your own tweaks to the games to make them harder or easier (adding clues, increasing number of cards in a set, etc.)

GAME 1: MINERAL BINGO

Goal of the game: get three (or four) in a row
Number of players: any (for larger groups use more than one set of cards)
Time allowance: can adjust the game to fit any amount of time
You will need: the playing cards, the list of clues, and tokens to put on the squares (Edibles are nice, such as cereal bits or candies, because they can be eaten at the end of the game)

How to play:
1) Distribute 9 cards to each player for a 3x3 square, or 16 cards for a 4x4 square.
2) Distribue tokens to the players (to mark the mineral squares).
3) Read the clues and play like standard Bingo. (If you are playing with a 4x4 grid, you could also play "postage stamp" where the winning shape isn't a line across the board, but a 2x2 square.)
4) Feel free to add your own clues.

BINGO CLUES:
1) Is made of just one element (sulfur, copper, silver, gold, graphite, anthracite coal) (Graphite and coal are made of carbon.)
2) Has cubic crystal shape (pyrite, galena, garnet)
3) Contains iron (hematite, magnetite, pyrite, hornblende, jadeite, heliotrope; possibly tourmaline, garnet, biotie, muscovite)
4) Is magnetic (magnetite)
5) Is clear (use pictures to determine this)
6) Is red or has red in it (use pictures to determine this)
7) Begins with the letter A (amber, afghanite, aragonite, agate, amethyst, anthracite)
8) Is organic (amber, anthracite coal)
9) Is purple (use pictures to determine this)
10) Contains carbon (graphite, calcite, anthracite coal)
11) Is green (use pictures to determine this)
12) Has special optical property (calcite-just the clear one, ulexite)
13) Has hexagonal shape (clear quartz, smoky quartz, amethyst, beryl, tourmaline; aragonite is only pseudo-hexagonal, but go ahead and count it)
14) Is yellow (use pictures to determine this)
15) Contains SiO_4 (olivine, topaz)
16) Contains aluminum (ruby, sapphire, feldspar, topaz, beryl, possibly some garnets)
17) Is blue (use pictures to determine this)
18) Begins with the letter F (fluorite, feldspars)
19) Luster is dull (use pictures to judge which ones have a dull, not shiny surface)
20) Have hardness of 8 or higher (topaz, ruby, sapphire)
21) Begins with the letter G (graphite, gypsum, galena)
22) Is a memeber of the cryptocrystalline group (agate, jasper, heliotrope, opal, chalcedony)
23) Contains calcium (fluorite, calcite, gypsum, afghanite)
24) Begins with the letter M (magnetite, muscovite, malachite)
25) Has rhombohedral shape (calcite)
26) Has metallic luster (pyrite, gold, silver, copper, galena, anthracite coal)
27) If you have two of the same mineral, you can put tokens on both of them (gypsum, calcite, feldspar, fluorite)
28) Cleavage is in sheets (biotite, muscovite)
29) Crystal shape resembles needles (gypsum, smoky quartz, clear quartz)
30) Contains sulfur (sulfur, cinnabar, pyrite, barite, gypsum, galena) (Actually, coal probably does, too.)
31) Is orange or has orange in it (use pictures to determine)
32) Doesn't have any flat sides (use pictures to determine)
33) Has vitreous luster (glassy) (use pictures to determine this—they must look really glassy)

GAME 2: MINERAL "DOMINOES"

Goal of the game: get rid of your cards
Number of players: any
Time allowance: can adjust the game to fit any amount of time
You will need: the playing cards

How to play:
  1) Distribute 5 cards to each player. The rest go in a draw pile, face down.
  2) Take the top card from the draw pile and put it on the table, face up, as the starting card.
  3) The first player looks at the cards in his hand to see if any of them have a similarity with the starting card. The criteria for similarity all start with the letter C:
     COLOR (must be the same or very close)
     CRYSTAL (same shape, such as hexagons, cubes, needles, or cut gemstones
     CHEMISTRY (same family group or contain same element, you must name the group or element)
  If the player doesn't have a matching card, he must take one from the draw pile. (If the draw matches, he can lay it down.)
  4) Play continues like this, with each player attempting to get rid of a card in their hand by matching to one already on the table. If they don't have a match they take a card from the draw pile. The cards on the table need not be in a straight line. However, the cards must match every card they touch. For example, orange calcite could be touching both another orange mineral and another calcite.
  5) To make the game harder, add a rule that says the cards must form a straight line and you can only lay down a card at either end. In this case, you might also want to allow players to lay down more than one card on a turn.

GAME 3: "MINERAL AUCTION"

IMPORTANT:
        This game gives the players a chance to use a lot of the information they've learned in the past two chapters. There aren't any rules for what makes a set. The players have freedom to think creatively and make their own choices. The goal is collect a set of three cards that have something in common. Simple criteria might be that they are all the same color, or have sharp crystal shapes, or are transparent, or even that they start with the same letter. Students can be encouraged to use criteria such as belonging to the same mineral classification group (carbonates, sulfates, etc.), containing a certain element such as iron or aluminum, being cryptocrytallines, having approximately the same hardness (within 2 points on Mohs scale) or being hexagonal or cubic. Two criteria that you can't use, however, are "ending in -ite" or "contains silicon," because too many of them would fall into these categories. You can have the adult in charge be the official judge of categories, or you might want to have a group vote and if everyone agrees then it's okay.

Goal of the game: to collect sets of three cards.
Number of players: 2 to 6
Time allowance: One game might only take 5-10 minutes, so you can play multiple rounds. You can require two or three sets to be collected in order to win. Adjust the rules to fit your time restrictions.
You will need: the playing cards, plus 5 pennies for each player

How to play:
  1) Shuffle the cards and give each player two cards, then put the rest in a draw pile, face down.
  2) Give each player five pennies.
  3) If you are playing with only two players, you won't have an auctioneer; you will both play every round. If you have more than two players, each player will take a turn being the auctioneer. That means when it is your turn to be the auctioneer, you are the only one who won't be playing during that round. Kind of backwards. The autioneer turns over the top card on the draw pile and asks if anyone wants to bid on it. The players look at the cards in their hand and decide if they'd like to have that card. This is when they need to think about possible categories and guess what they might like to collect. The goal is collect three of a kind.
  4) Any player who would like to bid on the mineral up for auction puts out a penny. If only one person bids, they get the card for one penny. The penny can either go back into the bank, or it can go to the auctioneer. (My 11-13 year olds liked the extra craziness of having the auctioneer get the money. Sometimes large numbers of pennies changed hands quickly and they thought this added a lot of fun. However, your group might like the bank option better.)

5) If two or more players bid, the autioneer asks if anyone would like to increase their bid to 2 pennies. If only one person increases their bid, they get the card and their 2 pennies go back into the bank (or to the auctioneer). Everyone else keeps their pennies.

6) If there is a tie, with two or more players bidding the same amount, then the auctioneer uses a coin toss to determine who wins the card. (If there are three people bidding, just have them choose a number between 1 and 10.)

7) After the auction is over, there will then be a chance to trade. Any player who would like to trade a card, puts that card out onto the table. No one has to trade; this is completely voluntary. If two players can agree to a trade, then they may trade cards. (One of my groups asked about trading pennies for cards. You can make your own rule about this if it comes up.)

8) After an auction and a chance to trade, that round is officially over, and it is the next player's turn to be the auctioneer. This sounds like it would make for very long turns, but, in fact, the game moves rather quickly. Often there will be only one bidder, or no one interested in trading.

9) Players may accumulate any number of cards in their hand. (I never ran into any problems with this. If you experience problems because of players having too many cards in their hand, you can add a rule about it, limiting the number.)

10) When someone has three of a kind, the round is over. Those three cards are then out of play and get set aside. The players who did not win are allowed to keep two of their cards if they wish, but have to put the others back into the draw pile. Or they may take two fresh cards. The player who won has to draw two new cards. Everyone gets their 5 pennies back again.

11) Play several rounds, maybe until someone gets two sets. Whoever gets the most sets of three is the winner. (But everyone wins if they had fun and learned something!)

NOTE about using LUSTER as a category:
 GLASSY: (or "vitreous") means it is really shiny, like glass (gemstones, quartz, agate, polished jaspers, etc.)
 SHINY: still shiny but not as shiny as glassy. (Fluorite or calcite would be in this category as well as others.)
 WAXY: still slightly shiny, but much less so than "shiny" (Unpolished chert, flint, jaspers have waxy luster.)
 METALLIC: shiny in a metallic way. (Galena and anthracite can look metallic, as well as the metals.)

OPTIONAL EXTRA:
 Just in case you feel the need to make tetrahedrons, either for demonstrating the science, or as a craft for the students, here is an easy pattern you can use.

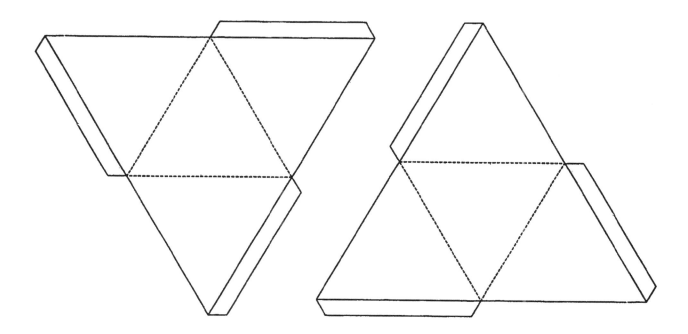

FELDSPAR

FELDSPAR

FELDSPAR

SAPPHIRE

RUBY

TOPAZ

AFGHANITE

ARAGONITE

ANTHRACITE COAL

BARITE
desert rose

GYPSUM

GYPSUM

AMBER

FLUORITE

FLUORITE

ROSE QUARTZ

QUARTZ

AMETHYST

TOURMALINE

CALCITE

TALC

GRAPHITE

MAGNETITE

ULEXITE

SULFUR

SILVER

GOLD

COPPER

BIOTITE

MUSCOVITE

LANDSCAPE JASPER

CINNABAR

MALACHITE

BLUE CALCITE

ORANGE CALCITE

PYRITE

187

HEMATITE

GALENA

HORNBLENDE

BERYL

HELIOTROPE

JADEITE

GARNET

OPAL

TIGER JASPER

SMOKY QUARTZ

AGATE

CHALCEDONY

2) CRAFT: PAINT YOUR OWN AGATE

sample of student work

Agates are made of silica, SiO_2. ("Agate" rhymes with "bag it.") Glass is also made of silica so it is not surprising that agates look smooth and glassy when polished. This name was given by Theophrastus, a Greek philosopher and naturalist who discovered this type of stone along the Achates River.

Most agates are found in volcanic rock, especially basaltic rock, although some are found in metamorphic rock. They started out as empty bubbles in the rock, then the space eventually filled in with SiO_2 that was mixed with hot water. Some bubbles filled in completely and others only partially, leaving an empty space at the center. Therefore, agates can be completely solid, or can have a hole in the center. Agates are not always round. They can also be found as stripes filling long gaps and cracks in volcanic rocks. These are known as banded agates or striped agates.

Agates usually come in shades of red and brown, gray, or blue. Bands of chalcedony often alternate with bands of crystalline quartz. If there is a hole in the center, it is often not smooth, but has large crystals around the edge. Agates are very resistant to weathering. In some places, they are found as hard, round balls, called "geodes" which must be cracked open with a hammer in order to see the beautiful minerals inside. One of the largest geodes ever found was in Brazil. It weighed 35 tons and the inside was lined with amethyst crystals (another form of silica, often light purple in color).

Agates are primarily used for decorative purposes such as jewelry, book ends, and other crafts. Two practical uses that have been documented (in just a few parts of the world) are burnishing leather, and using very thin agate slices as panels for stained glass window

You will need:
- 8.5" x 11" sheet of clear plastic, one per student (copier transparencies are ideal and can be purchased at the printing department of any office supply store)
- the following pattern pages copied onto white paper (each student will need one pattern and one title page)
- acrylic craft paints in various colors (use Google image search to get ideas for color schemes)
  NOTE: Paint markers work well for thin stripes.
- small brushes
- paper clips to hold the pattern pages in place
- white or clear craft glue (such as "Tacky Glue")
- a small amount of sand
- also have paper towels on hand, and some scrap paper to cover your work surface
- hairdryer to speed up drying time if you are working with a group that has limited work time

What to do:
  1) Copy (or print out) the agate pattern pages onto white paper. Each student will need one agate pattern and one copy of the title page that says AGATE at the top. The final project will get cut out and glued to the title page. (Card stock is nice for the title page if you have some available.) If you want to extend this project and make it more challenging, have the students draw their own black and white agate patterns instead of using these patterns.
  2) Place the sheet of plastic over one of the pattern pages. Use some paper clips or small pieces of tape to keep the plastic sheet in place while you paint.
  3) Use acrylic paints to paint colored stripes. (TIP: Limit the number of colors you use. Your agate will look more genuine if it is in shades of just blue, or just red and brown, or just gray, or just purple.) Bear in mind that the agate will be seen from the reverse side. You are painting on the back. If you paint a stripe on top of a stripe, only the first stripe will be seen on the front side. If you are using paint markers for thin stripes, do all the thin stripes first.
  4) Notice that the outer (crusty-looking) layer will be covered with sand. You can trace around this with a permanent marker or with a layer of paint if you are worried that they'll forget about this strip while cutting it out (and accidentally trim it off).

5) When the paint is dry, cut out the agate, including that unpainted strip (for the sand) on the outside.

6) Turn your agate over and put craft glue on the outer edge. Then dust with sand, making sure the sand is firmly embedded in the glue. Let it dry.

7) Glue your agate (shiny side up!) to the title page sheet that says AGATE at the top. sheet.

NOTE: The brightly colored agates you find in gift shops might have been dyed artificially. If you are interested in how this is done, here is a website that describes the process in great detail:
http://dyeingagate.com/dyeing-agates/

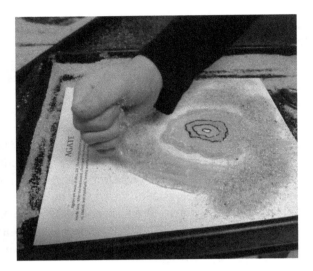

STEP 4   STEP 6

MORE STUDENT WORK:

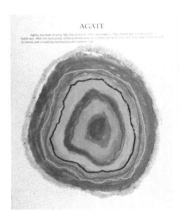
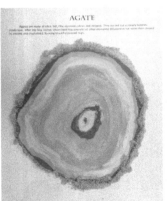

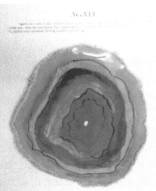

SAMPLES of real agates (to inspire student creativity!)

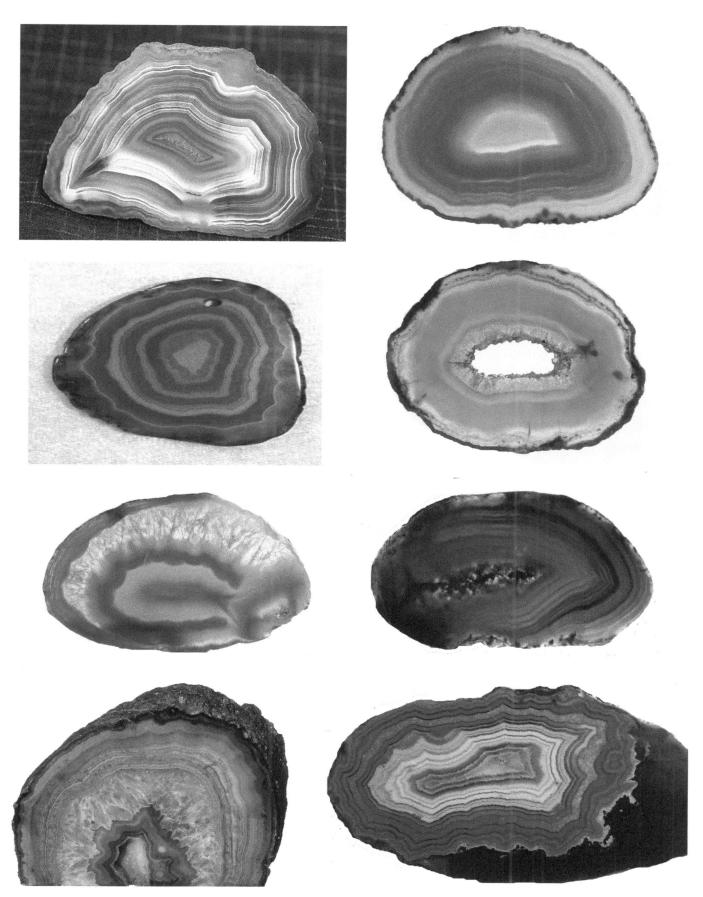

AGATE

Agates are made of silica, SiO_2 (the elements silicon and oxygen). They started out as empty bubbles inside lava. After the lava cooled, silicon (and tiny amounts of other elements) dissolved in hot water then seeped in, cooled, and crystallized, forming beautiful colored rings.

196

Place plastic sheet over this pattern.

Crusty outer layer
should not be
painted.

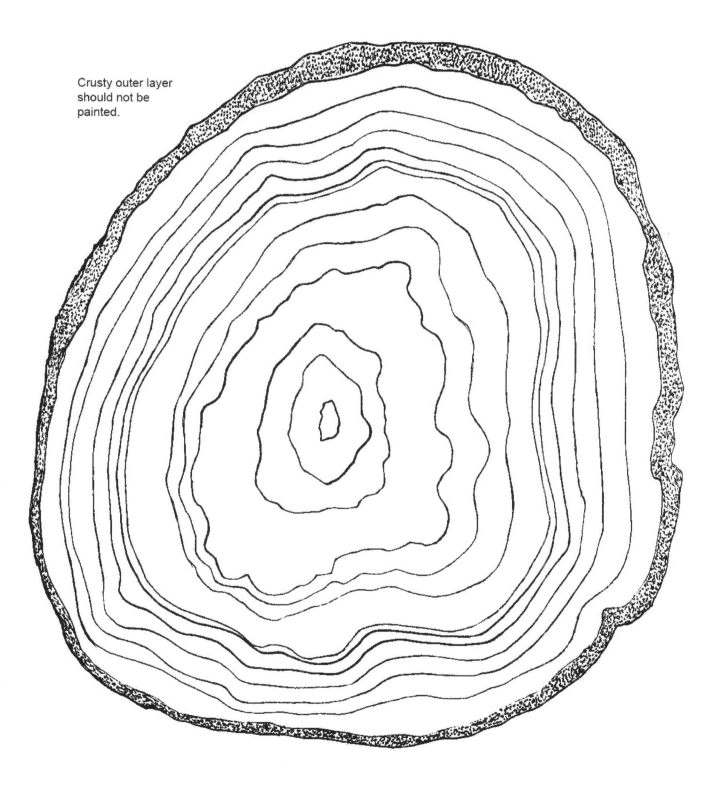

Place plastic sheet over this pattern.

NOTE: You don't need to use all the lines in this pattern. Feel free to simplify it a bit.

Crusty outer layer
should not be
painted.

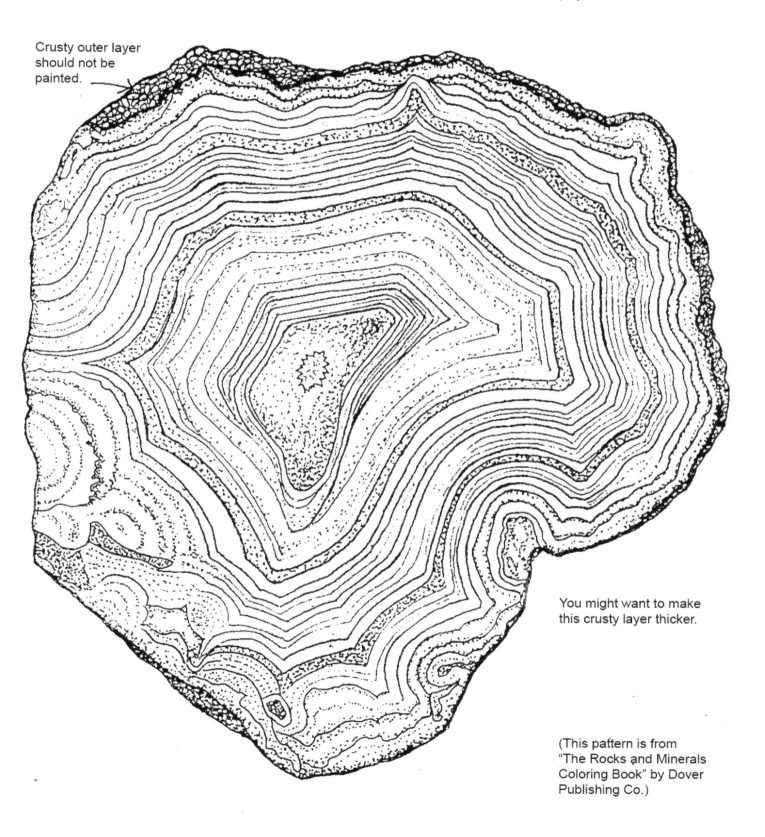

You might want to make
this crusty layer thicker.

(This pattern is from
"The Rocks and Minerals
Coloring Book" by Dover
Publishing Co.)

CHAPTER 4

1) REVIEW/QUIZ on chapters 1 to 4

This activity is optional. If you think your students need more review, or you would like to give them a quiz, you can make copies of the following review pages. (It is labeled as review, but you can use it as a quiz.)

If you would like to use it as a quiz, and want your students to study for the quiz, a study guide is also provided. Hand out the study guide the week before you give the quiz.

Answer key:
1) D   2) E   3) F   4) A   5) B   6) C   7) d   8) a   9) b   10) b   11-15) listed on page 7
16-17) amber, coal   18) F   19) E   20) G   21) salt   22) H   23) A   24) D   25) B   26) C   27) A   28) B   29) a   30) e
31) C   32) D   33) E   34) A   35) B   36) d   37) b   38) b   39) a   40) d   41) b   42) c   43) T   44) c   45) a
46) a   47) c   48) a   49) B   50) D
BONUS: 1) c   2) d   3) a   4) a   5) F

2) LAB: SIMULATE COOLING OF MAGMA

This lab uses salt water as a substitute for lava. Salt water has dissolved minerals in it, just like magma does. When magma cools, the minerals come out of the solution and form crystals. The same things happens with salt water. We can simulate minerals that are made by either slow-cooling or fast-cooling magma.

You will need:
• salt
• an eye dropper (not absolutely necessary, though)
• a glass or cup with a small amount of warm water
• an aluminum pie pan, or a piece of very stiff aluminum foil
• a small candle (very short "tea lights" are ideal)
• some small objects from which you can build a holder for your foil (blocks, soup cans, whatever is on hand that won't burn easily)
• magnifier

What to do:
  1) Put a spoon of salt into the warm water and stir well. The water does not have to be hot, though.
  2) Use whatever objects you have to make a holder for the tin pan or piece of heavy foil. It needs to stick out (like a diving board) about 5 or 6 inches (10-12 cm). The picture shows how we did this, but you can rig something different. Just make sure that the end of the foil will hang out far enough and be an inch or two above the candle flame.
  3) Put the small candle under the end of the foil. If the foil is not high enough, adjust your rig. The candle flame should not touch the bottom of the foil.
  4) Light the candle and let the end of the foil warm up. This end will cause the "magma" to cool very quickly. The end away from the flame will stay cool and will allow the "magma" to cool slowly.
  5) Put some single drops of salt water right over where the candle is, then begin to put small drops every half inch (1 cm) or so until you get to the base of the foil, as far away from the candle as possible.
  6) Watch how quickly or how slowly each droplet dries. The drops right over the candle will dry almost instantly.
  7) After watching for a few minutes, blow out the candle and let your experiment sit for a while. Go do something else and come back to it in an hour or more. (You can also let it sit overnight if necessary.)
  8) Observe the foil after all the drops have dried. Use the magnifier to look at the texture of the dried drops.

What should happen:
    The drops that dried quickly should have little or no texture. The salt will look like white powder. The drops that dried slowly should have some nice crystals in the middle or around the edges. You should see some very nice cubes! Making cyrstals take time, even for minerals. If a mineral is rushed, it does not have time to arrange the atoms into nice crystals. As we saw in chapter 3, simple shapes like chains form first, then later on you get sheets and framework.

STUDY GUIDE for QUIZ on chapters 1-4

1) What did these scientists do?

Jabir ibn Hayyan   Henning Brandt   Georgius Agricola

Dmitri Mendeleev   John Dalton   Neils Bohr

2) What determines the identity of an atom? Is it the number of electrons or protons or neutrons?

3) What are the five guidelines that determine whether a substance is a mineral or not?

4) Name a few substances that disobey these rules and are still included as minerals.

5) Make sure you know the chemical formulas for calcite (limestone), pyrite (fool's gold), salt (halite), quartz, and gypsum. Also, know the symbols for the native elements. All of these are given on page 9.

6) The only mineral crystal shapes you need to know are salt, quartz and calcite.

7) Be familiar with the Mohs hardness scale. Where is gypsum on the scale? quartz? corundum? diamond?

8) Be familiar with the characteristics of minerals, such as color, luster, cleavage, etc. You will not be asked to name them. You only need to recognize them when you see them.

9) Know the order of formation in silicate minerals. What happens first when the magma cools? Does a framework form right away or at the end? When is it stone soup? Do lines form before sheets?

10) Can you name a framework mineral? A stone soup mineral? A sheet mineral?

11) What does color tell you about the iron and magnesium content of a mineral? When a mineral contains a lot of iron and magnesium we call it m_____.

12) Feldspar is a very common mineral. It's the "field stone" in many countries. Feldspar is a silicate mineral so it has a lot of silicon in it. But it also has a lot of what three other elements? (These elements keep the color light.)

13) Quartz has a property that the other silicates do not. What is it?

14) Are rubies and sapphires made of the same stuff as diamonds?

15) What is the relationship between clay, shale and slate?

16) What is marble?

17) Which one is found on the ocean floor, granite or basalt? And on the continents?

18) Can you identify the pattern (from page 46) for granite, basalt, breccia, limestone, sandstone, and shale?

19) What is an evaporite?

20) What is a foliated rock?

Name _____

Match the person to what they did.

1) Georgius Agricola _____   A) Was an alchemist
2) Henning Brandt _____   B) Came up with "atomic theory" that everything is made of atoms
3) Neils Bohr _____   C) Invented the Periodic Table
4) Jabir ibn Hayyan _____   D) Studied mines in Europe
5) John Dalton _____   E) Discovered the element phosphorus
6) Dmitri Mendeleev _____   F) Proposed the solar system model of the atom

7) Which one of these substances is NOT on the Periodic Table?
   a) gold   b) bismuth   c) phosphorus   d) water   e) iron

8) When two or more atoms pair up, they form a: a) molecule   b) element   c) ion   d) nucleus

9) What determines which element an atom is?
   a) number of electrons   b) number of protons   c) number of neutrons   d) number of electron shells

10) What did alchemists think the philosopher's stone would do. besides turning things into gold?
   a) make you invisible   b) cure your diseases   c) make you wise   d) give you power and wealth

Can you remember the five rules that decide whether something is a mineral?
11) _____
12) _____
13) _____
14) _____
15) _____

Name two substances that disobey one of these rules and are still included in most mineral books.
16) _____ and 17) _____

Can you match the chemical formula with the name?
18) $CaCO_3$ _____   A) gold
19) FeS_2 _____   B) silver
20) SiO_4 _____   C) salt
21) NaCl _____   D) iron
22) Cu _____   E) pyrite (fool's gold)
23) Au _____   F) calcite
24) Fe _____   G) quartz
25) Ag _____   H) copper

Match the mineral with its crystal shape.
26) halite (salt) ___   A) hexagon
27) quartz ___   B) rhombohedron
28) calcite ___   C) cube

29) Which is the softest on the Mohs hardness scale? a) gypsum   b) quartz   c) corundum   d) apatite

30) Which one of these is NOT used to help identify a mineral?
   a) color   b) luster   c) cleavage   d) density   e) size   f) hardness

When magma containing silicon begins to cool, the tetrahedrons form patterns in a certain order. Put the patterns in the correct order.

31) First: ___
32) Second: ___
33) Third: ___
34) Fourth: ___
35) Last: ___

A) sheets
B) framework
C) "stone soup"
D) lines
E) double lines

36) Which of these is an example of a framework mineral? a) mica   b) olivine   c) asbestos   d) quartz

37) Which of these is a "stone soup" mineral?   a) mica   b) olivine   c) asbestos   d) quartz

38) Which contains more iron and magnesium? (HINT: look at color)   a) muscovite mica   b) biotite mica

39) Which has a lot of potassium, K, and sodium, Na? a) feldspar   b) quartz   c) olivine

40) What special property is unique to quartz?
   a) It has a hexagonal crystal structure.   b) Its formula is SiO_2.   c) It can form sheets.   d) It is piezoelectric.

These crystals are not related to any of the questions. They are just sitting here looking pretty. :)

41) The ocean floor is made mostly of: a) granite   b) basalt   c) gabbro   d) pumice

42) Rubies and sapphires are both made of this mineral: a) diamond   b) quartz   c) corundum   d) tourmaline

43) TRUE or FALSE?   Mafic rocks are darker in color than felsic rocks. ____

44) Which of these has a "frothy" texture? a) granite   b) obsidian   c) pumice

45) Which is thin and runny and is less likely to explode out of a volcano? a) mafic lava   b) felsic lava

46) Slate used to be shale, and shale used to be: a) clay   b) silt   c) sand   d) magma

47) What do field geologists use to differentiate clay from silt? a) eyes b) fingers   c) tongue   d) nose

48) Which is softer and easier to carve?   a) marble   b) granite

49) Which of these patterns represents limestone? _____
50) Which of these pattern represents sandstone? _____

A
B
C
D

BONUS QUESTIONS: (just a little harder)

1) This type of metamorphic rock always has stripes: a) mica schist   b) marble   c) gneiss   d) diorite

2) Which one of these is NOT an evaporite? a) gypsum   b) travertine   c) salt   d) coal

3) The word "quartz" comes from a German word meaning: a) hard   b) clear   c) silicon   d) hexagon

4) Foliated rocks have: a) lots of thin layers   b) fossilized leaves   c) stripes   d) quartz crystals

5) TRUE or FALSE? By definition, all clasts are made of silicon-based minerals. _____

3) ACTIVITY: COOKIE MINING

In this activity you simulate mining coal by digging chocolate chips out of cookies. Guaranteed to be a hit! You can find many versions of this activity if you Google "cookie mining." You might want to take a look at some other versions and see if there is one you like better than this one. Some websites offer fancier student printables, for example.

You will need:
- three types of chocolate chip cookies   (Choose a standard cookie plus at least one that advertises extra chips. You might also want a soft cookie to compare to hard.)
- flat toothpicks
- round toothpicks
- paper clips (the standard, smaller size, not the jumbo size)
- copies of the student hand-out (expense/profit sheet)
- a quarter of a sheet of plain paper for each student (have tape on hand to secure it to table if it slips around too much)
- pencils
- HELPFUL: at least two adults if you are working with a group. If the group is large, three is even better.

What to do:
1) The teacher/adult will need to look at the cookies carefully and decide which ones are likely to contain the most chips. Perhaps it might even be wise to sacrifice one of each kind of cookie ahead of time and count the chips in each one. The cookie that is likely to have the least amount of chips will be sold for 3 cents. The cookie that has the most chips will be 7 cents and the one in the middle will be 5 cents. The miners will have to purchase their cookie property. The difference in price simulates the reality mining companies face when purchasing land. Not all land is the same price. Land that is known to contain large seams of coal might be much more expensive than land that has less coal.
NOTE: These are the standard instructions you will find online. Personally, I found very little difference between any of the cookies that I purchased. I decided to charge a flat rate of 5 cents per cookie and did not bother with varying the prices.
2) Explain about start up costs and tell the students that they will have 19 cents to begin with. Go over each expense they will incur and see if there are any questions or comments. Make sure they understand they will have to pay that back with the profits they earn from mining.
3) Give each miner a quarter (or half) sheet of blank paper, and a pencil.
4) Have each miner buy their "property." Record cost on tally sheet. TIP: The temptation to nibble the cookies will be almost unbearable for some students. You may want to think about giving them a cookie to eat ahead of time. They may eat the crumbs of their mined cookie after they are done, but they really must NOT eat the mining cookie while they are mining it! Warn them strictly about this. Eating their mining cookie will invalidate the results of their mining.
5) Tell the miners to place their cookie on the paper and trace around it with the pencil. This outline will be used when it comes time for reclamation.
6) Explain to the miners that they will also have to purchase their tools, like real miners do. Their tool choice for this activity is a flat toothpick for 2 cents, a round toothpick for 4 cents or a paper clip for 6 cents. They may choose to buy as many or as few tools as they wish. If a tool breaks during mining, they will have to buy a new one. Should they buy several cheap tools? Buy just one expensive one? (TIP: Make sure each miner has at least two tools. If they have a free hand that doesn't have a tool in it, they will end up touching the cookie, guaranteed.)
7) Give the miners 5 minutes to mine. They will pay 1 cent per minute for labor costs. If some miners really need extra time, you may give them an extra minute, but they must pay for that extra minute.
IMPORTANT: The miners are not allowed to touch the cookies with their hands! The only thing that may touch the cookie are the tools they purchased.
8) While they are mining, supervise their work. If a tool breaks they must replace it. The adults supervising need to look out for miners who are cheating by using their hands. (You can institute a 1 cent penalty for cheating, if necessary.) Chips must be relatively free of cookie dough to be considered as mined.
9) After mining stops, the miners should put all chips, or pieces of chips, into a pile to be counted. For broken chips, combine them to make same volume as whole chips. Just estimate.
10) Finally, miners will need to do reclamation, which is repair of the land. Try to fit all crumbs into the original circle. Crumbs may not be stacked. Miners must use the tools to do this. Allow 2 minutes for reclamation but no charge for time.
NOTE: Other versions of cookie reclamation will tell you to use graph paper under the cookie and have the miners count how much cookie goes outside the outline by counting squares, including calculating partial squares that add up to whole squares. This is a bit tedious for younger students. Also, I found that almost all of my students were able to get the crumbs into the outline, so it would have added an unnecessary complication at the end.
11) Use the printed sheets to calculate investments and profits. Younger students might need help with the final calculation.

COOKIE MINING   **Start-up funds: 19 cents**

Land acquistion (purchase of property): _____

Equipment costs:
# of flat toothpicks _____ x 2 cents each = _____
# of round toothpicks _____ x 4 cents each = _____
# of paper clips _____ x 6 cents each = _____

Labor costs:
Number of minutes _____ x 1 cent per minute = _____

Reclamation costs: _____
2 cents if cookie remains are inside original outline
4 cents if cookie remains go outside original outline

TOTAL MINING COSTS: _____

Gross profit:
# of chips _____ x 2 cents per chip = _____

CALCULATE NET PROFIT:

Start up   19
Gross profit   +
                              _____

Minus mining costs   -
                              _____
        NET PROFIT: [   ]

Was this a successful mining operation?

COOKIE MINING   **Start-up funds: 19 cents**

Land acquistion (purchase of property): _____

Equipment costs:
# of flat toothpicks _____ x 2 cents each = _____
# of round toothpicks _____ x 4 cents each = _____
# of paper clips _____ x 6 cents each = _____

Labor costs:
Number of minutes _____ x 1 cent per minute = _____

Reclamation costs: _____
2 cents if cookie remains are inside original outline
4 cents if cookie remains go outside original outline

TOTAL MINING COSTS: _____

Gross profit:
# of chips _____ x 2 cents per chip = _____

CALCULATE NET PROFIT:

Start up   19
Gross profit   +
                              _____

Minus mining costs   -
                              _____
        NET PROFIT: [   ]

Was this a successful mining operation?

4) GAME: "FOLLOW THE ROCKY ROAD"

This game can be played with 2-6 players, but 3 or 4 players is ideal.

NOTE: If you happen to have a collection of rocks and minerals, you might want to make a set of cards that have actual specimens glued to them. This makes the game even more interesting!

<u>You will need</u>:
• copies of the pattern pages printed onto card stock
NOTE: If you would like to print directly from a digital file and you only have the paperback, go to www.ellenjmchenry.com/rocks-and-dirt, and you'll find a file you can download. If you have no way to make color copies, you can cut these pictures out of the book.
• scissors
• pencil, or a pen that does not bleed through paper

<u>Set up</u>:
1) Copy the pattern pages onto card stock.
2) Write the names of the rocks on the backs of the cards if you can't do double-sided copies. Handwritten labels are fine.
3) Cut the cards apart.

<u>How to play</u>:
1) Decide how many cards you will use in your first round. If you think the players know most of the rocks, go ahead and use all the cards. If using all the cards will seem overwhelming, set some of the cards aside. (Notice that mineral specimens have been included, as well, on two of the pages.)
2) Divide the cards evenly among all players. If the number of cards won't divide evenly, you can set a few cards aside and just use them at the end of the game if you need more cards.
3) Players should set all their cards face up so that the name on the back is hidden. It does not matter if other players see the cards.
4) The first player chooses one of his/her cards, and reads the name on the back of the card so that all players can hear. It is important that everyone hears the name of the rock. Then the player sets the rock in the middle of the table.
5) The second player must remember and repeat the name of the first rock. If he/she gets it right, then he/she may choose a card, say the name of the rock, and then set that card next to the first card.
6) The third player must now repeat the names of the first two cards. If correct, he/she then lays down a third card (saying its name) right next to the second card, to being forming a long line.
7) The play keeps going like this, with each player saying all the names of the rocks that have laid down already, then adding one of their own.

NOTE: This game can be played competitively or cooperatively. I have had both kinds of groups. If a group wants to help each other remember the names, that is okay. The personality of the group will determine how competitive the game will be.

TIP: After playing a couple of times, you could offer a small reward for anyone who can say the entire line of rocks.

<u>Igneous page</u>:
andesite, tuff, rhyolite
breccia, basalt, diorite
pegmatite, pumice, obsidian
granite, gabbro, scoria

<u>Sedimentary page</u>:
limestone, coquina limestone, siltstone
travertine, chalk, oolitic limestone
conglomerate, shale, sandstone
coal, flint, oolitic chert

<u>Metamorphic and minerals</u>:
mica schist, gneiss, marble
quartzite, tuff, aragonite
barite, cinnabar, hematite
muscovite mica, biotite mica, asbestos

<u>More minerals</u>:
calcite, fluorite, hornblende
malachite, olivine, feldspar
gypsum, ulexite, galena
talc, agate, anthracite

IGNEOUS ROCKS for "Follow the Rocky Road"   COPY ONTO CARD STOCK

rhyolite
(Lighter in color than tuff)

tuff
(Maybe be hard to tell apart from rhyolite. You can choose one or the other if you prefer.)

andesite
(Kind of a blend between granite and basalt.)

diorte

basalt

breccia

obsidian

pumice

pegmatite

scoria

gabbro
(Notice all the green. Gabbro is mafic and therefore darker than granite, which is felsic.)

granite
(Notice the pink feldspar.)

SEDIMENTARY ROCKS for "Follow the Rocky Road"   COPY ONTO CARD STOCK

siltstone

coquina
limestone

limestone

oölitic limestone

chalk

travertine
(This is a photo of Mammoth Hot
Springs in Yellowstone Park.)

sandstone

shale

conglomerate

oölitic chert

flint

coal

METAMORPHIC ROCKS and MINERALS for "Follow the Rocky Road"   COPY ONTO CARD STOCK marble

(Not all marble has
those colorful impurities.
Marble is often just white.)

gneiss

mica schist

aragonite
(a form of calcite, $CaCO_3$)

tuff
(made from volcanic ash)

quartzite

hematite

(There are other forms,
including one that is black.)

cinnabar
(HgS)
(the ore for mercury, Hg)

barite
the "desert rose"

asbestos

biotite mica

muscovite mica

MORE MINERALS for "Follow the Rocky Road"   COPY ONTO CARD STOCK

hornblende

fluorite

calcite

feldspar
(also comes in blue and white)

olivine

malachite

galena
(PbS)
(an ore of lead)

ulexite

gypsum
(non-crystalline form)

anthracite

agate

talc

5) EDIBLE ROCKS

This activity can be done with just one or two students, or with a large group. If you are doing it with a group, you can make adjustments to suit your situation. For example, if you do not have access to an oven, you can do more "no-bakes" and perhaps bring pre-baked cookies to hand out or to decorate. You could also do it as a cooperative "cookie exchange" and assign one type of cookie to each student (or family) and have them make the cookies at home ahead of time. On the day of the exchange, provide a box or bag for each student to collect one or two of each type of cookie brought in.

If you want to use this as a craft activity with a group, you might want to think about setting up "stations" and having the students rotate about the stations, making one type of cookie at each station. You will have the necessary ingredient for each cookie at each station, along with a picture of the rock they are simulating. You can have more than one station per table, and the pictures of the rocks will serve as markers for where each station is located. If you want to bake on site, you can give each student a piece of aluminum foil on a large paper plate or a piece of cardboard (even a flatten cereal box will do) and use a permanent marker to write each student's name on their foil. Each student's cookies will be placed on that foil until it is time to bake the cookies. For the baking process, just slide the entire foil sheet off the cardboard and onto a baking sheet.

You can choose to make just a few of these cookies, or you can do them all. Feel free to add your own creativity and make up some concoctions of your own, as well!

Students visiting various stations. Each station had the appropriate ingredients for a particular rock.

NOTE: You can adapt these recipes to be gluten-free, or low sugar. I attempted to keep sugar and candy to a minimum, but you can make further improvements if you want to. I tried to avoid candy as much as possible, saving its use for places where it is really needed, such as the Swedish fish in the "fossils."

You will need:
• cookie ingredients, as noted in set of instructions
• boxes or bags if doing a cookie exchange
• a printed picture of each rock if you want to set up a "station" for each cookie

GRANITE

This cookie is more of a "fruit leather." Use various types of chopped dried fruit to represent the small mineral crystals in granite. Quartz (white) can be represented by dried apples or finely chopped coconut (or marshmallow, if you don't mind the extra sugar). Biotite (black) might be raisins or chopped dates. Feldspar (pink or orange) could be apricot, peach, or cranberries, or any other colorful fruit.

Provide students with a disposable muffin paper (the kind you put into the holes of muffin trays) and let them put their own prefered assortment of dried fruits into it. Also provide a cup of thick liquid "glue" to drizzle over the fruit bits, to help bond together while in the oven. For the "glue" try one of these options: fresh egg white, egg white substitute, egg white powder, or tapioca starch. For all of these options, mix with a little water until the consistency is good for drizzling. After the glue is drizzled over the fruits, press them down, smashing them together a bit to encourage them to stick together. The muffin paper will sit on the tray that will go into the oven.

This is the unbaked version.

If you don't have an oven available, don't use any drizzle glue. Just put the fruit bits into a piece of plastic wrap or a plastic bag, then press hard on the fruit bits, kneading them together as best you can. They'll stick together long enough to look like granite. Pieces of white marshmallow might also be used, as they are very sticky.

VARIATIONS YOU CAN MAKE:

WHITE GRANITE: Use lots of white and a little dark.
PINK GRANITE: Use lots of color, very little white.
PEGMATITE: Add one or more pieces pretzel stick.
DIORITE: Use only dark-colored pieces and white pieces.

BASALT

This is basically a chocolate cookie dough. You can use a standard "sugar cookie" dough recipe and then add some cocoa powder. Since basalt is a fine-grained rock, you don't need to add any grit to it. You can let the students put a chunk or raw dough on the foil that will go into the oven. OR, if you'd rather skip baking these, you can just bring in a large, very flat and thin pan of brownies designed to look like a section of ocean floor. I let my students cut a piece of ocean floor, then add a gummy shark stuck onto a toothpick, so the shark looks like it is swimming above the ocean floor. This reinforces the concept that the ocean floor is made of basalt. (You could also use Swedish fish from the limestone recipe.)

OBSIDIAN

The best option I could come up with for obsidian was to smash some shiny root beer-flavored candies. I put the candies into a plastic bag, then tapped them with a hammer. The result was clear brown shards that look just like obsidian. The resemblance to real obsidian was quite amazing, actually. Real obsidian does come in red and brown, not just black, so even the color was reasonably accurate. If you feel adventurous, you can let your students smash their own obsidian, or you can bring a bowl of prepared obsidian and just let them collect a few pieces.

PUMICE

Pumice can be simulated with a hard meringue cookie. Crunchy meringues start out as a frothy mixture with lots of air mixed into it, which is exactly how pumice starts out, too. Both harden as they cool. The key to the crunchy texture of these cookies is the high sugar content, just something to be aware of. If you want to surf the Internet for a recipe, search for "French meringue cookies."

A good low-sugar alternative might be a crispy "crackle bread" cracker. (Wasa is a popular brand.) These often have a very scratchy texture, too, quite a bit like pumice; you have to be careful when eating them. Real pumice is sometimes used to make abrasive pads, so these crackers are a good simulation in that regard!

MUSCOVITE (clear mica)

This one can't be done gluten free because you need phyllo dough, which relies on gluten for its delicate texture. Purchase a packet of phyllo dough (usually in the frozen dessert section) and allow to thaw. Layer the paper-thin sheets of dough in a baking dish, brushing a mixtureof melted butter and honey between the layers. Bake according to package directions.

SHALE/SLATE

Slate is shale that got a bit harder because of metamorphic processes, and sometimes it can be hard to tell the difference. This recipe idea can be used to represent either one.

Choose a cookie dough recipe that can be rolled out with a rolling pin. I used a gingerbread cookie recipe, just to add a little variety to my dough flavors. Roll the dough very thin, or have the students press out a ball of dough until it is very thin.

THE FOLLOWING COOKIES USE A STANDARD "ROLLED" COOKIE DOUGH. You can make a huge batch of dough and then divide it to make these cookie options. Use your favorite "rolled sugar cookie" recipe, but reduce, or even omit, the leavening agent (baking soda or baking powder). You don't really want these cookies to puff up and expand a lot. You want them to remain fairly flat or square. (These are edible rocks, not gourmet cookies!)

If you don't have a cookbook or Internet access, you can use this basic recipe:

1 cup butter or shortening, 1 cup sugar, 5 cups flour, 4 eggs, 2 teaspoons vanilla

You might need to adjust the texture by adding flour or water. The dough should be like play dough.

MARBLE

Provide students with plain dough and a little dough to which red food coloring has been added. (You can use regular food coloring or all-natural, as you prefer.) Student will take a small ball of plain dough and a small ball of red dough and roll them into long cylinders. Twist the cylinders together, then form into a ball. Press or twist the ball a bit until the red color looks like it is swirling through the dough. Flatten the ball and put onto baking sheet.

LIMESTONE

This will be fossiliferous limestone, because that's a lot more fun that plain old limestone. Press out a thin piece of dough, then press a Swedish fish into it. (I tried other gummy animals, but they all melted into goo. Only the Swedish fish survived the oven.) You might also try pressing an animal cracker into the dough. I recommend leaving the fish or animal in place while baking. If you pull it out ahead of time, leaving just the impression in the dough, the baking process will puff up the dough enough that the imprint will likely become almost unrecognizable.

OÖLITIC CHERT

Provide a bowl of poppy seeds that can be mixed into the dough by hand. Then flatten the dough or shape it into a rectangle. Also, if you want to vary the flavor, add a few drops of almond flavoring.

SANDSTONE

Provide a bowl of (raw) hot cereal, such as white grits or Cream of Rice. These are found near the oatmeal in most grocery stores. They are designed to have boiling water poured on them to soften them, but we will use them "as is" with their gritty texture. Let the students mix a large amount of grit into their dough, until it starts reminding them of sandstone. Then form into rock shapes and put on baking tray. (Remember, these are edible rocks, not gourmet cookies!)

BRECCIA

Provide bowls of sharp, angular pieces of cereal or cracker. By definition, breccia contains angular shards of rock. Broken bits of breakfast cereal or hard crackers work well. Use whatever you can find or that you have on hand. The shards in this picture (which look a lot like real breccia!) came from a blue corn chip.

CONGLOMERATE

This looks similar to breccia, but, by definition, the shards must be rounded, not angular. Provide bowls of smooth objects such as peanuts, round cereal pieces (such as Kix), or small round candies (such as mini M&Ms).

PEBBLES and COBBLES

Use the dough to make balls of various sizes. Pebbles are small, cobbles are medium sized. Make either or both. Set finished balls onto baking tray. (Chocolate powder may be added to the dough if you want brown pebbles.)

MAKING THE LABELS

Pictures are provided on the following pages if you want to make labels for your cookie stations. Cut out the pictures and stick them to cards that have been folded in half to make them stand up. You can also make the cards larger, if you want to add some written instructions or other notes.

PINK GRANITE (lots of feldspar)

GABBRO

PEGMATITE

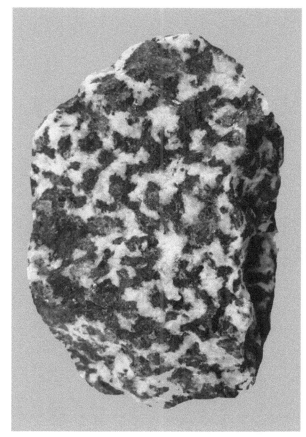

DIORITE

221

BASALT

OBSIDIAN

PUMICE

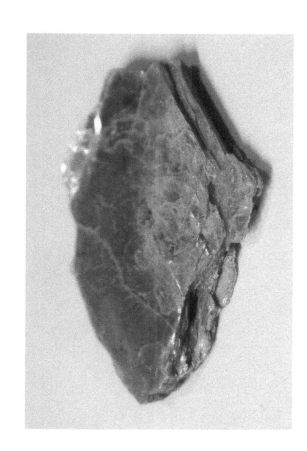

MUSCOVITE (mica)

223

SHALE or SLATE

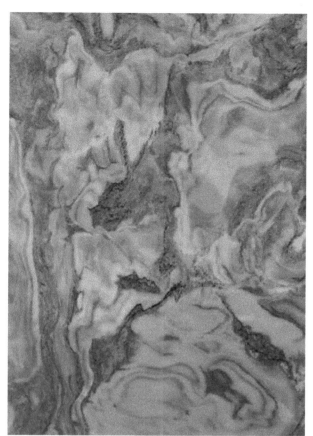

MARBLE

OÖLITIC CHERT

FOSSILIFEROUS LIMESTONE

SANDSTONE

BRECCIA

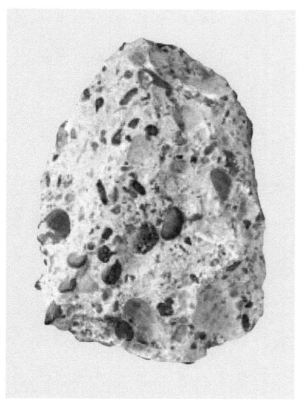

CONGLOMERATE

PEBBLES and COBBLES

227

CHAPTER 5

1) LAB: THE "FIZZ TEST"

NOTE: If you don't have limestone but would still like to do this lab, there is a video version on the youtube playlist.

You will need:
• one or more pieces of limestone
• some rocks that are not limestone
• a piece of chalk
• a piece of coral or sea shell if you have them
• vinegar or lemon juice, or both
• water
• small plate or cup

What to do:
1) Put a few drops of water on your rock specimens. Observe what happens. Look for any bubbles that might form, or a fizzing sound. (Put your ear up close.)
2) Now try the same thing with the vinegar or lemon juice. Look for tiny bubbles or the sound of fizzing. (Students often find that listening for fizzing is easier than watching for bubbles.)
3) Try both the water and the vinegar on the piece of chalk, and on any other items you've collected.

What should happen:
When you apply acid (vinegar or lemon juice), you should see and hear bubbles on the limestone and the chalk, but not the other rocks. The acid is dissolving the $CaCO_3$ and releasing molecules of CO_2. Coral and sea shells also contain $CaCO_3$ so technically, these should work also. If your shell is very shiny, though, it may not work, as the coating may prevent the acid from touching the $CaCO_3$.

2) LAB: PRECIPITATING $CaCO_3$

You will need:
• calcium hydroxide, sold as "pickling lime" or agricultural lime (TIP: pickling lime is easy to purchase on Amazon.)
• a large clear glass and a spoon
• a straw
• water
• OPTIONAL: An alternative source of CO_2, such as dry ice or yeast

What to do:
1) Put a few spoons of lime into a glass of warm water. Stir well. You are making "lime water."
2) Allow the water to sit several hours or overnight until the water looks clear. The excess lime will be at the bottom of the glass. The water looks clear but it contains dissolved calcium and carbonate ions, the ingredients for limestone.
3) Blow into the water using the straw. There is enough carbon dioxide in your breath to precipitate some $CaCO_3$.

What should happen:
You should see some tiny white dots forming in the water and slowing sinking to the bottom. It's possible that the dots will be so small that it will simply look like the water is getting cloudy.

NOTE: If you would like a more potent source of carbon dioxide, you can put a piece of dry ice into a bag and insert the straw. Put a rubber band around the straw to make the seal airtight. Another alternative would be to put some water and yeast and a little sugar into a bag and let it sit for a few minutes. The yeast will start to metabolize the sugar and will create carbon dioxide as a waste. Again, make sure your straw has an airtight seal so that all the gas will go out the straw and into your glass of water. In both cases you should see bubbles coming out of the straw.

3) LAB: WEIGHING CO$_2$

NOTE: If you don't have a digital scale but would still like to do this lab, there is a video version on the youtube playlist.

As you learned in the text, when limestone is heated, carbon dioxide is driven off, so that only CaO remains. Even though carbon dioxide is a gas, that doesn't mean it doesn't weigh anything. You should be able to drive out enough CO$_2$ to see a difference in mass, before and after heating.

You will need:
• a piece of limestone
• a digital scale
• a propane torch

What to do:
1) Weigh a piece of limestone. Write down the exact number; don't round off the numbers. Hopefully, your scale will give you two decimal places. For example: 21.83 grams
2) Heat the limestone with the torch for 5 to 10 minutes. Get it right in the flames.
3) Let the limestone cool for a few minutes. Do NOT put it into water to cool it faster. Keep it dry.
4) Weigh the rock again. Has it lost weight? Why?

4) LAB: FUN WITH CO$_2$

You will need:
• a tall, clear glass
• a can of seltzer water (soda water) or clear soda beverage such as ginger ale or 7-Up
• a few raisins

What to do:
1) Open the fizzy drink and pour into the clear glass.
2) Add the raisins.

What should happen:
You should see the raisins sink, then begin to collect bubbles at the bottom. Enough bubbles will form that they will lift the raisins to the top again. At the top, the bubbles will burst. The raisins will be heavy again and sink to the bottom. This process will repeat itself over and over again.

Notice how the CO$_2$ bubbles seem to appear out of nowhere. Where are they all coming from? The carbon dioxide gas is actually dissolved in the water. As long as it stays in solution, you can't see it. When it comes out of solution and turns into a gas again, a bunch of molecules will come together to form a bubble. The bubble is less dense that the solution around it, so it rises.

5) ACTIVITY: MAKE SOME STALACTITES

You will need:
• two jars or glasses
• a piece of soft string or yarn
• epsom salt
• water
• a piece of cardboard

What to do:
1) Fill the jars or glasses 2/3 full of warm water.
2) Stir several spoons of epsom salt into each glass. (Stir well.)
3) Place the glasses about 15 cm (6 inches) apart.
4) Put one end of string into each glass. The middle of the string should be stretching from jar to jar, but loosely so that the middle hangs down just a bit. Put the piece of cardboard under the middle of the string.
5) Let this sit for several days, perhaps even a week and watch what happens.

What should happen:
The epsom water solution should travel up the string and keep the entire string wet. In the middle, some water should be dripping occasionally. As the water drips down, a mineral formation should start to form on the string (a stalactite). You might even see a stalagmite start to form on the cardboard. (FYI NOTE: This lab is not foolproof. It doesn't always work well.)

6) OPTIONAL LAB: EFFECT OF ACID RAIN ON LIMESTONE BUILDINGS

If you happen to be looking for a more involved project to work on, here is a "science fair" project about limestone. It takes more time and materials than the other experiments in this chapter.

http://www.all-science-fair-projects.com/print_project_1356_118

7) LEARN MORE ABOUT CAVES

Considering purchasing or borrowing some books about caves or caving. For ages 8 to 13, you might want to start out with "One Small Square: Cave." by Donald SIlver and Patricia Wynne. A used copy can be purchased for only pennies plus shipping on Amazon. It's a picture book, but with high information content, and even adults will learn things.

8) CRAFT: MAKE A COCCOLITHOPHORE ORNAMENT

NOTE: Images of coccolithophores are mostly copyrighted and therefore can't be printed in this book. Use Google image search to find images of various coccolitho-phores. Not all of them look like they are made of plates. One species looks like it has trumpets all over! The circles can be fairly flat, or they can be cup-shaped. You might be able to think of your own method for making one.

You will need:
• a copy of one of the pattern page, printed onto white card stock
• a Styrofoam® ball, exact size not important, around 2-3 inches (5-7 cm)
• a box of straight pins
• a bottle of white craft glue (Aleens's "Tacky Glue" works well)
• a piece of string or yarn

What to do:
1) Look at the pattern pages and choose the appropriate size circle for your ball.
2) Copy the pattern page onto white card stock. (Digital pattern pages that you can print out can be found at www.ellenjmchenry.com/rocks-and-dirt.) (If you really can't make copies and you have the paperback version of this book, you can also cut the pattern pages out of the book and use them.)
2) Cut out the circles. If you run out of circles, make another copy of the pattern page.
3) Put a dot of craft glue in the center of the back of a paper coccolith. Stick it to the ball and put a pin through the center of the circle (right where the glue is) and into the ball.
4) Pin circles all over the foam ball. There will be a lot of overlapping.
5) If you want to add a hanger, make the string or yarn into a loop and secure it to the ball.

EXTRA INFORMATION:

Coccolithophores are living cells! Underneath those hard limestone plates there is a squishy living cell. This cell is classified as a protozoan, but it can do photosynthesis like a plant cell. The cell has the basic cell parts: nucleus, mitochondria, chloroplasts, Golgi bodies, endoplasmic reticulum, etc. (The chloroplasts are brown, not green like those found in plants.)

The formation of the plates starts with the process of transcription and translation of DNA by mRNA and ribosomes. A protein pattern is formed, to which the calcium carbonate crystals will stick. The crystallization mainly occurs inside the cell, and the plate is then transported inside a vesicle (made by the Golgi body) to the outer wall of the cell. Therefore, the newest coccolith plates are the ones that appear to be on the bottom (underneath the others) when we look at a photograph of a coc-colithophore. Some coccolithophores only make new plates when the reproduce, but others shed plates and grow new ones all the time.

The name, coccolithophore, comes from the Greek words "coccus," meaning "grain or seed," "lithos," meaning "stone," and "phoros," meaning "to carry or bear."

Coccolithophores are found in every ocean of the world. When the population grows very quickly, it is called a "bloom." There was a huge bloom in the Bering Sea in 2011.

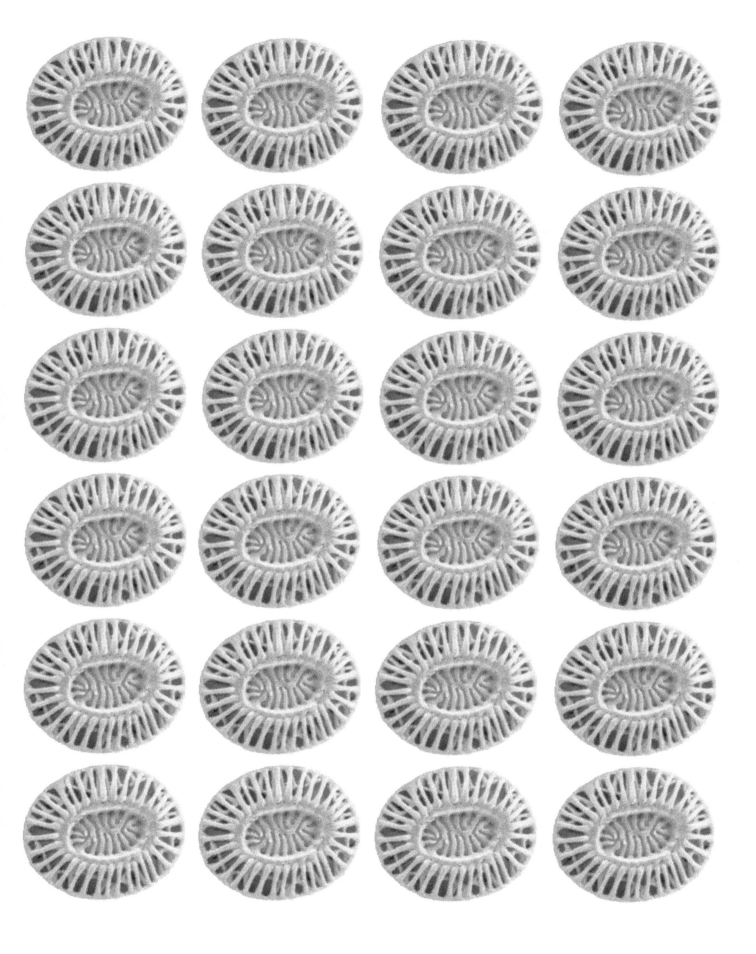

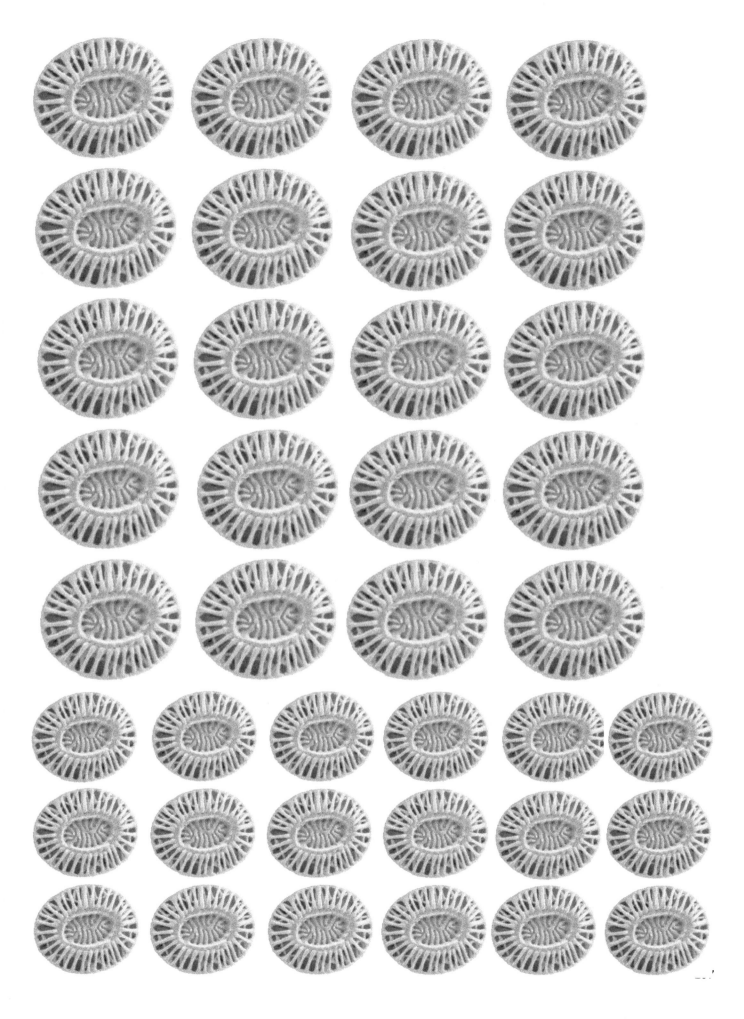

9) "THE LIMESTONE GAME"

This is a board game for 2 to 4 players. If you have more than 4 players, consider making multiple copies.

You will need:
- copies of the pattern pages copied onto card stock (or regular paper if you can't get card stock)
  (If you purchased a hard copy of this book and would like to be able to print pages directly from your computer's printer, or would like to get them printed at a print shop, use this web address to download digital copies of the pattern pages: www.ellen-jmchenry.com/rocks-and-dirt)
  NOTE: If you have a hard copy, are unable to do color printing and need only one copy of the game, you can cut the following color pages out of the book and use them.
- pair of dice
- scissors and glue stick (or white glue or clear tape)
- small piece of limestone for each player (or another rock if you can't get limestone)
- small piece of chalk for each player

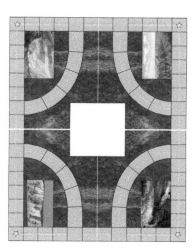

Set up:
1) Cut and paste the four board pieces and assemble them to make the playing board, as shown. (TIP: If you leave a strip of white on some of the inner edges, you can use them as glue tabs, thus avoiding use of tape.)
2) Cut out the cards, shuffle them well, and put them into a draw pile.
3) Cut out the bumper stickers for the cave and canyon and put them on top of the photo.
4) Put 1 piece of limestone on the quarry picture. The other pieces will be set on one at a time as the BLAST cards are drawn.
5) Put 4 small pieces of chalk on the chalk cliff picture.
6) Cut out a car for each player and assemble as shown on that page.

How to play:
1) Each player chooses a car and sets it on the middle square of one of the curved road pieces. It doesn't really matter where you start, as you will eventually need to cover all roads on the board. Players should keep their cars on one side of the road so that they can pass each other. Cars may move in any direction, but only one direction per turn. (No U-turns.)
2) The object of the game is to collect four items before the limestone cycle equation is completed. You will need to collect one piece of limestone from the quarry, one piece of chalk from the cliff, and a bumper sticker from the cave and the canyon. As soon as the last piece of the equation is set into place, the game is over. Whoever collects their items wins. (So this game is not player versus player, but players versus the equation!)
3) A turn always consists of first drawing a card, then rolling the dice (unless the card tells you not to). If the card has instructions on how to move, just do what the card says.

   If the card says BLAST! this means that blasting has occurred at the quarry and there is now a fresh supply of limestone. If you draw a BLAST card, set one piece of limestone on the quarry, then roll the dice and move.

   If the card says SINKHOLE, you take a small sinkhole token (not the actual sinkhole card, which is too large) and set it right on the square where your car is now, but after you have rolled the dice and moved on. If you draw a SINKHOLE card, you do NOT get stuck in the hole. This sinkhole remains on the road for the rest of the game, effectively creating a detour. Players can't travel on that section of road. The only way to get rid of a sinkhole is for someone to draw the sinkhole repair card.

   If the card is one of the equation cards, the player sets that card in one of the squares at the center of the board. Once these cards start to accumulate, match them up so they make the limestone cycle. When the last card is put into place, the game is then over.

   If the cards says that anyone can move during that turn, the player then reads the question out loud and the other players put up a number of fingers that corresponds to which of the answers they think is correct. Putting up fingers makes the players commit to an answer and stick with it. Anyone who gets the question right will then be able to move the same number of spaces as the player whose turn it is. The original player then rolls both dice.
4) The players take turns even though on many turns more than one player will move. Cars can move in any direction, but only one direction per turn.
5) To collect an item, simply pass or land on the star near that location. Note that the quarry's supply of limestone will be generated as the game goes on. The quarry starts with only one piece available at the beginning of the game.

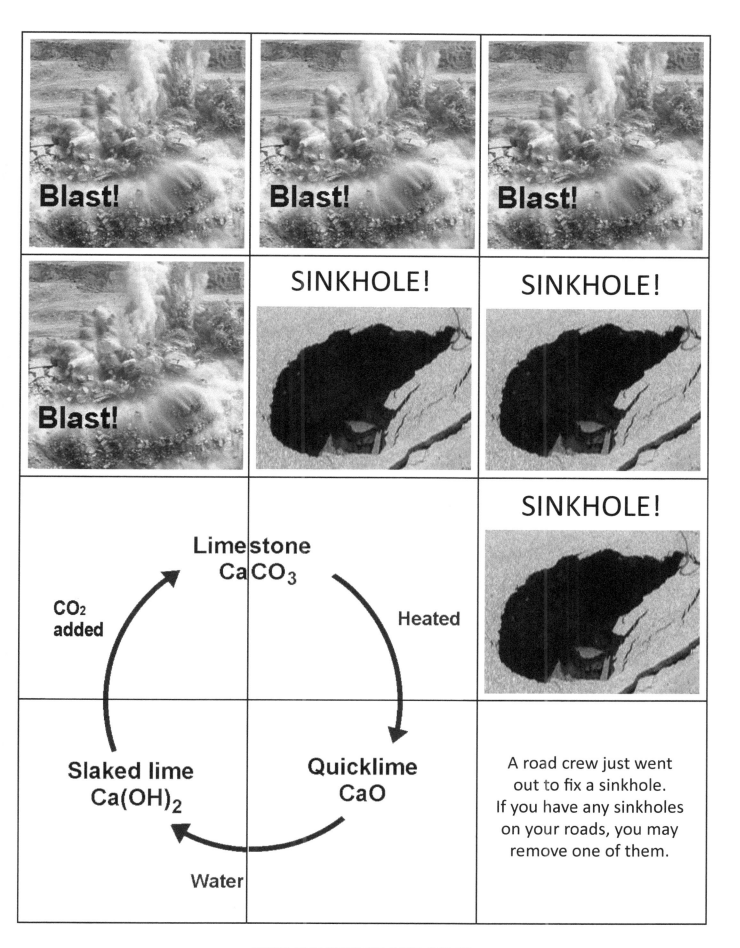

Blast!

Blast!

Blast!

Blast!

SINKHOLE!

SINKHOLE!

SINKHOLE!

Limestone
CaCO₃

CO₂
added

Heated

Slaked lime
Ca(OH)₂

Quicklime
CaO

Water

A road crew just went out to fix a sinkhole. If you have any sinkholes on your roads, you may remove one of them.

Anyone who can answer this question correctly can move on this turn.

Which one of these is NOT a type of limestone?
1) shale   2) chalk
3) travertine   4) dolomite

Show your answer by putting up fingers.   (Answer: 1)

Anyone who can answer this question correctly can move on this turn.

Which one of these is classified as an impurity in limestone?
1) calcium   2) iron
3) oxygen   4) carbon

Show your answer by putting up fingers.   (Answer: 2)

Anyone who can answer this question correctly can move on this turn.

Which one of these sea creatures would not contribute to the formation of limestone?
1) clam   2) oyster
3) jellyfish   4) snail

Show your answer by putting up fingers.   (Answer: 3)

Anyone who can answer this question correctly can move on this turn.

What do you need to convert $CaCO_3$ (calcium carbonate) to CaO (calcium oxide)?
1) water   2) chemicals
3) CO_2 4) heat

Show your answer by putting up fingers.   (Answer: 4)

Anyone who can answer this question correctly can move on this turn.

How many bonds does an oxygen atom want to make?
1) 1   2) 2
3) 3   4) 4

Show your answer by putting up fingers.   (Answer: 2)

Anyone who can answer this question correctly can move on this turn.

How many bonds does a calcium atom want to make?
1) 1   2) 2
3) 3   4) 4

Show your answer by putting up fingers.   (Answer: 2)

Anyone who can answer this question correctly can move on this turn.

Where might you find slaked lime?
1) food   2) sewage treatment
3) sports field   4) all of these

Show your answer by putting up fingers.   (Answer: 4)

Anyone who can answer this question correctly can move on this turn.

What is karst?
1) a landscape 2) a sinkhole
3) a mineral   4) a cave

Show your answer by putting up fingers.   (Answer: 1)

Anyone who can answer this question correctly can move on this turn.

When a solid comes out of a solution, we say it:
1) evaporated   2) consolidated
3) precipitated

Show your answer by putting up fingers.   (Answer: 3)

This question is for the person to your left. If correct, he or she may move on this turn.

Which of these elements may have speeded up the formation of many caves?

1) carbon   2) sulfur
3) silicon   4) magnesium

(Answer: 2)

This question is for the person to your right. If correct, he or she may move on this turn.

True or false? You can predict when and where a sinkhole will appear.

False. Even if you know an area might have sinkholes, you can never predict when and where.

This question is for the person to your left. If correct, he or she may move on this turn.

True or false? It is okay to break off a small piece of stalactite in a cave and keep it as a souvenir.

False. If everyone did this, there would be nothing left. Take nothing but pictures and memories.

Anyone who can answer this question can move on this turn. The Grand Canyon has layers of sandstone, limestone and shale. Sandstone and shale have an element that limestone does not. Is it: 1) sulfur 2) carbon 3) magnesium 4) silicon Show your answer by putting up fingers. (Answer: 4)	This question is for the person to your right. If correct, he or she may move on this turn. In dolomite, about half the calcium atoms have been replaced with: 1) silicon 2) carbon 3) magnesum 4) iron (Answer: 3)	Anyone who has visited a cave may move on this turn, IF they can remember the name of the cave. If the cave did not have a name, tell where it was and who you went with.
A road crew just went out to fix a sinkhole. If you have any sinkholes on your roads, you may remove one of them.	Instead of rolling the dice on this turn, switch places with another car.	Instead of rolling the dice on this turn, switch places with another car.
You stopped to take pictures at a scenic overlook. The pictures are wonderful but it slowed you down a bit. Roll only 1 of the dice.	You car looks like it might overheat. Better go slow for a while! Roll only 1 of the dice on this turn.	You had to stop for a family of ducks that was crossing the road. This slowed you down. Roll only 1 of the dice on this turn.
The weather is perfect for traveling and you are making good time. Roll the dice as usual and move your car along.	That skunk on the side of the road sure smelled terrible but it didn't slow you down. Proceed normally, rolling both dice.	It is pouring down rain and you must reduce your speed. Roll only 1 of the dice.

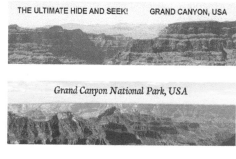

Cut out these miniature bumper stickers and put them on the canyon/cave photos.

These sinkholes are the ones that actually get put onto the road. (The SINKHOLE cards are too large to fit onto the road sections.)

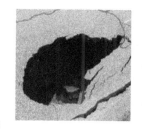
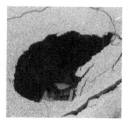
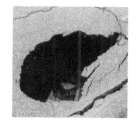
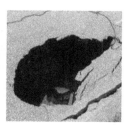

Cut out a car for each player and fold as shown. You can glue the two sides together, or just leave them folded. Either way is fine.

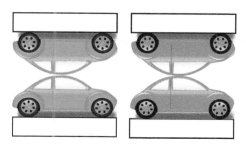

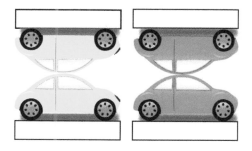

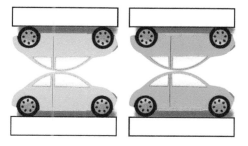

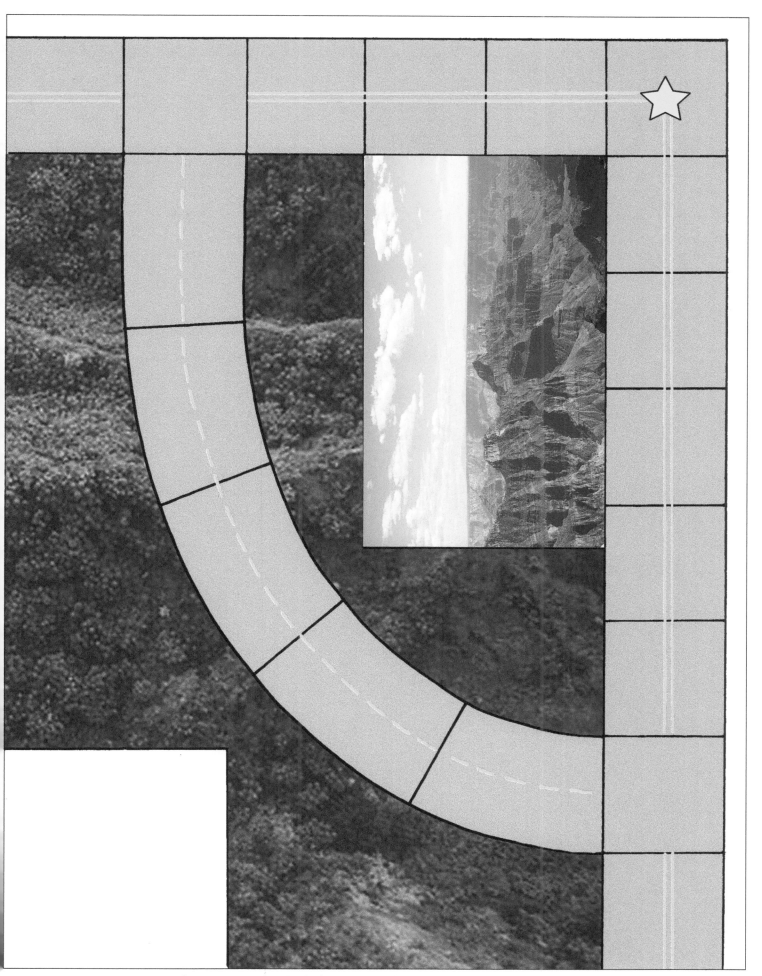

If you want to use glue tabs, don't trim this side off and use it as a tab.

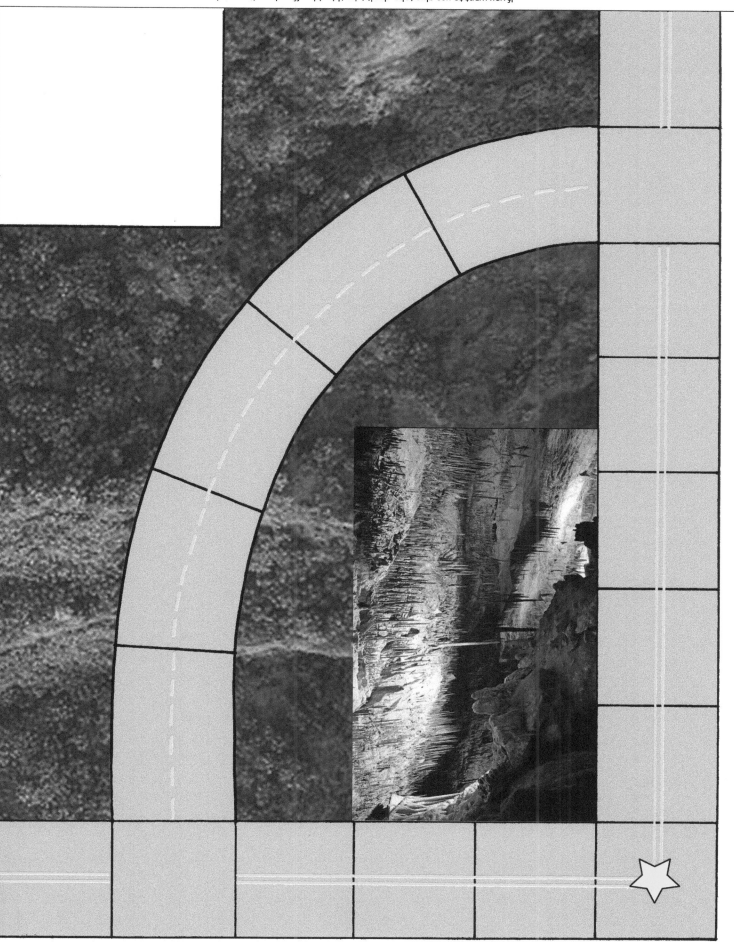

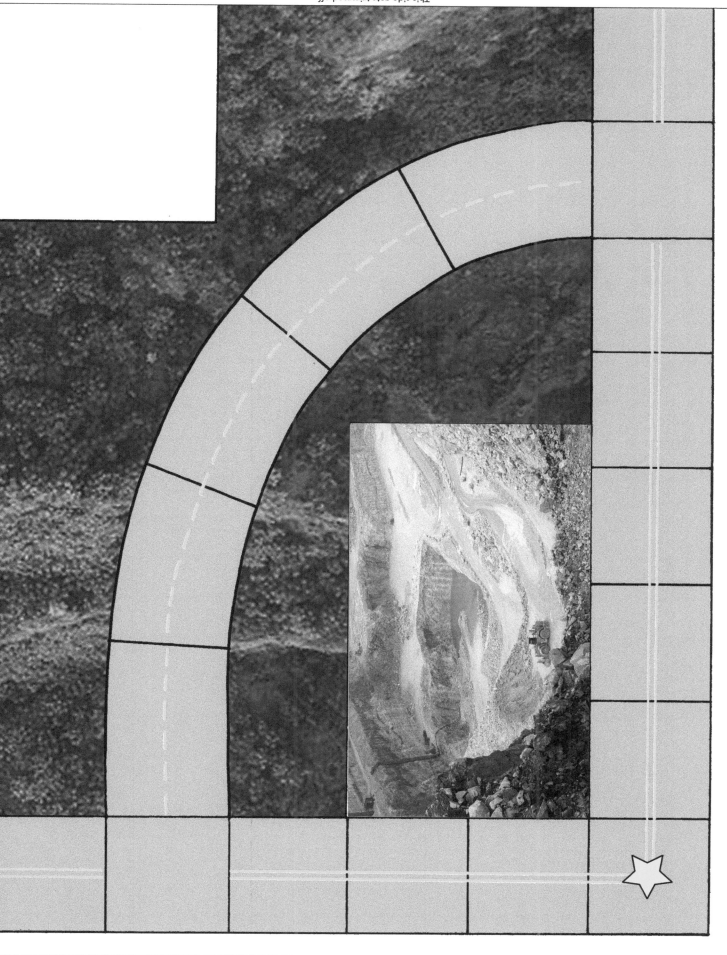

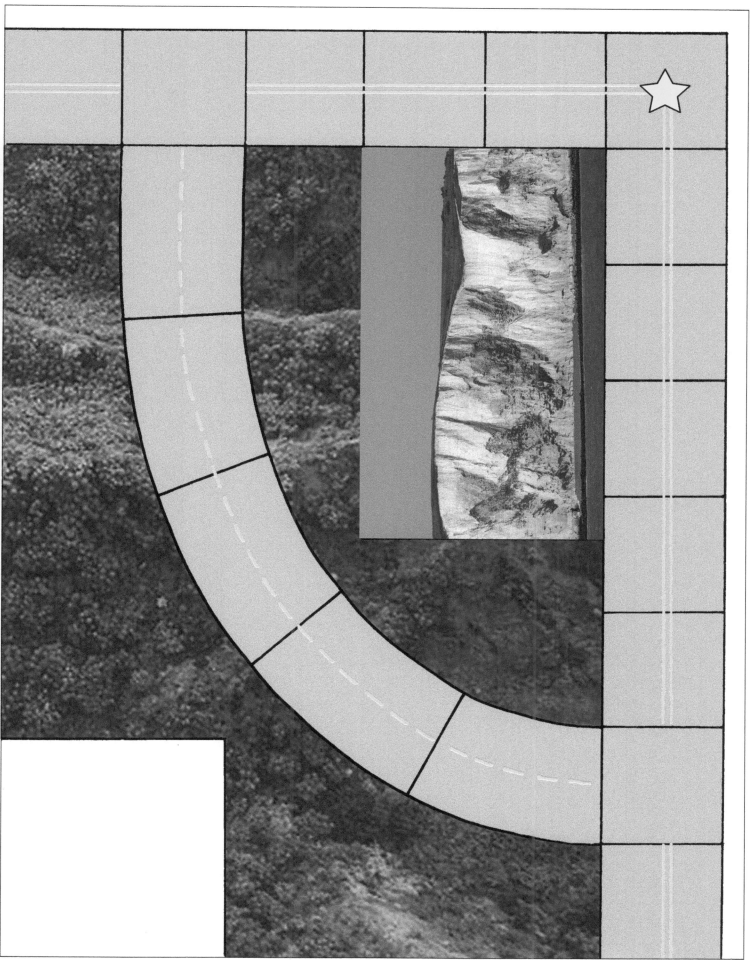

This side gets trimmed off.

CHAPTER 6

1) CRAFT: SOIL PROFILE BOX

There are two options for this project, so there are two pattern pages. Both boxes have re-closable flaps on the top. The blank pattern also has a slot that can be cut to turn it into a penny bank. (Sort of a pun on the word "save.") You don't have to cut and use the slot, however. These boxes can be used for anything. Kids never seem to have enough places to put their miscellaneous stuff. Or they can be filled with dry (non-greasy) snacks.

OPTION 1: Use watercolor and/or colored pencils to make the pre-printed box.
OPTION 2: Use acrylic paints to make your own soil profiles.

For option 1, you will need:
• copies of the pattern page (pre-drawn design), printed onto white card stock
• watercolor paints and/or colored pencils
• scissors and X-Acto knife
• white glue (or high quality glue stick)
• ruler and something to score with (compass point, dead ball point pen, nail, scissor tip, etc.)

For option 2, you will need:
• copies of the pattern page (blank template), printed onto white card stock
• acrylic paints (chocolate brown, reddish brown, tan, gray, white) (optional: acrylic varnish, sold alongside the paints)
• a selection of small brushes
• scissors and X-Acto knife
• white glue
• ruler and something to score with (compass point, dead ball point pen, nail, scissor tip, etc.)

What to do:
OPTION 1:
  1) Color or paint the various profiles. If you are using watercolor paints, remind the students to keep the color VERY light so that they don't obscure the horizon letters. Watercolor and pencils can be combined if you do the watercolor first, let it dry, then add your pencil on top. You could do a background color on horizon C, for example, then go back in and color the stones. Again, remind the students to keep the watercolors very light.
  2) Cut out the box. Cut down into the top flaps so they will bend in. (This is fairly obvious to an adult, but kids can get confused.)
  3) Cut the slit with the X-Acto knife.
  4) Score on the fold lines.
  5) Use white glue (not too much!) on the side glue tab. TIP: You can flatten the box so that you can press and hold the flap and make it stick really well.
  6) Fold and glue the flaps on the bottom. Then fold down the top flaps so that they close.

What to do:
OPTION 2:
  1) You can paint first, or assemble the box first. (Acrylic paints dry very fast, and even faster with a hair dryer.)
  2) The horizons can be any thickness. Google image search can show you some examples to work from. Just make sure the A horizon is brown, as that is almost always the case. The B horizon is often redder, but sometimes is more gray or yellow or purple. You can put in an E horizon, or not. Also, you can do 4 different profiles or you can make the box all one profile. To make the C horizon, I suggest painting the background first, letting it dry, then adding the stones on top. The students might want to do a little bit of blending at the horizon interfaces. This is a great project for experimenting with paint. You really can't mess up because, after all, you are just painting rocks and dirt.
  NOTE: You can also label the horizons if you'd like. Make sure the paint is completely dry, then label using a fine point permanent marker. Also, if you want the box to have a glossy sheen, you can add some acrylic varnish as the final coat. (Varnish is sold along with the acrylic paints, almost like a color option.)
  3) Assemble as described in option 1. Make sure slit is cut before you begin assembly. (Though, in a pinch, you can cut it afterwards if you are careful.) TIP: Scissors do a terrible job. You really want to use an X-Acto or razor.

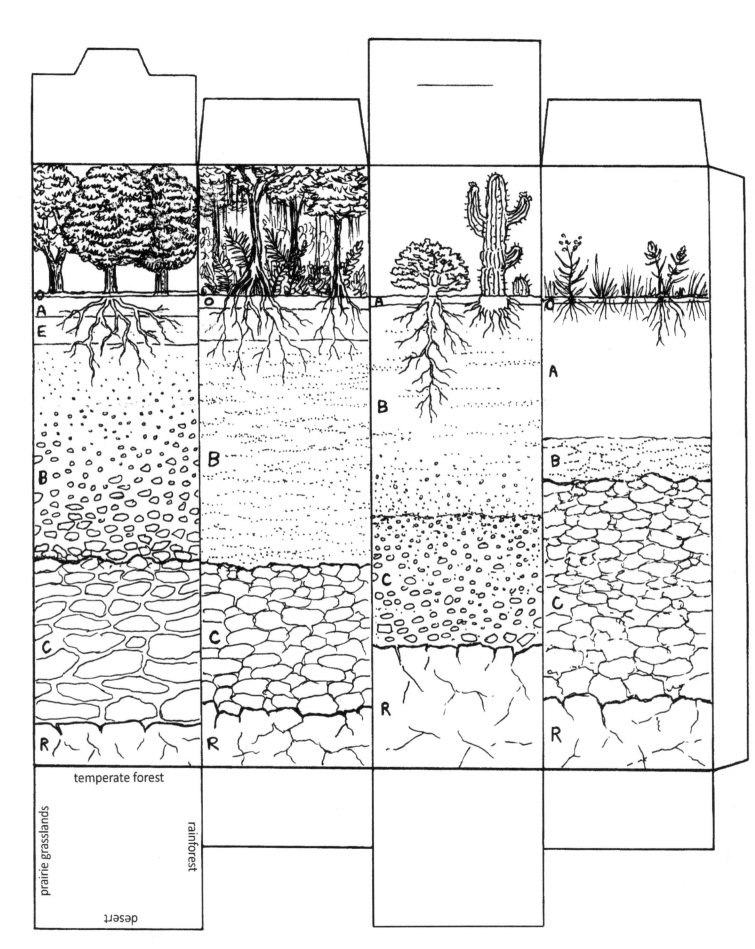

temperate forest

prairie grasslands

rainforest

desert

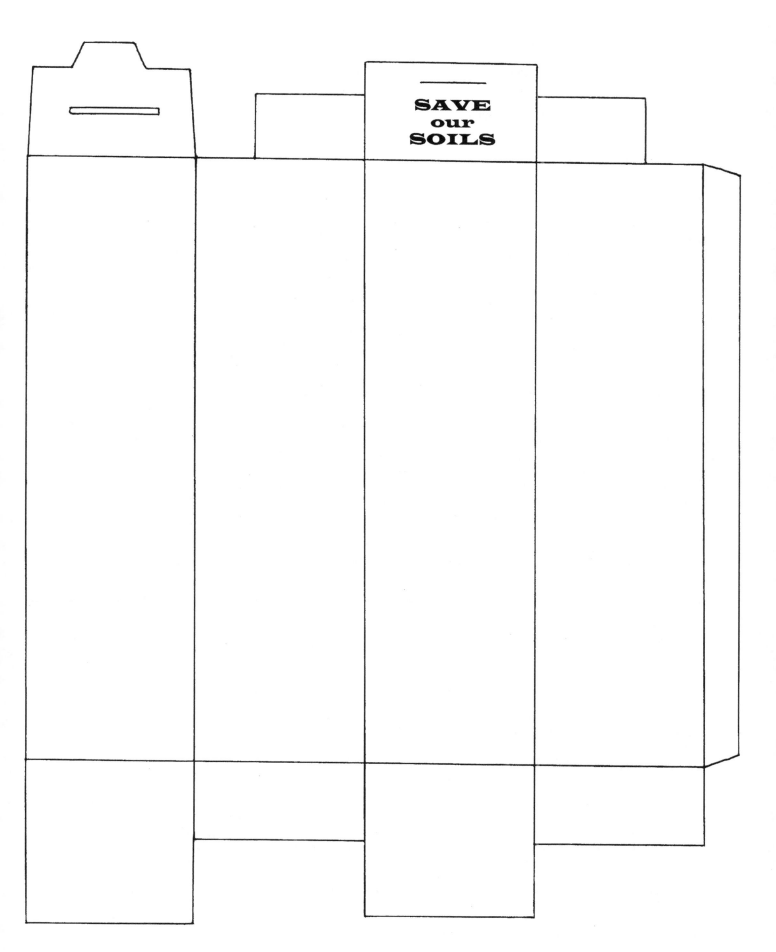

SAVE
our
SOILS

259

2) LAB: SOIL TEXTURE ANALYSIS

In this lab, students will determine three aspects of soil texture. Two of them relate to the soil triangle on page 63.

<u>You will need</u>:
- tall glass jars with lids (one jar per soil sample)
- soil samples (as many as you want to do)
- water
- digital scale (You might need a larger scale, more like a bathroom scale, if your soil samples are large.)
- oven (to dry soil quickly)

The reason we care about soil texture is because it gives us clues about the soil's potential. Is the soil good for farming? Would you be able to dig a pond there? Could it turn into a forest? Would it be a good place for a housing development? If a soil is too sandy, it won't hold enough water and nutrients for growing good crops. The air spaces between the particles are not small enough to retain nutrients and water. Sandy soil would also be a terrible place to try to dig a pond. The pond would never hold water. A clay soil has extremely small spaces between the particles, so it can have just the opposite problem and hold water too tightly. Soils need air as part of their composition, and the tiny spaces in clay are always filled with water, never letting air get in. Clay soils make great ponds, but are terrible for farming. A loamy soil has just the right blend of air spaces and water, so plant roots can access water, nutrients and oxygen. They are great for farming, and therefore this use should be prioritized. Housing developments might not be the best use for soil that is good for growing crops. Rocky or shallow soil would be terrible for farming, but might be excellent as a building base for houses or apartment buildings.

Soil scientists take many soil samples from a given area, then analyze them. They use tools that look like long metal straws, poking them deep into the ground to get a complete section of soil horizon. When they compare the core samples, they find certain patterns that can be positively identified, so they give them names. These patterns are called "series." A series gets its name from the first place it is discovered. For example, the "Hublersburg series" soils were first discovered near this town, but after that they were also discovered in many other places across that state. The Hublersburg series has a written description that describes the thickness of each soil layer, along with its colors and textures. Any soil found to have those same horizons will be labeled as Hublersburg soil, no matter where it is located. State agricultural agencies keep computerized databases of all their soil information, and anyone can go online to access it. A soil scientist who thinks he or she has discovered a new soil series must first search online to see if their soil has already been described.

This information on Hublersburg series soils was copied from an official government database, so you can see how precise soil descriptions are. The B horizon is divided into six sub-layers. Friable means the soil will crumble nicely in your hand.

HUBLERSBURG SERIES SOILS
The Hublersburg series consists of very deep, well drained soils. They formed in residuum weathered from impure limestone. They are on convex upland slopes of 0 to 35 percent. Permeability is moderate. Mean annual precipitation is 41 inches. Mean annual temperature is 52 degrees F.

TAXONOMIC CLASS: Clayey, illitic, mesic Typic Hapludults
TYPICAL PEDON: Hublersburg gravelly silt loam, on 3 to 8 percent northwest facing convex slopes in an apple orchard. (Colors are for moist soil.)

Ap: 0 to 8 inches; dark grayish brown (10YR 4/2) gravelly silt loam; moderate very fine granular structure; friable, slightly sticky, slightly plastic; 15 percent rock fragments; moderately acid; abrupt smooth boundary. (5 to 12 inches thick)
E: 8 to 14 inches, light yellowish brown (10YR 6/4) gravelly silt loam; moderate fine subangular blocky structure; friable, slightly sticky, slightly plastic; 15 percent rock fragments; slightly acid; clear wavy boundary. (0 to 10 inches thick)
BE: 14 to 21 inches, yellowish brown (10YR 5/8) silty clay loam; moderate medium subangular blocky structure; friable, slightly sticky, plastic; 10 per-cent rock fragments; moderately acid; gradual wavy boundary. (0 to 9 inches thick)
Bt1: 21 to 35 inches; yellowish brown (10YR 5/8) channery silty clay; moderate medium subangular blocky structure; firm, sticky, plastic; common faint clay films on faces of peds; 15 percent rock fragments; strongly acid; gradual wavy boundary. (5 to 17 inches thick)
Bt2: 35 to 46 inches; yellowish brown (10YR 5/6) gravelly silty clay; moderate coarse blocky structure; firm, sticky, plastic; common faint clay films and common distinct black coatings on faces of peds; 15 percent rock fragments; strongly acid; diffuse wavy boundary. (5 to 18 inches thick)
Bt3: 46 to 55 inches; yellowish brown (10YR 5/8) silty clay; moderate coarse blocky structure; firm, sticky, plastic; many faint clay films and common distinct black coatings on faces of peds; 5 percent rock fragments; strongly acid; clear wavy boundary. (5 to 20 inches thick)
Bt4: 55 to 65 inches; strong brown (7.5YR 5/6) silty clay; moderate coarse blocky structure; firm, sticky, plastic; many faint clay films and common distinct black coatings on faces of peds; 5 percent rock fragments; very strongly acid; gradual wavy boundary. (0 to 20 inches thick)
Bt5: 65 to 74 inches; strong brown (7.5YR 5/6) clay; moderate coarse blocky structure, firm, sticky, plastic; many prominent clay films and common distinct black coatings on faces of peds; 5 percent rock fragments; very strongly acid.

Another important reason to record all this data is to allow future generations of soil scientists to be able to track changes that might be occurring in the environment. If farmers suddenly begin having trouble growing their crops years from now, soil samples at that time can be compared to those taken years earlier, and perhaps this will provide clues as to what might be happening.

Soil doesn't change quickly, which is why it is so important for a farmer to know what he is getting when he buys a piece of land. You can increase the organic matter through management, but the main soil characteristics will remain the same. If you were looking to buy farmland to grow grain crops and you were not allowed to actually go see the fields, but were allowed to examine a bucket of soil from each field, you would probably be able to choose the best field. The soils' textures would tell you a lot about the potentials of those properties.

<u>What to do</u>:

DETERMINING WATER-HOLDING CAPACITY OF SOIL

1) The trickiest part of this experiment is getting a good starting sample. You need a soil sample that is maximum moist but without any water trickling out. Imagine a garden a few hours after it has rained. All the puddles are gone but the soil still has maximum moisture. If you pick up a handful of soil it will be very wet, but you won't actually have water running out of it. Add water to your soil samples until you think you have achieved this level of water saturation. Remember, no dripping!
2) Weigh each wet soil sample.
3) Put each soil sample onto a separate baking tray and spread them out a bit so they will dry faster.
4) Put the trays into the oven and set the oven on low heat. You want the samples to dry out but not burn. Leave them in the oven for a few hours, or until the soil is completely dry. You can mix and crush the samples a bit while baking.
5) Take the trays out and let them cool. Very carefully gather ALL the pieces and crumbs from each sample, taking care to keep them separate and not to lose any dust, if possible. You want to weigh everything.
6) Weigh each dry soil sample.
7) Put the weight of the dry sample over the weight it was when it was wet. This fraction represents the amount of actual soil, minus the weight of water. Treat the fraction as a division problem and type it into a calculator. Move the decimal point to the right two places and that is your percentage of soil. If you subtract this number from 100, that will give you the percentage of the sample that was water. If you compare a sandy soil to a clay soil, these percentage numbers will be hugely different. Clay soils should show a very high water content percentage.

DETERMINING SOIL TEXTURE PERCENTAGES (i.e. how much sand/silt/clay)

1) Put soil into tall glass jar, filling one half to one third of the jar.
2) Add water until the jar is almost full. Leave a little air space at top.
3) Shake the jar. You can also use a spoon to crush and mix the soil a bit to encourage it to dissolve into the water.
4) When you've got a well-mixed water and soil solution, let the glass jar sit for a while.
5) After just several minutes, you should start to see particles collecting at the bottom. This will be your sand content. Sand settles out of water very quickly. In just five minutes or so, most of the sand content will have collected.
6) After the jar sits for an hour or two, you will see another layer collecting on top of the sand layer. This will be the silt layer.
7) Let the jar sit overnight or until the water on top is clear. The clay particles are extremely tiny and will stay suspended in the water longer than any other particle. (However, you might have some organic matter that floats and won't ever settle. Don't count that into your calculations.)
8) Once the water on top is clear, take a ruler and carefully measure the number of millimeters in each layer.
9) Do a few simple math problems to figure out the percentages of each type of grit. For example, to find the percentage of sand, put the number of millimeters of sand over the total number of millimeters of all three layers to make a fraction. Then treat this fraction like a division problem and type it into a calculator. Then move the decimal point to the right two places and that is your percentage. The three percentages (sand, silt, clay) should total 100 percent.
10) Once you have the percentages, you can find your soil on that soil triangle on page 63. If you want a better image to work from, just type "soil texture triangle" into Google.

NOTE: My attempts at this experiment were not very successful. I found that the divisions between the layers were very hard to see, or the divisions were very blurry making it hard to decide where to put the ruler. It all looks so simple when you see the demo on youtube, but when you try it yourself it is not so easy! A soil scientist told me that she suspects that the soil samples they used on the video were a "pre-mix" of sand, clay and silt. They took a handful of each and mixed them together. This guarantees a better result for settling out. If you want to make sure this lab works for the students, you might want to consider this "cheat" method and pre-mix the soil.

DETERMINING SOIL TEXTURE SIMPLY BY FEEL (i.e. how much sand/silt/clay)

This isn't just an experiment. This is a real technique that soil scientists use out in the field. Sending soil samples to labs for analysis gives a more accurate assessment, but labs take weeks to return results. Often, soil scientists need an analysis immediately, and something that is approximately accurate is good enough. Part of soil science education involves training students to do this in-the-field analysis. Their final exam is to analyze soils that have been pre-analyzed by computerized lab machines, and then to compare their personal results to the computer analyses. The computers do something similar to what you did in part one, with the jars of water and soil, but in a much more accurate way. The end result that is important is to determine where the soil belongs in the soil triangle.

1) You will need a soft ball of soil, about the size of a golf ball. It should fit in the palm of your hand.

2) Knead the ball a bit to see if you need to add water. The ball should be soft and workable, like putty. If it is too dry, add a little water and knead it into the ball.

3) Use the flow chart on the opposite page, and do what it says at each step. The chart will ask you questions about your ball of soil and the answers to those questions will determine which way you go on the chart. In the end, you should arrive at a conclusion about your soil. This chart was taken from a government website, so is public domain and can be copied.
(If you make copies for your students, you might want to consider putting them into plastic sleeves so they don't get covered with smudges from clay-covered hands and fingers. Just a tip from experience.)

If you have students who are interested in soils and would like additional information, here is a related topic that might interest them. Sports fields are a very practical application of soil science. This website gives detailed information about the nature of sandy soils that are used on ballfields. **http://www.ultimate-baseball-field-renovation-guide.com/dirt-mix.html**

3) OPTIONAL LAB: SOIL pH

If you have older students who have the patience for more chemistry, you might want to purchase a soil pH kit at a garden center or online. The kit will come with instructions and explanations of why soil pH is important and how to test your soil sample. Some plants, such as blueberries, prefer a more acidic soil. Others, such as ivies and some blub flowers, prefer the opposite—alkaline soil. The acidity or alkalinity of the soil affects the roots' ability to exchange mineral ions with the surrounding soil. An acid substance creates lots of hydrogen ions ($H+$) and an alkali creates hydroxide ions ($OH-$). These ions can make it harder, or easier, for roots to absorb certain mineral ions.

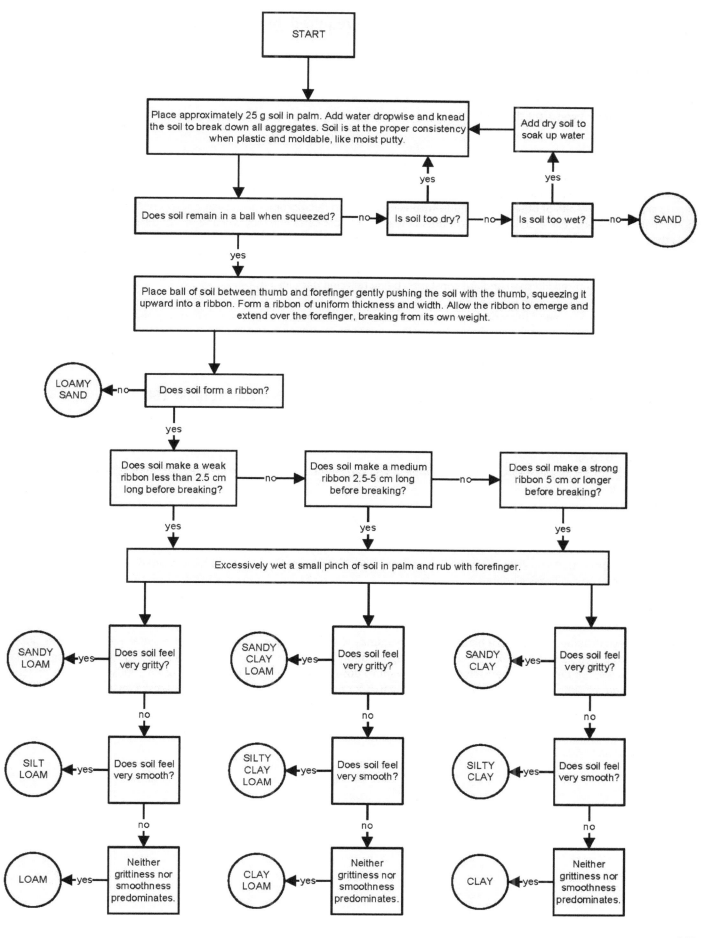

4) LAB: SOIL SCAVENGER HUNT (no microscope required, but magnifier recommended)

You will need:
• soil samples that are likely to have lots of interesting things living in them
• a high-power hand lens (over 10x, if possible)
• some paper plates and paper towels for mess control
• pencils and paper if you want the students to record their results

What to do:
   1) Give each student a soil sample (on a paper plate) and a magnifier. If you don't have magnifiers, you can still do the activity, you just won't be able to see things like very small quartz crystals or worm eggs.
   2) Tell the students to record everything they find in their sample. Look for minerals, plants and animals. They may find tiny quartz crystals or tiny bits of other rock. (Quite surprisingly, I found microscopic-sized clear quartz crystals in the soil at the bottom of my compost pile!) Look for bits of roots or leaves. Almost certainly, they will find bits of decomposed plant materials that has become almost unrecognizable. The decomposers (bacteria) will be too small to see. Look for the animals that were listed in the chapter. However, don't be limited to just these; they may find other things, as well. Don't forget that soil creatures have several stages in their life cycles. Soils are full of eggs and larva, not just adult stages. There might also be dead bugs. They have a limited life span and are constantly being recycled back into the soil.

NOTE: If you live in an area of the world that is not a temperate climate, you will not be able to rely on the information in this chapter and will need to do your own research into what might be found in your soils.

5) MICROSCOPE LAB (if you have access to a microscope that has at least 100x)

You will need:
• compound microscope
• glass slide and cover slips
• soil samples (including a garden or compost sample if you can get one because they are great for finding tiny critters)
• water and a bowl
• eye dropper
• optional: diatomaceous earth (found in natural food stores or online)

What to do:
   1) This lab assumes that you already know a bit about working with microscopes. If you need more directions about using a microscope, you can easily ask Google.
   2) Place a few crumbs of soil into the bowl and add a few spoons of water. Gentle crush the crumbs and stir the mixture a bit. (NOTE: You need the water to remain relatively clear. A muddy paste will not be transparent enough and you won't be able to see anything.)
   3) Use the eye dropper to draw up a bit of the slightly muddy water and place a drop or two on a slide. Put on a cover slip.
   4) Observe the sample under 100x power. At this magnification you will be able to see things like worm eggs and large nematodes but you will not be able to see bacteria. If you want to observe bacteria, you will need to use a 1000x oil immersion lens. If your scope has a 1000x lens and you want to use it, search for online instructions on using this lens. (You will need immersion oil.) Patience is required in order to find things. You are only seeing a very small portion.
   5) Diatomaceous earth is very interesting under a microscope. You can see all the broken bits of diatom shells. You'll need to add a little water to the white diatomaceous powder, and then cover it with a cover slip. (Magnification of 400x might give better results than 100x.)

6) ART PROJECT about soil organisms (more than just an art project-- lots of new info)

You will need:
• the following pattern page ("Microscopic View of Soil") copied onto regular paper, one for each student
• optional: copies of the page with the bugs (students can share)
• good drawing pencil and eraser
• Internet access so you can watch the step-by-step video showing how to do the drawing
• optional: ink pen if you want your drawing to be ink instead of just pencil

NOTE: If you do not have access to the Internet in the place where you will be doing this drawing, you may be able to download the video from my website. Go to www.ellenjmchenry.com/downloads/soildrawing.mp4. Your computer may be able to save the file onto the hard drive.

What to do:
  1) Use the video drawing lesson to do the drawing. The video is almost exactly an hour long and does not include any time for putting macroscopic creatures on the reverse side, so you might want to break this into two sessions. At minimum, take a break in the middle. Total time allowance should be close to 90 minutes. Younger kids might want to pause the video at various points, also, which could add time.
  2) After the "microscopic view" drawing is complete, fold the page on the two vertical lines, so that the flaps cover the drawing. This will give you a front cover of sorts on which you can draw macroscopic features. "Macro" means "large" so macroscopic features are anything you can see without a microscopic, such as ants, centipedes, etc. Make the creatures approximately life-size. Springtails and mites are very small, smaller than ants.

SAMPLES OF STUDENT WORK, ages 9 to 12:

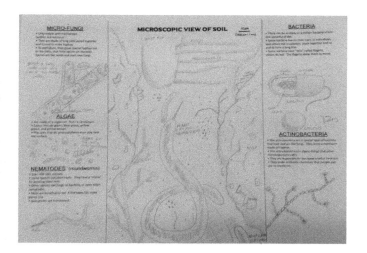
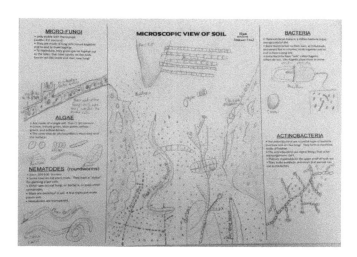
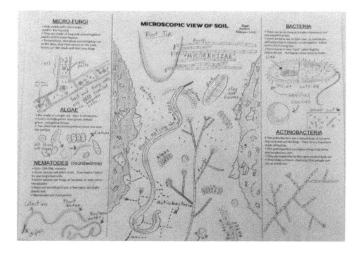
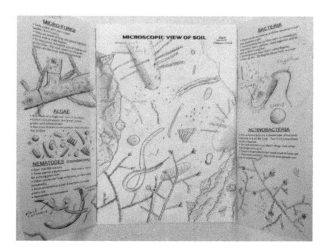

MICROSCOPIC VIEW OF SOIL

$$\underset{(1000\,\mu m = 1\,mm)}{\underline{10\,\mu m}}$$

BACTERIA

- There can be as many as a million bacteria in just one spoonful of dirt.
- Some bacteria live on their own, as individuals, and others live in colonies, stuck together end to end to form a long line.
- Some bacteria have "tails" called flagella, others do not. The flagella allow them to move.

ACTINOBACTERIA

- The actinobacteria are a special type of bacteria that look and act like fungi. They form a mycelium made of hyphae.
- The actinobacteria can digest things that other microorganisms can't.
- They are responsible for the sweet smell of fresh soil.
- They make antibiotic chemicals that people can use as medicines.

MICRO-FUNGI

- Only visible with microscope. (width= 4-6 microns)
- They are made of long cells joined together end to end to make hyphae.
- To reproduce, they grow special hyphae out to the sides, that have spores on the ends. Spores act like seeds and start new fungi.

ALGAE

- Are made of a single cell. Size= 5-10 microns
- Colors include green, blue-green, yellow-green, and yellow-brown.
- The ones that do photosynthesis must stay near the surface.

NEMATODES (roundworms)

- Size= 200-500 microns
- Some species eat plant roots. They have a "stylus" for piercing plant cells.
- Other species eat fungi, or bacteria, or even other nematodes.
- Most are benefical in soil. A few types can make plants sick.
- Nematodes are transparent.

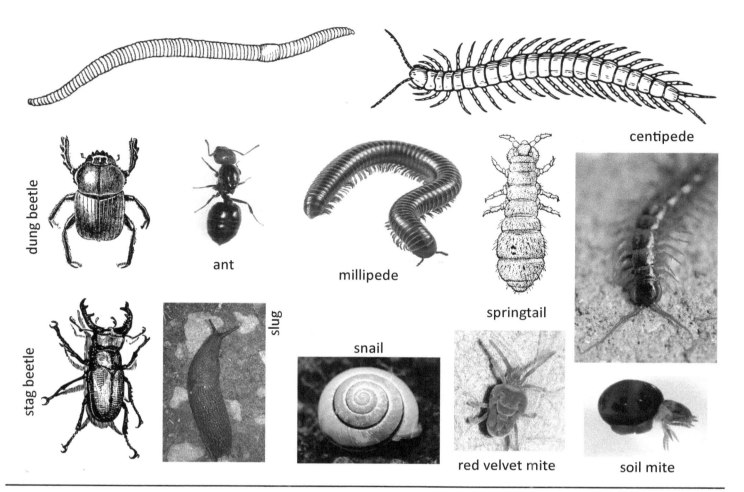

centipede

dung beetle

ant

millipede

springtail

stag beetle

slug

snail

red velvet mite

soil mite

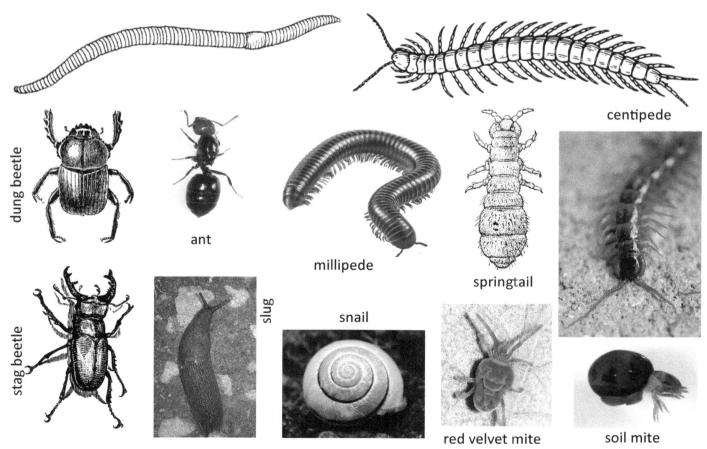

centipede

dung beetle

ant

millipede

springtail

stag beetle

slug

snail

red velvet mite

soil mite

CHAPTER 7

1) WEBSITE/VIDEO: An amazing true story showing the importance of aquifers

   In 1969, David Bamberger purchased over 5,000 acres of the worst property he could find in southern Texas, in the hopes that with application of sound environmental practices, that land could be restored to productivity. That same property is now an award-winning nature preserve with abundant natural vegetation and wildlife. How did David restore the land? Simply by understanding how aquifers work, and the part that native plants play in maintaining those aquifers. Developers had ripped out all the native plants and brought in non-native juniper trees, trying to turn the land into something it could never be. As a result, the aquifers dried up, plants died, and animals moved out. David got rid of all the trees that did not belong there, and planted the native prairie grasses that had once thrived there. The results have been seemingly miraculous. You can either read or watch the rest of the story at their website.

   http://bambergerranch.org/video-library

2) MAPPING THE OCEAN FLOOR: A SIMULATION

NOTE: If you have used a lot of free downloads from the author's website, you might have already done this activity while studying oceans. Feel free to skip it, in that case.

   In this activity, students map a piece of ocean floor that they can't see. It will sort of be like a "slice" of ocean. They will push measuring rods down into their slice of ocean and note the depth when they touch bottom. In real life, sonar is used, of course, but this activity still gets the point across.

 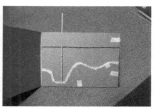

Students poke measuring rods down into the ocean.

The cover is then opened so they can check their results.

   There is some adult preparation necessary ahead of time since the students must not see the shape of the ocean floor they will be exploring. The prep time is about 15-20 minutes per ocean slice once you have the materials gathered.

You will need (for each slice of ocean):
- 3 pieces of 8.5" x 11" (22cm x 28 cm) corrugated cardboard
   (They don't have to be exactly this size, but all the slices should be the same size.)
- scissors and/or X-acto knife
- masking tape or duct tape
- 2 medium-sized binder clips
- a bamboo skewer (sold in grocery stores for making kebabs)
- a copy of the graphing page
- pencil

Preparation:
   1) Two of the pieces of cardboard will remain as they are, 8.5 x 11. The third piece will be cut to make the section of ocean floor. Here are some ideas for floor shapes.

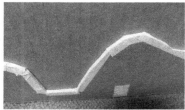

The dotted portion is the top you will cut off. BUT DON'T DISCARD IT.
You will cut off the top strip and use it in the interior.

   2) Put masking tape along the top edge of your contoured piece. The tape will prevent the measuring stick from going down in. Make sure there are not any holes in the tape where the measuring stick might get stuck.

This picture shows both steps 2 and 3 accomplished. Tape has been applied to top edge contour piece, and the contour piece has also been mounted to a whole piece of cardboard.

3) Glue your contour piece to the bottom of one of the remaining rectangles.

4) Cut a strip from that "discard" piece (the dotted area in the pictures from step 1) and glue it to the top, as shown in the picture.

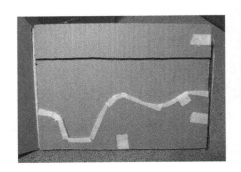

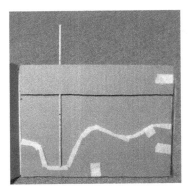

This picture shows how the measuring stick will go down through that top strip, and then hit the bottom down below.

5) That last rectangle will be the front cover. You can either just clip it in place with binder clips, or you can use tape to make it hinge open.

When the front cover is in place, the top will look like this. It will be three pieces thick. The middle piece is where the measuring rod will go down.

I glued a piece of blue paper to the front, and labeled it SLICE OF OCEAN. You can leave yours blank if you want to.

6) Make sure both ends of the rod are blunt (not sharp). Use a fine point marker to make evenly spaced marks on the bamboo rod. I made my marks at one centimeter intervals. Every fifth mark I made darker than the rest so that the eye will easily see groups of 5. This makes counting the marks much faster and easier.

7) I also added numbers along the top. I put a number in front of each corrugate hole in the cardboard. These numbers correspond to the numbers along the top of the chart that they will fill in. The numbers are just points at which you take a reading. These number do NOT represent any actual increment such as miles or meters. **Again, note that the measuring rod slides down through the MIDDLE piece.**

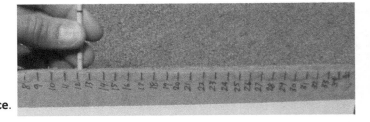

How to do the measurements:

1) Start with hole number 1. Push the measuring rod down into the hole that is directly behind number 1. Push it slowly until it hits bottom.

2) Once it hits bottom, put your thumb or finger on the rod right along the top, to mark how deep the rod went.

3) Holding your thumb/finger in place, pull the rod out and count how many units are below your thumb/finger.

4) Go over to your graph paper and find the number 1 along the top. Count down the number of spaces that you just measured. Put a pencil dot there.

5) Now go to slot 2 and do the same thing. Record the depth at number 2 on your graph paper.

6) Keep doing this until you get all the way to the end.

7) Now connect the dots to what shape they make.

8) Take the front cover off (or just open the lid) and see how accurate your graph is compared to the real thing.

9) Answer the questions at the bottom of the graph page.

NOTE: A common mistake students make is to count the number of units on the measuring rod above the cardboard, instead of below. This will result in an inverse graph, an upside down version of the bottom shape.

MAPPING THE OCEAN FLOOR

Use your wooden measuring rod to take depth measurements in your slice of ocean. After you are done, connect the dots to make a profile of your sea floor. When you are finished, open the cardboard to see how your map compares to the real profile.

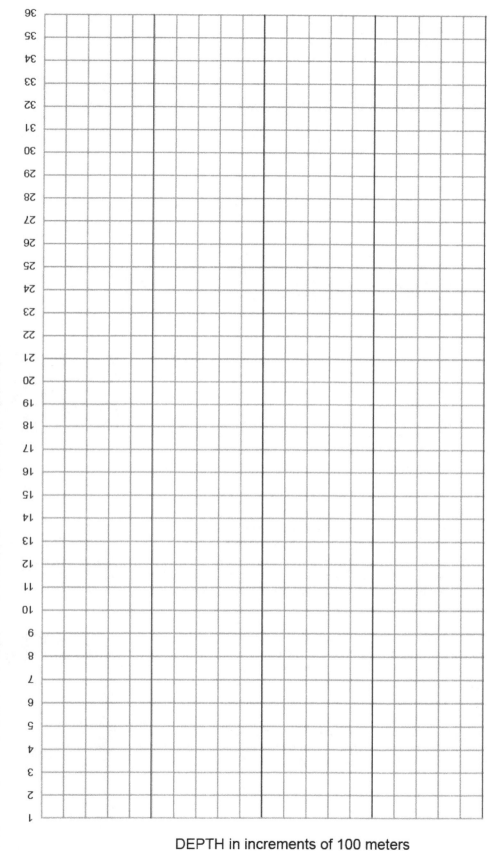

DEPTH in increments of 100 meters

QUESTIONS:

1) If you had taken more measurements, would your drawing have been more accurate? _____

2) In real ocean mapping, they often use _____ waves to measure depth.

3) If each block represents 100 meters of depth, how deep is the deepest part of your slice of ocean? _____

4) Were any of your measurements hard to interpret? Which numbers were hardest to record? _____
   Why? _____

3) CRAFT: INSIDE THE EARTH "SHRINKIE" (clear plastic ornament)

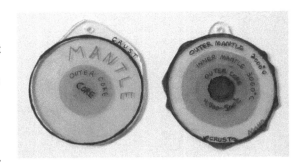

This craft is oddly appropriate for this chapter because it uses a type of plastic that shrinks substantially when heated. Of course, this shrinking plastic phenomenon is not exactly the same as the shrinking that magma does under extreme heat and pressure, but at least the craft can remind us that sometimes heat does surprising things to materials, and, in some cases can cause considerable reduction of volume.

The main point of this craft is to review the layers that we think are found inside the earth. (Remember, this has never been directly observed!) Students who struggle with crafts can keep their design simple. Students who love detail may want to add surface details or written labels.

You will need:
• a package of shrinking plastic (a well-known brand name is "Shrinky Dinks" but there are other brands, too)
• permanent markers (or whatever is recommended on the package—some shrinky plastic will accept colored pencil) (Make sure to include a thin black marker if you want them to label their diagrams.)
• scissors
• OPTIONAL: paper hole punch (an easy way to add a hole for hanging the ornament)
• a copy of the concentric circles for each student to put under their plastic as a drawing guide
• OPTIONAL: copies of the full-color reference page showing layers of earth
• baking trays and access to an oven

What to do:
1) Provide each student with a copy of the concentric circles template (note that there are two templates on the page), a piece of plastic that is larger than the circles, and a supply of permanent markers of various colors (including brown). The pattern page gives you two templates.

2) The students will place their piece of plastic on top of the template so that the circles are clearly seen under the plastic. They will then proceed to draw a colorful diagram of what is inside the earth. (We have never seen the inside of the earth, so colors can be creative, if they prefer, not "realistic.") Be generous with the amount of color applied.

NOTE: You may choose to provide a copy of the following page showing colorful illustrations of the inside of the earth, or you may choose to have students do their own research on the Internet or from books. Make sure students know that they are not limited to the color schemes found in any drawing, but may choose their own colors. You can also vary the requirements for labeling depending on the age and ability of your students. The drawing showing the compositional layers versus the texture layers will be very helpful for understanding the differences in labeling that you will see in illustrations online or in books.

3) If they want to add labels, provide very thin black markers so the words can be written with small letters. The black marker should write on top of the colors if they are dry. (Of course, the shrinking process will make the letters much smaller.)

4) When finished, they will cut out their earth ornament. If a hanging hole is desired, a hole can be punched with a metal hole punching device. When completely finished, place the ornament (marker side up) on a baking tray.

5) Bake according to instructions on package. They only take 2 to 3 minutes. If you are concerned about the surface of your baking pan (because of the plastic cooking on it) you can use the back side of the pan, or you can place a sheet of baking parchment on the pan.

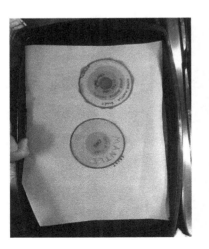

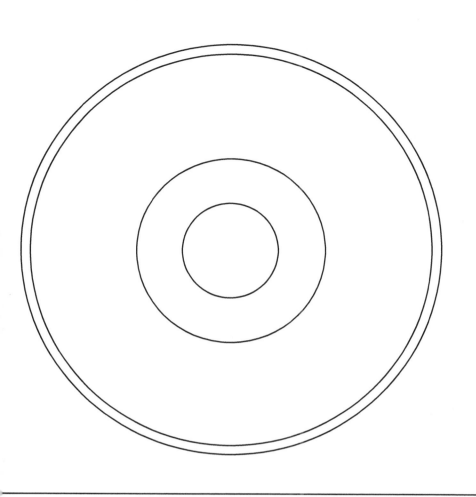

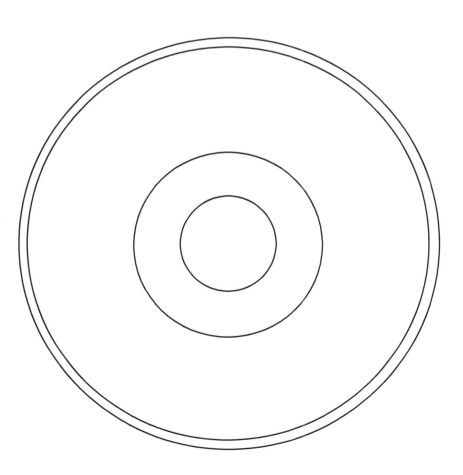

REFERENCE PICTURES THAT MAY HELP YOU

(You do not have to include all these layers. Only the basic 4 are required: crust, mantle, outer core, inner core)
Notice that some layers are defined by what they are made of, and other layers are defined by texture. This helps us understand why we see different labels of various drawing in books and on websites.

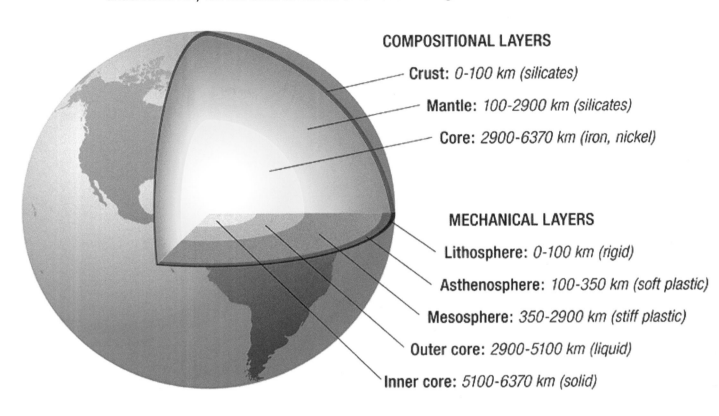

COMPOSITIONAL LAYERS

- **Crust:** *0-100 km (silicates)*
- **Mantle:** *100-2900 km (silicates)*
- **Core:** *2900-6370 km (iron, nickel)*

MECHANICAL LAYERS

- **Lithosphere:** *0-100 km (rigid)*
- **Asthenosphere:** *100-350 km (soft plastic)*
- **Mesosphere:** *350-2900 km (stiff plastic)*
- **Outer core:** *2900-5100 km (liquid)*
- **Inner core:** *5100-6370 km (solid)*

APPROXIMATE
TEMPERATURES
OF LAYERS:

Crust: 0° to 1,000° C
    (0 to 1,800° F)
Outer mantle: 2,000° C
    (3,600° F)
Inner mantle: 3,000° C
    (5,400° F)
Outer core: 4,000-5,000° C
    (7,200-9,000° F)
Inner core: 6,000° C
    (11,000° F)

The MOHO is the dividing line between the crust and the mantle.

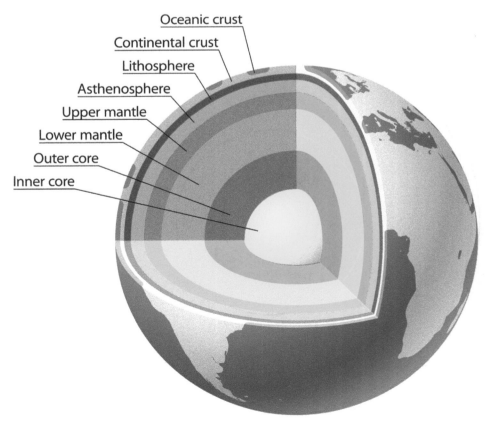

Oceanic crust
Continental crust
Lithosphere
Asthenosphere
Upper mantle
Lower mantle
Outer core
Inner core

4) CORE SAMPLING ACTIVITY

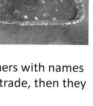

This activity attempts to simulate core sampling, where geologists drill long tubes into the ground to bring up cylindrical samples of dirt and/or rock. (See picture on page 97.) We will use a clear plastic straw as our drill, and an edible snack as our landscape. Students will also simulate the process of interpreting core sampling data and using it to create a chart that shows wide horizontal layers.

This activity requires a lot of preparation, but the activity itself will take up quite a bit of class time, so the prep time will be well spent. If you'd like the students to participate in making the landscapes and you want to do the SWEET option, set aside 30-40 minutes at least several hours before you want to do the core sampling. You could even do it one or two days before. The sandwich option doesn't need to be prepped far in advance, and can be done immediately beforehand. For any of the options, have the students assemble their landscapes "secretly" (as well as that can be done in a group). You can mark the containers with names or group numbers, and then make sure that no person or group receives the landscape they created. If they all trade, then they all should end up with a "mystery landscape" to investigate.

You will need:
- clear plastic straws (can be ordered on Amazon if you can't find them locally)
  NOTE: If you really can't get clear straws, use regular ones and either cut them open or carefully expel the core samples by blowing them out (in a controlled manner so the core stays together for the most part).

- copies of the data recording sheet that best matches the shape of your container
- colored pencils for recording data
- a variety of baking supplies, as discussed below
- opaque (not clear) containers, such as small loaf pans (If all you have are transparent containers, line them with foil.)
NOTE: You can make these any size. An ideal size to work with is those tiny loaf pans used for holiday gift breads. However, you can use other things that you may have on hand.

The overall goal in creating the landscape is to make 6 to 10 identifiable layers that a straw can poke through.

Edibles:

You will need to make some decisions about what type of edible materials you want to use. If you have health concerns, you can make the edible landscape gluten-free, sugar-free, preservative-free, fat-free, or whatever works best in your situation. You could do a meat, cheese and bread landscape as long as you use firm cheese and medium-firm bread. If you decide to use the sweet option, bear in mind that the layers don't have to be eaten afterwards. You can just recycle them into compost or rinse them down the drain. Don't feel guilty about not eating them.

If you are using any type of pudding or gelatin in your layers, create the landscapes the day before, or at least several hours before. Make sure the layers are firm enough to hold their shape reasonably well. With any of these materials, you will find that the straw will compact the layers as it pokes downward.

SWEET
- fruit-flavored gelatins ("Jello") with added unflavored "Knox gelatin" to make them extra-firm
  NOTE: Jello brand now makes an "all-natural" powdered gelatin mix that contains no artificial colors, flavors or preservatives.
- applesauce with gelatin and blue food coloring, to represent an aquifer
- stiff pudding (If you use a box mix, put in less liquid than the instructions call for)
- graham crackers (which will get soggy if they are in the middle)
- Oreo crumbs, moistened with a little applesauce
- anything else you can think of that might be fun (colored sprinkles in places for gemstones?)

SANDWICH THEME
- bread in light and dark colors (pumpernickel is usually very dark and might be used for coal)
- cheese slices, both white and yellow
- processed sandwich meats that are homogeneous in texture (bologna, salami, liverwurst, etc.)
- pickle slices
- zucchini or cucumber (peeled, cut into slices, and patted dry)
NOTE: Blue food coloring might be added to zucchini, cucumber or pickles to represent an aquifer (soak in blue ahead of time, pat dry)

ULTRA HEALTHY
- cooked carrots or orange squash (still firm, not mushy)
- raw or cooked zucchini (still firm, not mushy)
- raw cucumber
- raw or cooked green and red peppers (raw would need to be in with other rather stiff things, like cucumber and raw zucchini)
- cooked potato or sweet potato
- tofu
- raw cantaloupe or honeydew melon
- raw kiwi (green) (This might represent an aquifer?)
- raw apple or pear (Apple soaked in [natural] blue coloring could also make an aquifer.)

<u>Assembly notes:</u>

FOR ALL PROJECTS:
> Again, the overall goal is to create 6 to 10 layers that a straw can poke through. Remember that in real rock strata, you often find just a few types of rock that are arranged in repeating layers, such as limestone, shale, sandstone, limestone, shale, sandstone, etc.
> You don't have to assign rocks to materials ahead of time, but you can if you want to. Do whatever works best in your situation. If you have a large group, it might be confusing to have each student or team calling the layers different things, so you might want to set up a "key" ahead of time. (For example: red Jello= sandstone, Oreo crumbs= coal, green Jello= shale)
> Remember to keep track of how many layers there are. The students will need to know the total number of layers.

SWEET
1) Add a package of unflavored gelatin to each packet or box of regular gelatin that you use. Also, use a little less water than the recipe calls for. I added applesauce, as well, to make it less transparent.

2) Vanilla pudding mix can be used either as a light yellowish color, or as a tan if you add a little bit of chocolate pudding mix to it. I found that a little chocolate powder goes a long way, so don't add too much. Remember to add less milk than the recipe calls for. You want the pudding to be pretty stiff. If it ends up too stiff, just thin with a little water or milk. This is a science project, not a baking contest, so it won't matter if it is lumpy.

3) Make at least 6 layers, but not more than 10. It will be surprisingly challenging to identify the layers inside the straw. Make sure you count how many layers there are so that the students know what they are looking for. Knowing the number of layers will help to interpret what they are seeing in the straws.

4) You can make a layer two different things. For example, layer 3 could be an aquifer on one side and limestone on the other. You can even divide the layers diagonally, not just straight across. (For older students, you could even try putting three things into one layer, or making a small round aquifer or a coal seam hidden in the middle somewhere.)

5) If your materials are still a little soft, put the containers in the freezer for a few minutes after the addition of each layer.

SANDWICH THEME
1) White bread and white cheese will be almost impossible to tell apart in the straw core samples. Keep white bread and light colored cheese between darker layers.

2) You can cut and patch your materials to make the layers. For example, if you have small round meats, use several slices and trim them to fit as a layer as well as you can. It'll be good enough!

ULTRA HEALTHY
This is the option I did the least experimenting with. However, if you are the type of person who would choose this option, I'm betting that you probably have a lot of initiative and creativity when it comes to working with food and therefore can figure out a method that works well for your chosen foods.

<u>Taking core samples:</u>
1) Choose the data sheet that most closely matches the shape of your container.   Gather colored pencils that can be used to record the core samples you will bring up.

2) Look at the data sheet and plan where your sample holes will go. Take the first core sample by pushing a clear straw down into the landscape, all the way to the bottom. It may help to twist the straw as you push down.

3) Put your mouth on the top end of the straw and apply some suction, as though you were going to suck up the sample, although you won't be able to do so because of immense friction. Sucking on the straw will keep the core sample from falling out when you lift the straw out.

4) Gently lift out the straw while still applying suction. As soon as you have the core out, stop sucking.

5) Record each core sample on the data sheet. Use colored pencils to record each layer going down. If you have less than 10 layers, just use the numbers that you do have and leave the rest on the bottom blank. For example, if you have only 6 layers, fill in numbers 1 through 6. Also, don't worry that the layers on the data sheet are not the same thickness of your edible layers. Just record the color or texture of each layer, and don't worry about the size or thickness. Yes, geologists do carefully measure each of their layers, but I found that many students have trouble figuring out what to do if there are no lines or numbers on the data charts. The numbers will correspond nicely to the layers down below.

NOTE: You will most certainly run into problems along the way. These problems are likely to be similar to the problems that geologists face as they take their core samples. For instance, you may have trouble distinguishing one layer from another. Geologists can also have trouble with this. You may find that your layers shift or get blurred on the way back up. This is also something geologists deal with. Layers can be thicker in one place and thinner in another. Occasionally, a core sample might come up unusable and another core has to be drilled nearby. Just do the best you can. That's what geologists do.

EXAMPLES OF HOW CORE SAMPLES MIGHT LOOK:

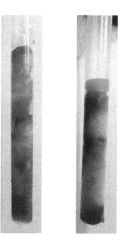

The blue layers are obviously meant to be aquifers. The other colors can be assigned to whatever types of rock you want to imagine them to be.

Notice that the layers are not always easy to tell apart. It really helps to know how many you are looking for.

NOTE: When students are done looking at the core sample and have recorded it, they can blow it out the end into a bowl or a paper cup.

6) When all your core samples are recorded, proceed to the bottom half of the data sheet. Those rectangles (or squares) represent the layers, starting at the top and going down. This means that the students must translate vertical data into horizontal views. Some students find this easy and obvious and others struggle to "get it."

Layer 1 will be very easy because you can see it. In real life there would be plants and dirt on top covering the rock, but in this activity you can see the rock. Color rectangle number 1 to match your top layers. Then, go down to the KEY, and color one of those rectangles that same color and write what it represents.

For the second layer, look at all your number 2s in the core samples above. What is layer 2 made of? Is it the same all the way across, or does it change somewhere in the middle? Color rectangle 2 to show what that layer is made of. Also color another rectangle in the KEY and label what it represents.

Do the same for the third layer. Look at all the 3s in the core samples above and figure out what layer 3 is made of. If it is different from layers 1 and 2, add that color to the KEY, also.

Keep doing this until you get to the bottom layer. Don't forget to finish the KEY, too.

CORE SAMPLING and STRATA MAPPING

Your assignment is to create a geological map of your property. You will need to take core samples of at least 9 different sites, but you can take more samples if you want to. (If your core sample has less than 10 layers (strata), just use the numbers you need and leave the bottom ones blank.) After you have your core samples recorded, figure out what each horizontal layer is made of and color it appropriately.

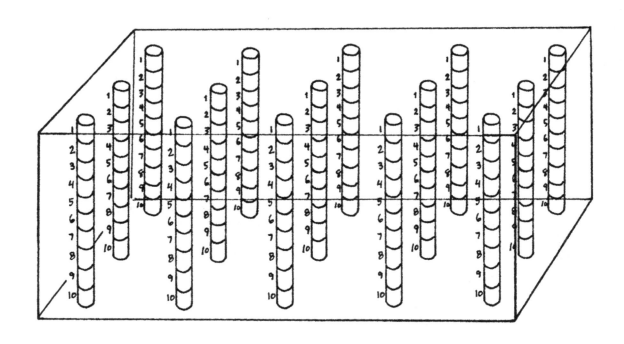

EACH LAYER AS SEEN FROM THE TOP:

1

2

3

4

5

6

7

8

9

10

KEY:

CORE SAMPLING and STRATA MAPPING

Your assignment is to create a geological map of your property. You will need to take core samples of at least 9 different sites, but you can take more samples if you want to. (If your core sample has less than 10 layers (strata), just use the numbers you need and leave the bottom ones blank.) After you have your core samples recorded, figure out what each horizontal layer is made of and color it appropriately.

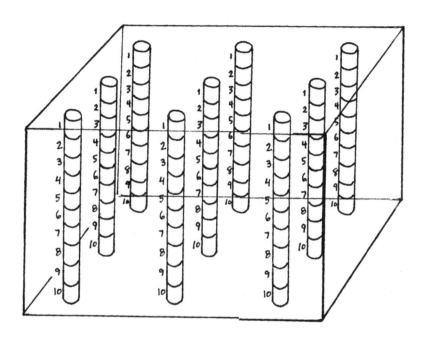

EACH LAYER AS SEEN FROM THE TOP:

1 [   ]

2 [   ]

3 [   ]

4 [   ]

5 [   ]

6 [   ]

7 [   ]

8 [   ]

9 [   ]

10 [   ]

KEY:

5) ART PROJECT: Scientific illustration of trilobites

This drawing project is video-based, like the layers of the canyon drawing and the microscopic view of soil. The video can be accessed on the YouTube channel, or by direct streaming at: www.ellenjmchenry.com/downloads/TrilobiteDrawing.mp4 (You might be able to download the video from the site if you need to use it in a place that does not have Internet access.)

<u>You will need:</u>
• a copy of the following pattern page printed onto "toned" paper, such as tan, light gray, light brown, light blue, etc.
• colored pencils: white, black, and one that is a darker version of your paper (brown if you are working on tan or light brown paper, medium or dark gray if you are working on gray paper, blue or gray if you are working on blue paper)
• a pencil with eraser
• a fine point pen

<u>What to do:</u>
The video will tell you everything you need to know. Just watch and draw along.

SAMPLES OF STUDENT WORK (ages 9-10):

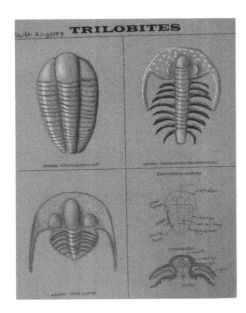 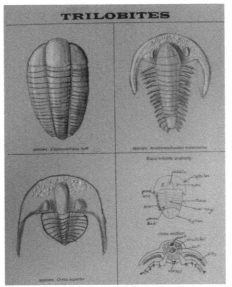 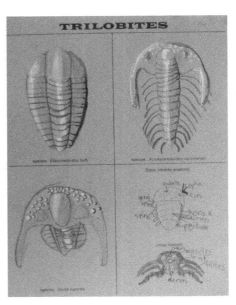

ADDITIONAL CRAFT IDEA:
If you'd like another paper craft project, you can download this (free) pattern page for making a paper trilobite.
http://www.hgms.org/Paleo/TRILOBIT.PDF

There is also a pattern for a paper nautilod:
http://jclahr.com/alaska/aeic/taurho/nautiloi.pdf

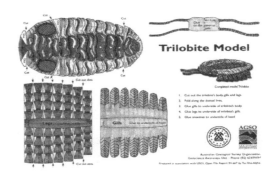

280

TRILOBITES

species: *Ellipsocephalus hoffi*

species: *Acadoparadoxides mureroensis*

species: *Onnia superba*

Basic trilobite anatomy:

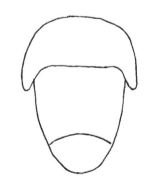

cross section:

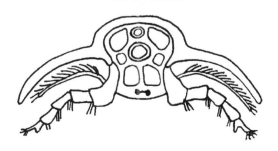

6) DEMONSTRATION: Natural springs

Here is something else you can do with your clear straws if you have some left over from the core sampling activity. You can demonstrate the physics principle by which natural springs operate. Gravity will always cause water to be at the same level inside a tube or in a water system such as an aquifer. A natural spring (or Artesian well) is a place where one end, or one part, of a water system is substantially below the rest. The water is forced out of the well or spring simply by the fact that it is lower than the rest of the system.

You will need:
• a few clear straws
• a few regular straws (the kind with the bendable part towards one end)
  NOTE: If your clear straws happen to have bendable places, then you won't need the regular straws.
• clear tape
• scissors
• water (possibly with a little food coloring to make it show up better)

What to do:
  1) Cut a few of the bendable places from the straws that have them. Leave a centimeter or two of the straight area on either side of the bendable part.
  2) Assemble the straws to form a long U, as shown. A technique you can use to join ends of two straws is also shown. Pinch the end of one straw, then insert it into the other straw. Push until the pinched part disappears. If the straws are exactly the same diameter, this will make a very tight joint.
  3) Wrap clear tape around each joint, trying to make it water tight.
  4) Pour a little water down into one end. (If you want the water to show up well, add a little food coloring.)
  5) Notice where the water level is in each straw. The lines should be even. You can check this by holding the U tube in front of something perfectly horizontal, such as a shelf or window frame.
  6) Raise one end of the U. In a split second, the water level will adjust and the levels will be even (at the same horizontal mark) even though the ends are uneven. Alternate raising and lower the sides of the U, and watch the water levels always stay equal, no matter how high or low the straws are.
  7) You can make an Artesian well by lowering one side low enough that the water comes out the top. You will run out of water very quickly, but imagine that the high end has a funnel attached to it, providing a generous and constant supply of water. This is like collected rain water in a watershed that feeds into an aquifer. As long as rain water keeps filling the aquifer, the Artesian well or natural spring will keep flowing.

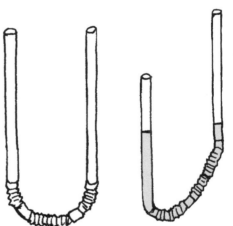

7) CRAFT/DEMO: Make your own seismograph

There are a number of ways you can build a simple seismograph. It can be as simple as a paper cup filled with marbles hanging inside a box, or it can involve electronic parts. If you'd like to make a very simple one, you could try these ideas from popular websites. or do your own Internet search.

http://www.kfvs12.com/story/307729/make-your-own-seismograph
http://pbskids.org/zoom/activities/sci/seismometer.html

If you'd like to try something more difficult, and perhaps something that really can detect distant earthquakes, use your Internet search engine with key words such as "seismometer, seismograph, DIY, make, homemade, hobby, educational." (You might also try the MAKE magazine archives. The link was unavailable at the time this book went to press, but perhaps you may be able to find it.)

8) CRAFT: Make a paper globe and mark ocean features

Some earth features, such as the Mid-ocean Ridge, are hard to understand when looking at a flat map. The earth is a sphere, of course, so a flat map can't ever be as accurate as a globe. The following page has a very easy to assemble paper "globe" (an octahedron) on which you can draw ocean features such as the Mid-ocean ridges, the deepest trenches, and the "ring of fire." If you want a more spherical paper globe, you can find patterns on line using key words "paper globe." The more sides there are, the closer to a sphere it will be, but the harder it will be to assemble.

If you want a lot of choices, including a globe that is in color and already has tectonic plates marked on it, check out this website: **http://www.3dgeography.co.uk/make-a-globe**

If you want to make a dodecahedron globe (12-sided), you can either use the pattern on the following page, or you can use this link: **http://www.progonos.com/furuti/MapProj/Normal/ProjPoly/Foldout/Dodecahedron/dodecahedron.html**

If these links are all broken by the time you are using this book, just use an Internet search engine set on "images" and use key words "dodecahedron paper globe."

<u>You will need:</u>
- the following pattern printed onto heavy card stock paper (or your Internet pattern)
- scissors and glue
- pencils and/or colored pencils or markers
- optional: short piece of yarn or string as a hanger

<u>What to do:</u>

1) Cut out the octahedron pattern.

2) Score all the fold lines by putting a ruler right along the lines and gently running the point of a compass (or a nail or dead ball point pen) along the line. The idea is to scratch the line without cutting through the paper. You might want to practice on the scrap paper around the globe, before trying it on the globe. If you are tempted to skip this step, try folding a line without scoring and compare that to a scored line. You will see how much easier it is to fold when the lines are scored first.

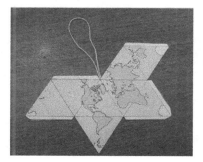

3) If you want to add a string as a hanger, glue the ends of the string to the inside of the globe before assembling it.

4) It doesn't matter which glue flap you begin with. Just do them one at a time, pressing and holding each joint for at least 10 seconds before you move on to the next. (If possible, let the joint rest for half a minute before you move on.) Use just enough glue to put a very thin film across the surface of the flap. If glue oozes out of the joint, you have used too much. You should not have any glue oozing out.

5) When the globe is dry enough to be handled, take a pencil, pen or marker and draw on the Mid-ocean Ridges, and any other features you just learned about. The magnetic poles are always interesting to note, as they are so different from the geometric poles. Also, you could label the Kola peninsula, Mount Everest, the Mariana Trench (deepest trench), Ring of Fire, Grand Canyon, and Surtsey Island. Consult a map to see where to draw all these features. If you want them to stand out, you can use red or orange when drawing them.

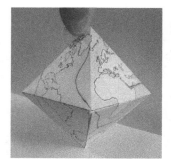

Mid-Atlantic Ridge on octahedron

You can add the fracture zones on the sides if you want to.

Mid-Atlantic Ridge on dodecahedron

ALTERNATIVE: You can also use a real globe if you have an old one, or if you use an erasable marker of some kind. Also, craft stores sometimes sell inexpensive rubber or foam balls printed to look like globes. (Amazon sells earth balls for just a few dollars each if you buy a package of one dozen.) Balloons printed with globe designs would work, too.

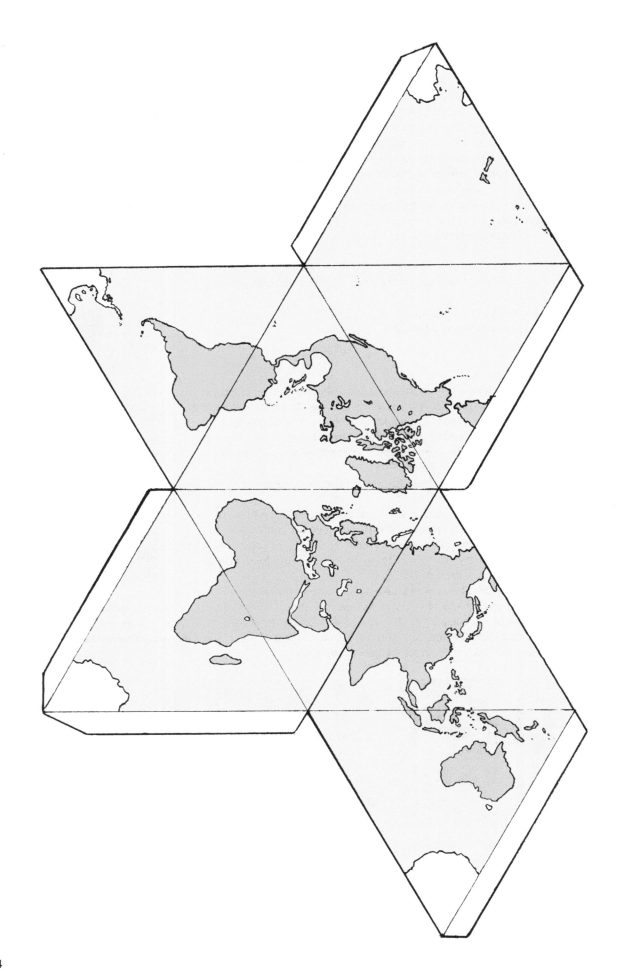

284

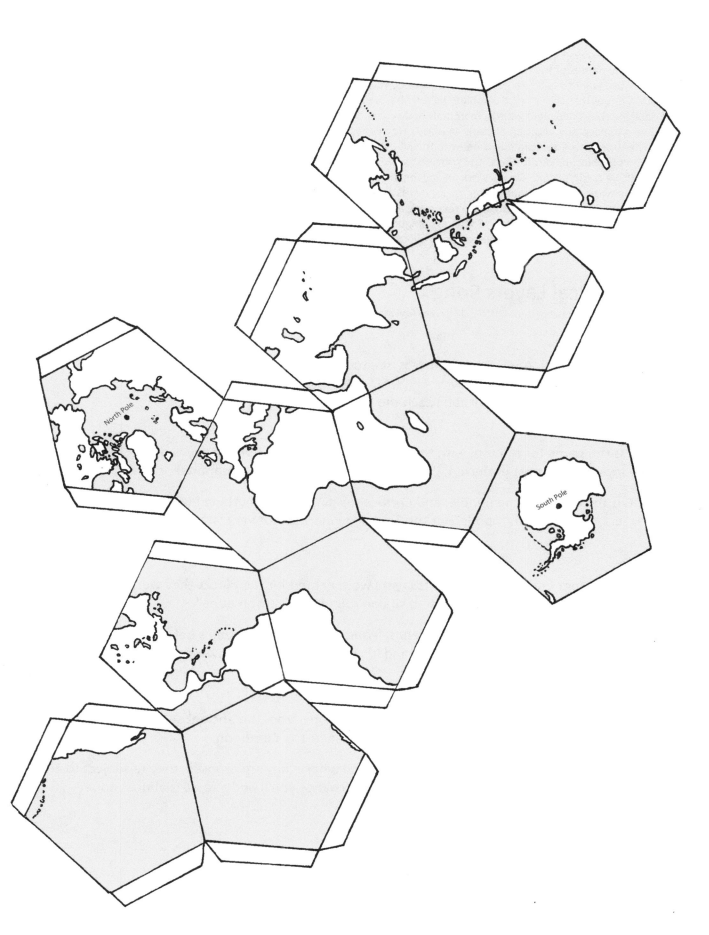

9) MUSIC: Sing a song about the geological layers

This song doesn't tell you anything about the age and origin of the geological layers. It's just a way to learn geology vocabulary. The names of these layers are used in geological charts more frequently than the words sandstone, limestone and shale. If you want to talk to a geologist about rocks, you have to know these layer names and have a clue what is found in them. Don't forget that this is not a complete list of what is in each layer! For example, the "dinosaur layers" have much more than dinosaurs; they have plants, birds, mammals and sea creatures. (Disclaimer: The top of the Precambrian layer in some places seems to have tiny fossilzed bacteria or algae. However, it does not have anything that kids would recognize as a fossil.)

The tune is an American folk tune from the late 1800s, sung by railroad workers out west as they dug holes in rocks in preparation for blasting out tunnels. The "tarriers" were Irish workmen. The original chorus goes like this: "Drill, ye tarriers, drill, and drill, ye tarriers, drill! Oh it's work all day for the sugar in your tay, down beyond the railway, so drill, ye tarriers, drill." Since the song was originally about drilling rocks, it was a natural for a song about digging down into layers of rock.

Unfortunately, there isn't an audio recording available yet. However, the tune is easily found on YouTube, or on web-sites that catalog American folk songs. To help you figure out the timing of the words, the stressed beats are marked with bold letters. Imagine that those bold syllables are said every time the hammer or shovel hits the ground.

The Geological Layers Song

(to the tune of "Drill, Ye Tarriers, Drill")   Lyrics (c) 2017 by Ellen McHenry

Chorus:
Dig, you ge**olo**gists, **dig**! And you **pale**on**tolo**gists, **too**!
Go and **dig** all day, stripping **layers** away,
Get **down** below the **Cam**brian, till you **reach** the basement **rock**,
And **then**, you can **stop**.

CENOZOIC is the **name** for the top, with the *Quaternary* layer at the **very** tippy-top.
This is the layer that is **most** preferred, if you're **looking** for a fossilized **mammal** or bird.

MESOZOIC is the **name** for the middle, and *Cretaceous* dinos were **anything** but little!
Jurassic layers have **huge** sauropods, but the *Triassic* fauna was **not** very large.

Chorus

PALEOZOIC is **deeper** underground, and its **layers** were named for the **places** they were found.
The *Permian* layer is **easy** to forget, 'cause it **sits** on top of the "**carbon** duet."

The **two** *Carboniferous* **layer**s have coal: *Pennsylvania*'s on top, *Mississippi*'s be**low**.
The *Devonian* lakes were **good** fishing sites, and it's **here** you'll find the **low**est ammonites.

Chorus

Silurian rocks have **foss**ilized leaves, and **also** many creatures from the **ancient** seas.
The **crin**oids start in the *Ordovician*, and **under**neath that is the *Cambrian*.

The *Cambrian* rocks have been **called** an explosion, and **where** they are exposed, they're **subject** to erosion.
On the very bottom we have **PRE**CAMBRIAN, and the **chance** you'll find a fossil is **almost** none.

Last chorus:

Stop, you ge**olo**gists, **stop**! You're **down** to **igneous rock**!
Your **dig**ging all **day** stripped the **layers away**;
You're **down** below the **Cam**brian, and you've **reached** the basement **rock**,
And **now** you can **stop**!

10) DEMONSTRATION: Cross-over depth

Here is a class demonstration you might want to do, showing the concept of cross-over depth. In the text, the context was magma surrounded by mantle. Here we have a balloon instead of magma and water instead of the mantle, but the principle of cross-over is the same. Pressure causes decrease in volume, which causes higher density compared to surrounding material, which eventually reaches a point of no return where the magma/balloon will sink and never be able to rise again. Of course, in this demo you can make the balloon rise again by pulling it up with the hook on the end of your wire.

NOTE: If you don't have the time, energy, or budget to do this demo but would like to see it in action, there is a video posted on the YouTube playlist. There is also a "how to" video showing how to do the steps listed in the directions below.

You will need:
- a clear plastic fluorescent tube protector (available at home supply stores that sell fluorescent tubes)
- a tube of caulk (or you could try substituting a soft, waterproof glue such as Quick Grip or Goop)
- a plastic cap that is about the same size as the end of the tube, such as a cap from a milk jug or a glass coffee drink bottle
- a wire coat hanger (a used one from your closet is fine)
- a few water balloons (or other small balloons-- not large balloons)
- a few small binder clips (3/4 inch and 1/2 inch)
- a twistie tie or rubber band
- a 1/4" diameter eye bolt
- a wing nut or regular nut that fit the eye bolt (optional-- I used just binder clips)
- paper clips to adjust weight of bolt
- a pan to catch any water that might leak out of the tube
- a pitcher of water to fill the tube
- possibly a pair of pliers to help you bend the coat hanger

supplies you will need (plus the plastic tube)

If you want to make a holder, you will also need:
- a piece of corrugated cardboard
- some strong tape, such as duct tape

holder for tube, made from a piece of cardboard

How to assemble the tube:
1) Take the plastic plugs out of both ends of the tube. Use the caulk to stick the cap onto one end of the tube. Make sure the seal is watertight! Let it dry.
2) If you would like to have a holder for the tube so that you don't need a person to hold it during the demo, cut a hole in the piece of cardboard that is the same size as the tube. Tape the cardboard to a chair seat.

How to assemble the sinking balloon:
1) Blow up the balloon just a small amount, so that it will fit into the tube without touching the sides. You don't want the balloon to get stuck.
2) Attach the balloon to the eye bolt using the twistie tie or a rubber band.
3) Clip the top of the balloon with a small binder clip.
4) You will use the nut and possibly the washers to adjust the weight of the balloon as you try the experiment.

What to do with the coat hanger:
The hanger will be used to make a device that can both push your balloon down until it reaches cross-over depth and also to retrieve your balloon from the bottom of the tube. Retrieving the balloon with the hook allows you to repeat the experiment without having to empty and refill the tube.
1) Straighten out the hanger so you have one long piece of wire.
2) Bend one end to make a small hook. Look at the picture to the right to see relative size of the hook compared to the balloon.
3) Be prepared to make some adjustments to the hook if you have trouble slipping it into the top binder clip. You will not use the hook on the way down. The bottom of the hook will be used to push the balloon down to cross-over depth. When the balloon reaches the bottom, use the hook to snag one of the metal loops on the top binder clip and pull the balloon up.

<u>What to do</u>:   (Remember, there is a video of this on the YouTube playlist.)

1) Fill the tube with water, but not quite to the top.

2) Put the tube into the holder if you made one. If not, you might want to recruit someone to hold the tube.

3) Put the balloon into the top of the tube. The balloon should float. If not, take off some weight.

4) Push the balloon down a little bit at a time with the coat hanger tool you made. Every few seconds, let the balloon go and see what happens. You want to find the point at which the balloon will stop rising to the top. If you push the balloon all the way to the bottom and it still rises, you need to add more weight.

5) When you have adjusted the weight so that the balloon achieves neutral buoyancy (goes neither up nor down) at close to the middle of the tube, you are ready to do the actual demo.

6) Explain that in this demo, the balloon will represent magma being created in the mantle and the water column will be the mantle. The magma is compressible. That means if you squeeze it, the atoms can move closer together so the magma takes up less space. In the demo, we will see compression of the air inside the balloon. The molecules of nitrogen and oxygen (and other gases) will move closer together as they are squeezed so that the air will take up less space. As it takes up less space it becomes more dense. When the density of the air inside the balloon is the same as the density of the water, the balloon will stay in one place, not rising or sinking. Now push the balloon down until it reaches the neutral point.

7) Now explain that as we continue to force the balloon down, eventually the density of the air inside will become greater than the water surrounding it and as a result, the balloon will sink. Things that are more dense will always sink. Give the balloon a final nudge, forcing it down below the cross-over depth. Watch as the balloon sinks (without being pushed) to the bottom of the tube. Then use the hook to retrieve the balloon and bring it back up. Repeat the demo several times, allowing students to push the balloon themselves.

What was the point of this demo?

To remind us that cross-over depth is a real phenomenon and applies to magma, not just balloons. Based on the findings of magma studies, magma has a cross-over depth that prevents it from coming up through the mantle. Diagrams like the one shown here (from Wikipedia) supposedly "show" magma plumes rising from a depth of 2500 km. The computer was programmed by someone who apparently did not know the chemistry of magma and decided to interpret the dense areas deep in the mantle as rising magma.

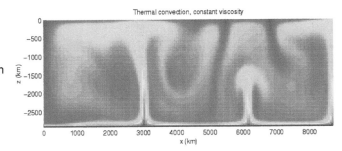

11) CRAFT: Make a collage of a hydrothermal vent

This activity is for those of you who are particularly interested in the ocean floor. If you don't want to take the time to study the ocean floor, feel free to skip this activity. We talked about hydrothermal vents on pages 94 and 95. These are places where hot, mineral-rich water is seeping up out of the oceanic crust. They were first discovered in the 1970s. There are some very nice advanced picture books (good for middle school ages, not just elementary) about deep sea vents, such as *Diving to a Deep-sea Volcano* by Kenneth Mallory. Youtube can also provide free videos. Try key words "black smokers" in addition to "hydrothermal vents."   Directions for this project can be found here:

http://ellenjmchenry.com/hydrothermal-vent-multimedia-collage/

Samples of student work:

CHAPTER 8

1) LAB: HALF-LIFE DEMONSTRATION

This activity is a hands-on way to explore the concepts of radioactive decay and half-life. The half-life of an isotope needs to be determined before it can be used for radiometric dating.

The idea of using pieces of straw to represent atoms was developed by a teaching instructor at Flinn Scientific. If you type "Flinn Scientific straw half life" into YouTube, you can watch their own video about the lab. I have not actually seen any of their printed lab materials, so beyond the basic idea of using straws, this lab is my own work. However, they deserve credit for the idea.

If you search the Internet for half-life labs, you can find simpler ones that use pennies or M&M candies, but these labs are very simple—you just put the candies or pennies into a bag and spill them out several times, each time removing any that come out with a certain side facing up. There is no way to vary the half-life to show that different atoms decay at different rates. If you use straws, you can simulate short and long half-lives, a much better simulation of reality. **However, if you have very young students and would like a simple lab, just Google "half-life lab M&Ms pennies."**

If you are working with a group, this is a great opportunity to show how an increase in available data can increase accuracy of results. The more data you have to work with, the better!

<u>You will need</u>:
• plastic straws (colored is great, but you can use clear if you still have extras)
NOTE: If you only have paper straws, go ahead and use them
• scissors
• ruler
• very fine permanent marker (or any thin pen that is able to make lines on the straw)
• small box with lid (can be plastic or cardboard as long as the bottom is flat)
• optional: tweezers to help remove pieces of straw from box
• copy of graph page
• pencil, and a few colored pencils

The box does not have to be transparent. A cardboard jewelry box will work, too.

<u>What to do</u>:
1) Cut two straws into pieces that are exactly 1 centimeter long. You will need 40 pieces total. (You can use more if you want to. The number 40 will make the math easier at the end.) Cut carefully and accurately; they need to be exactly the same length. You might need to use a thin marker to mark off the sections before you cut.

2) Cut another two straws into 40 pieces that are half a centimeter long (5 mm). Set these pieces aside to use later.

3) Cut another two or three straws into 40 pieces that are 1.5 centimeters long. Set these pieces aside to use later.

4) Put the 40 pieces that are 1 cm into the box. Carefully turn all the pieces so that they are on their sides.

5) Explain to the students that the straws lying on their sides represent radioactive atoms that are unstable. Unstable atoms have a nucleus with an awkward number of either protons or neutrons. As time goes on, the nucleus will try to fix itself by ejecting particles of some kind. At some point, enough particles will have been ejected that the nucleus will become stable. This means it will no longer be radioactive. If you wait long enough, radioactive atoms stop being radioactive. The process might be short and simple, or it might be complex and take a very long time. For example, it takes 14 steps for uranium to become stable. It ends up becoming an atom of lead, Pb, and ejects several helium nuclei along the way. Some atoms take only seconds or minutes to become stable. Atoms that become stable very quickly are used in the medical industry for all kinds of tests and therapies. Because the atoms stopped being radioactive in a short amount of time, there is little long term damage to the patient's body.

In this demo, the passing of time will be represented by shaking the box. One shake will represent one year. We will graph the results of several "years" of data.

6) After you make sure all the straws are lying on their sides, close the lid and give the box a shake. Put the box down and carefully open the lid. Remove any straws that have landed with their open end pointing up. Count the number of straws remaining in the box and record that number on your data chart.

7) Give the box another shake. Again, remove any straws that land open end up. Count the number remaining and record it on the data chart.

8) Continue shaking the box and removing straws until all the straws have landed open side up and there are therefore no straws left in the box. NOTE: When you get down to just a few straws, it is possible to have zero straws land open side up. This is valid data—record it. Your data sheet will show that there was no decay during that "year." Even if this happens two or three times in a row, keep recording it as valid data.
NOTE: The data chart gives you spaces for 20 shakes, but you might not need all of them. Shorter pieces might finish by 10.

9) After you are done recording data, transfer that data to the graph. Remember, one shake equals one year. Choose a color to represent this size straw and record your choice on the KEY. Make all your dots this color.

10) After you have your dots on the graph, draw a "best fit" curve. This means drawing a nice, smooth curved line right through the middle of the data points. (Use the same color.) Some dots might be a little under or over the line.

11) If you are working with a group, this might be a good time to pause and compare data. Let the students compare their results with others. Does everyone's best fit curve look the same? (Yes, they should.) You can compare curves by putting one paper on top of the other and holding them up to a light so you can see through both. You can also combine results numerically by averaging each trial. For example, if you have 6 students, add their number of straws left after the first "year" then divide by 6 to get the average. Does this averaged data point lie on the best fit curve line? (It should. If it is way off, check their math.) They should find that averaging their results gives points that lie almost exactly on the best fit curve line.

12) Repeat the experiment with the smaller pieces of straw. Since these pieces are shorter, they will be more inclined to fall with their open end up. When you put your dots onto the graph, choose a different color for this size straw and record the color in the KEY.

13) Repeat experiment with the longer pieces of straw. Since these pieces are longer, they will be less likely to land open end up, but some will still do so. If you still have straws left lying down after the maximum number of shakes in the data table, that is okay. Again, when you put your dots onto the graph, choose a different color for this size straw and record it in the KEY.

14) OPTION: You could cut some even longer pieces of straw and see what happens. How long do they have to be before none of them will flip to open side up? You can do as many sizes as you want to. Just choose a different color for each when you graph the points. NOTE: There is some extra space below the last line on the data chart. You can add another line if you want to.

15) Now it is time to figure out the half-life for each size of straw. The half-life of a radioactive element is the time it takes for half of the atoms in a sample to become stable. For example, if you have 100 atoms of an isotope of iodine called iodine-131, after 8 days half of them will have decayed and only 50 radioactive atoms will be left. After 8 more days, half of that remaining half will have decayed and only 25 radioactive atoms will be left. In other words, after 16 days, 3/4 of the original atoms will have decayed. After another 8 days (total of 24 days), another half of that second remaining half will have decayed so that 7/8 of your original atoms have now decayed and you only have about a dozen radioactive atoms remaining. After 32 days, 15/16 of your original atoms will have decayed and you will only have about 6 left.

In the text, we looked at Carbon-14. It has a half life of 5,730 years. After about 11,460 years, 3/4 of the original C-14 in your sample will have decayed (turning into nitrogen). Uranium-238 has a half-life of 4.5 billion years. For these very long half-lives, the graph was obviously extrapolated from a small amount of observable data in a short time period. A uranium sample would have billions upon billions of atoms in it, and enough of these decay often enough to be able to make a graph.

To calculate the half-life of a straw size, put a pencil point on the number 20 on the left side of the graph. We started with 40 pieces in the box, so half of 40 is 20. Slide the pencil point horizontally straight across the graph until you bump into the best fit curve line. (You don't have to draw a line, but a light line is okay.) Now slide the pencil point straight down until you get to the numbers (years) on the bottom line. What number of years is the pencil point indicating? It might be on a number, but it might also be between two numbers. If it is between numbers, estimate a decimal number for where the pencil point is. For example, if the point is sitting exactly half way between 1 and 2, the estimate would be 1.5. If the point is very close to the 2, you might estimate 1.8 or 1.9. Write this number down. Now go back and start on the number 10 on the left. (10 is half of 20.) Do the same thing and arrive at an estimated number on the bottom row. Now subtract the first number from the second. This will give you the number of years that the second half life took. It should be similar to the first time period. Then do 5 (because 5 is half of 10). This time you have to subtract the second number instead of the first to find how many years it took to get from 10 to 5. Then go back and do it again for 2.5.

Now find the average of these numbers. Let's say your first number was 1.8, the second was 1.6, third was 2.0 and the fourth was 1.2. Add 1.8 + 1.6 + 2.0 + 1.2 = 6.6. Now divide by 4 to get 1.65. The half-life of that size straw is 1.65 years.

Now find the half-life for the other sizes.

HALF-LIFE LAB

RECORD YOUR DATA: Write the number of remaining straws in the box after each shake.

shake number

straw size	1	2	3	4	5	6	7	8	9	10	11	12	13	14	15	16	17	18	19	20
.5 cm																				
1 cm																				
1.5 cm																				

GRAPH YOUR DATA:

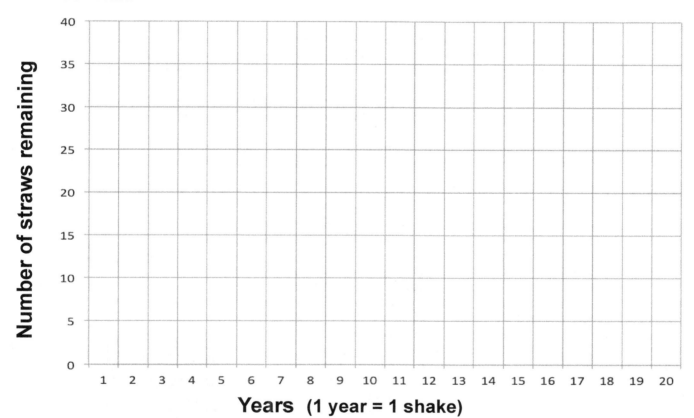

Years (1 year = 1 shake)

COLOR KEY:

☐ = .5 cm   ☐ = 1 cm   ☐ = 1.5 cm   ☐ =

HALF-LIFE CALCULATIONS: WORK SPACE:

1) Half-life of .5 cm straws = _____ years
2) Half-life of 1.0 cm straws = _____ years
3) Half-life of 1.5 cm straws = _____ years
4) Half-life of _____ cm straws = _____ years

2) LAB: EXTRAPOLATION ACTIVITY

This is sort of a follow-up to the extrapolation activity at the end of the chapter. In this activity, students will gather their own data and make their own graph, then do some extrapolating. The actual experiment really has very little to do with rocks and dirt. Yes, you could find some connection if you tried hard enough, but the main point of this activity is it understand what extrapolation is, because it is used so much in science.

You will need:
• a ramp of some kind (a ruler or yardstick with edges taped on [unless it has an indent in the middle, as some do], a cardboard tube cut in half the long way, an oblong piece of poster board folded in half to make a V-shape, or whatever you have around the house or classroom that would lend itself well to being a ramp)
NOTE: The exact length of the ramp is not important. Longer ramps will be more exciting, but also more work to set up.
• some objects to prop up one end of the ramp and adjust its height (use things you have lying around, like books or blocks)
• three small balls of different weights (small and large marbles, golf ball, ping pong ball, super ball, wooden ball, etc.)
• a paper cup
• scissors and tape
• three copies of the page with the lines on it
• a copy of the graph page
• pencil and a few colored pencils

Set up:
   1) Set up your ramp, putting some objects under one end. You will need to adjust the height of the ramp several times, so have some extra objects in reserve. Make sure the ramp is sturdy enough to hold the heaviest ball. If not, reinforce it a bit with extra cardboard or tape. Make sure all the balls will roll down freely. If the balls get slowed down by the edges of your ramp, make adjustments. Tape the ramp in place if it slides around.
   2) Trim the three lined pages and tape them together so that they form a continuous long sheet of 1 centimeter lines. Then tape the lined paper onto the table right at the end of the ramp. The lines will be used to measure how far the balls will be able to push the paper cup.
   3) Cut a U-shape in the paper cup, as shown. This hole will be where the ball goes in and pushes the cup. The hole will help to prevent the balls from bouncing off the cup. The cup will still tend to go off track a bit now and then, but it will be better.

What to do:
   1) Weigh each of the balls. If you don't have a scale, you can use these figures.
      small marble= 4 grams   regular marble= 6 grams   large "shooter" marble= 20 grams
      small (2.5 cm) super ball= 9 grams   large (3.5 cm) super ball= 18 grams
      ping pong ball= 2.5 grams   golf ball= 45 grams

   2) Decide on a color to represent each ball on your graph. Put that color in a block on the key, where it says BALLS, and then write the mass (weight) of the ball next to the color block. The mass of the ball is more important than the size. We could do another lab showing that this is true, but this lab will be lengthy enough as is.
   3) Put the paper cup right at the end of the ramp.
   4) Measure how many centimeters high the end of your ramp is. Start with a low number, like 5. (If you don't have a ruler marked in centimeters, you can use a small strip cut off the lined paper since the lines are one centimeter apart.)

   5) Choose one ball to begin with. Set it at the top of the ramp. Let it go and watch how far the cup is moved. Count how many lines backward the cup went, and record this number on the graph. It doesn't matter whether you measure from the front or the back of the cup, but measuring from the front is easier if you start by lining up the front of the cup with the first line on the paper. Put a colored dot (color for that ball) at the correct height, listed on the left side.
NOTE: Roll the ball down several times and watch the cup. Each trial might move the cup slightly differently. Repeat the roll enough times so that you get a feel for what the average result is. In other words, once in a while the ball might accidentally knock the cup off track. You would not want this to be your only data number!
   6) Raise the ramp by at least 5 cm. It can be more than that, as long as you record the height correctly on the graph.
   7) Reset the cup and then roll the ball down again. Count how many lines backward the cup went and record it on the graph.
   8) Raise the ramp again. Reset the cup and roll the ball down again. You can repeat this as many times as you want to. The more data points on your graph, the better.
   9) Now switch to a different ball. Reset the ramp to its lowest position, reset the cup, and roll the ball down. Mark the data point on the graph.

(continued on page 295)

EXTRAPOLATION LAB

Ramp height in centimeters (Each space is 1 cm.)

Distance in centimeters that the object moved   (Each space is 1 centimeter)

BALLS: [   ] =   [   ] =   [   ] =   [   ] =   [   ] =

Interpolate:
1) Choose a mass between two of your balls. _____ How far would that mass have moved the cup if the ramp was at its lowest setting? _____ At its highest setting? _____
2) Choose two of your ramp heights and calculate what height would be right between them. _____
How far would your heaviest ball have moved the cup at this height? _____ Your lightest? _____

Extrapolate:
1) How far would the cup have moved if you had used a 1,000 gram ball? _____ A 10,000 gram ball? _____
2) How far would the cup have moved if you had used a height of .5 centimeter? _____
3) How far would the object have moved if you had used a height of 100 centimeters? _____
4) Is there a limit to how high the ramp can be if you keep the length the same? _____

Factors that might have affected the data:
1) Did all the balls roll down the ramp the same way? Did they experience similar friction?
2) Did each ball always move the cup exactly the same distance every time you tried it?
What factors might have caused differences in these measurements?

10) Repeat this process for at least one more ball.

11) After you have all your data points on the graph, connect the dots with smooth, best fit curves. Keep drawing the curve far beyond your actual data points, following the trend of how the curve looks. Imagine that your data points are just one part of a larger curve. What does that larger curve look like?

12) After you've drawn all your curves, answer the questions below the graph. These questions will ask you to extrapolate. Allow time for discussion of these questions.

3) THE "ROCK HOPPING" GAME

This is a very straightforward game. You just hop around on the map board and "visit" these sites. It is targeted to younger students, but can be played by any age. You can make the game more challenging by giving players more sites to visit. You can add your own cards to the game, also. Just remember to make a location dot on the map and connect it to the trails.

If you want to increase content for older players and one or more of them has access to an Internet-connected device, you can also require that when they land on a site they must give one interesting fact about that site. The name of the site can be typed into a search engine and you can have your fact in about 10 seconds. For an additional layer of complexity, the player with the Internet device can ask a question instead of just giving a fact. Anyone who can answer the question correctly can win an extra roll on their next turn (or use two dice instead of one).

You will need:
• copies of the following pattern pages.
• scissors
• glue stick and/or clear tape
• tokens (to represent players on the board)
• number cube (die)

Set up:
1) Print out the following pages onto white card stock, if possible. (The map can be printed onto regular paper. Sometimes regular paper folds for storage more easily than card stock.)
2) Cut apart the cards.
3) Trim the edges of the map pages and tape them together so that the edges match. (You can leave some edges untrimmed and use them as "glue flaps" if you'd like, rather than putting cut edge against cut edge and taping.)

How to play:
1) Shuffle the cards and deal one card to each player. That will be the players' starting point. Return those cards to the deck.
2) Deal a certain number of cards to each player. For a short and quick game, use 3 cards. For a longer game, use 5 or 6. If you have a lot of time and older students, you can use the whole deck and divide the cards between the players. If the cards do not distribute evenly, leave out the extras.
3) The method of play is very straightforward. Players take turns rolling the die and moving their tokens around the board, following the paths. They must land on the sites that are on the cards in their hand. After they have "visited" a site, they turn that card over so that they can keep track of where they've been and where they still need to go. You can choose to require that players stop when they get to a destination (and not using any remaining "hops") or you can rule that players may continue moving after landing on their site.

4) OPTIONAL SUPPLEMENTAL VIDEO: "Puzzle on the Plateau"

If you want more information on the Grand Canyon and theories about how it might have formed, consider purchasing a DVD called "Puzzle on the Plateau." Host Mike Snavely takes you on a tour of this amazing geological feature, and points out key features that give clues as to how the canyon might have formed. You'll learn the significance of the Kaibab Plateau, for example, and of the geological event called the "Great Denudation" that must have preceded the formation of the canyon. You'll find out how people could have been living in the rock walls of one section of the canyon and why they finally left.

Running time is 53 minutes. DVD can be found on Amazon ($20) if you type in both title and name of host. Be advised that the last few minutes of the video present Christian ideas (general, not denominational). For those of you of this persuasion, this will be a positive attribute. For those not of this persuasion, the video would still be very worthwhile, as there are over 45 minutes of well-researched canyon geography and geology, and animations of some of the geological ideas discussed in this chapter. You will also find out the connection between the canyon and many other interesting geological features in that part of the country. It's a fascinating look into an out-of-the-box theory, and sometimes these crazy "pioneers" turn out to have some very valid points. If you want to show the video in a public school setting, just stop the video before the last few minutes.

Sugar Loaf Mountain
in the harbor of Rio de Janeiro, Brazil

Uluru (Ayer's Rock), Australia

Arches National Park, Utah, USA

Devil's Tower, Wyoming, USA

Giants' Causeway, Northern Ireland

World's largest meteorite, Namibia

Monument Valley, Arizona, USA

Mushroom Rock State Park, Kansas, USA

Goreme Valley Stone Chimneys
Cappadocia, Turkey

The White Cliffs of Dover, England, UK

"The Twelve Apostles," Victoria, Australia

Mammoth Hot Springs
Yellowstone Nat'l Park, Wyoming, USA

"Half Dome," Yosemite National Park
California, USA

"The Cheesewring"
Bodmin Moor
Cornwall
UK

"Devil's Marbles"
in the "outback" of Australia

Rock of Gibraltar

"Maltese Cross"
Western Cape, South Africa

"James Bond Island"

Phang-Nga Bay
Thailand

Stonehenge, England, UK

"Pierced Rock," Gulf of St. Lawrence

"The Wave," Matsushima, Japan

The Externsteine, Germany

Blarney Stone, County Cork, Ireland

Stone Forest, Yunnan Province, China

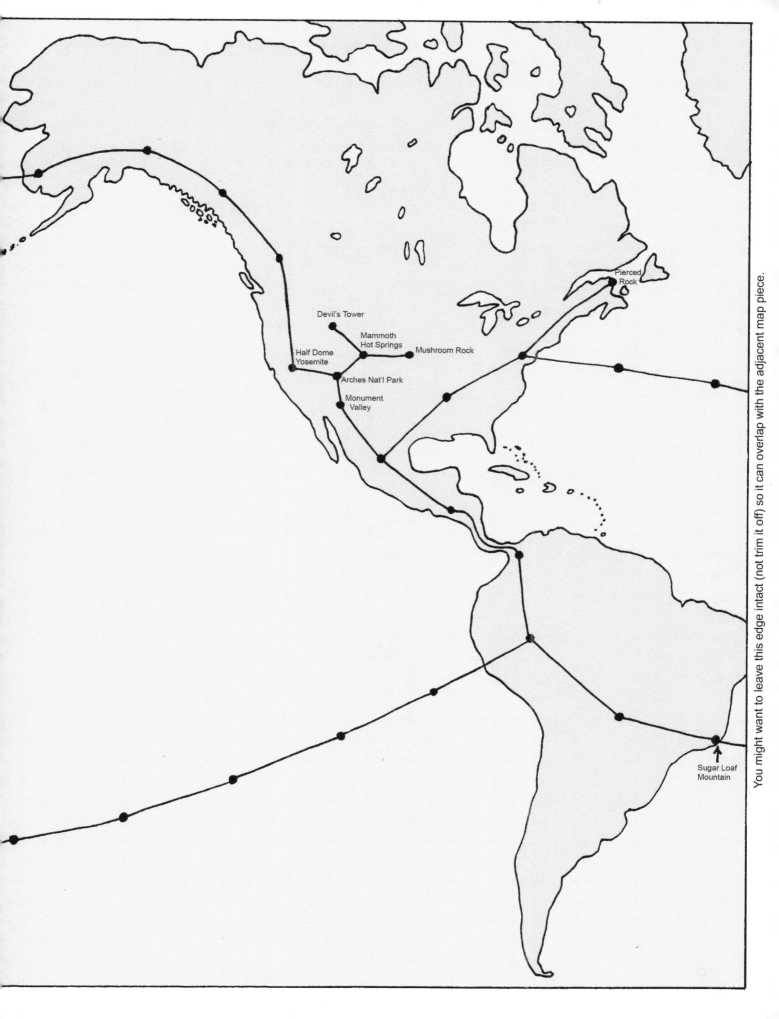

Devil's Tower

Mammoth
Hot Springs

Half Dome
Yosemite

Mushroom Rock

Arches Nat'l Park

Monument
Valley

Pierced
Rock

Sugar Loaf
Mountain

You might want to leave this edge intact (not trim it off) so it can overlap with the adjacent map piece.

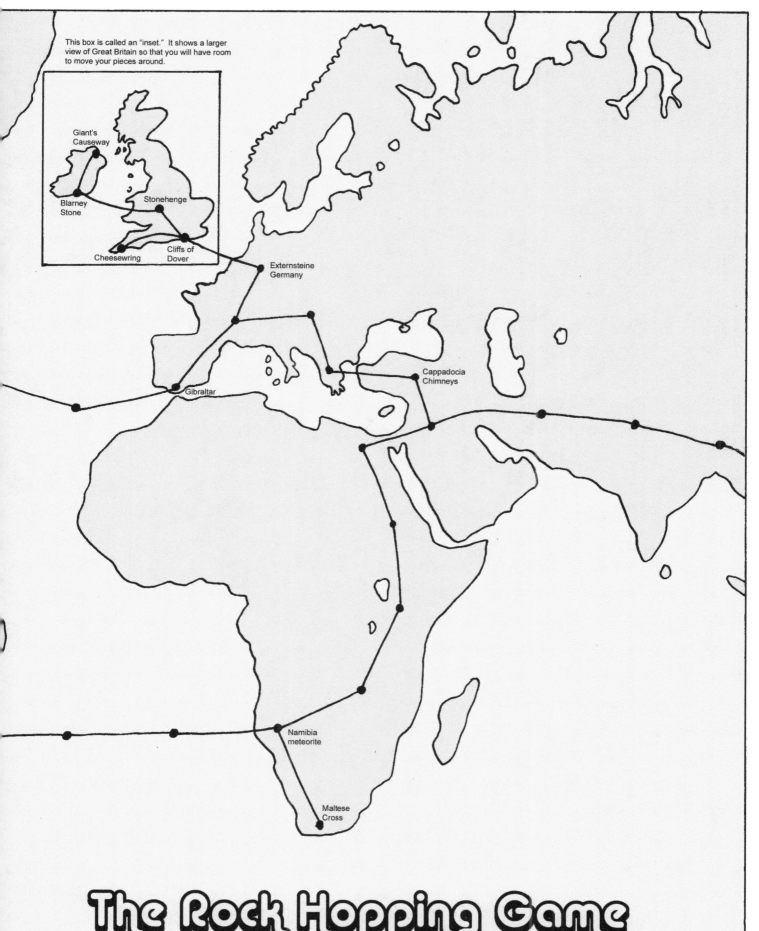

This box is called an "inset." It shows a larger view of Great Britain so that you will have room to move your pieces around.

Giant's Causeway

Blarney Stone

Stonehenge

Cheesewring

Cliffs of Dover

Externsteine Germany

Gibraltar

Cappadocia Chimneys

Namibia meteorite

Maltese Cross

The Rock Hopping Game

You might want to leave this edge intact (not trim it off) so it can overlap with the adjacent map piece.

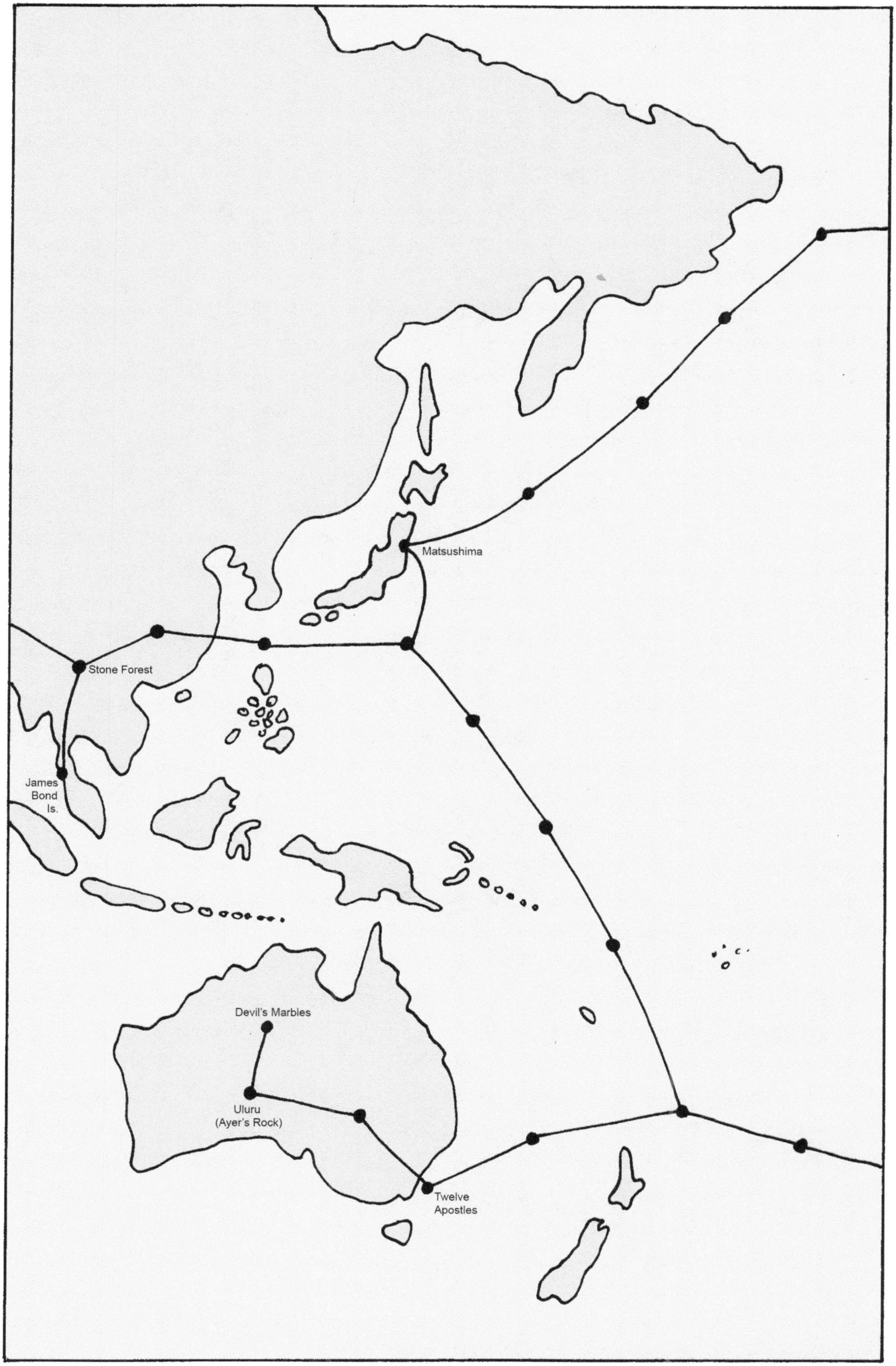

Matsushima

Stone Forest

James
Bond
Is.

Devil's Marbles

Uluru
(Ayer's Rock)

Twelve
Apostles

5) QUICK DEMO: Twist a piece of chalk to show shear faulting

In this activity, you can experience shear lines (shear faults). Shear lines often result when materials are stressed to the breaking point. Shearing is a type of breaking. Structural engineers have done thousands of experiments with many materials and found that shear lines aren't random, but follow a predictable pattern, always being at about a 45 degree angle to the surface. In this activity you'll see this for yourself.

<u>You will need</u>:
• pieces of chalkboard chalk

<u>What to do</u>:
Hold a piece of chalk firmly, and <u>twist</u> it until it breaks. Look at the broken middle. Are the broken edges straight or beveled? Was that just an accident? Twist another one. Do the broken edges look the same? Twist a few more and see if the pattern continues.

<u>What is happening</u>:
The twisting motion is meant to simulate the forces that would have occurred in the outer mantle if the inner earth really did shift and melt, as the hydroplate theory suggests. This is one explanation for the Benioff zones, which occur mostly around the outside of the Pacific Ocean. (This explanation was worked out by several professional engineers, not me.) The Benioff zones descend at about 45 degree angles, exactly the angle the theory would predict. Does this "prove" that Benioff zones are shear faults? No. It's simply a suggestion in that direction.

6) MORE INFO: SIDE TOPICS RELATED TO HPT

For those of you who are intrigued by the hydroplate theory and would like to know more, here are a few more interesting tidbits that would not fit into the text. If you are already on information overload, feel free to skip this section. This is only intended to be a fun extra for those who are interested.

This pallasite is from the author's personal collection. It is a very thin slice of the "Brahin meteorite" that landed in Russia in 1807. The meteor had apparently broken up into several small pieces, each weighing up to 80 kg (180 lbs). The district of Brahin was contaminated by Chernobyl in 1986, so very few meteorite hunters have been able to access this site since then. In 2002, another chunk was found underground but at a distance from the original site. There is a Wikipedia article on this meteorite if you want to know more.

The only explanation astronomers can think of is that pallasites came from the inside of certain asteroids where metal touches silicate rock. However, calculations show that there are not enough asteroids to produce this number of pallasites.

PALLASITE METEORS

Pallasites are meteors that are made of iron and nickel, but have "inclusions" of olivine. This means that they look like shiny pieces of metal with little green gemstones embedded in them. You can see in the picture on the left that the metal surface is as shiny as a mirror. The picture on the right show the transparent olivine crystals. Pallasite meteorites are relatively rare, less than 2 percent of all meteorites.

We learned about olivine in chapter 3. To make olivine you have to melt a mafic rock until it becomes a liquid, cool it a little bit until "stone soup" has formed, and then chill it quickly to prevent it from turning into another type of silicate further down the series (e.g. pyroxenes).

Astronomers are baffled as to how these meteorites could have formed in outer space. Pallasites are as surprising as finding an iceberg with ping pong balls stuck in the middle. (What kept the balls from floating to the surface while the water froze?) Olivine is much less dense than iron and nickel and should have collected in one place. And is it a coincidence that these meteorites are made of exactly the same things that we think are inside the earth?

The hydroplate theory ventures to offer an explanation. These meteors were chunks of hot rock from deep in the earth. Extreme heat had allowed iron and nickel to settle down into the bottom of the pillars that supported the granite crust. When the crust cracked open, the pillars crumbled and the chunks of hot pillar (iron and nickel with some olivine mixed in) were jettisoned into space where they cooled very quickly. Cooling in zero gravity allowed that "ping pong ball effect." The meteors went into an elliptical orbit and after several thousand years returned to the solar system and to earth, falling as meteorites.

CRATERS ON THE MOON

Did you know that the moon has many more craters on the side facing the earth? In particular, there are five very large, dark craters that have no match on the other side. The moon also experiences deep earthquakes (moonquakes) only on the near side. The dark patches seem to be ancient lava flows that cooled. The lava even filled in some craters. The moon does not have a detectable large, liquid core like the earth does. There may be a small liquid core, but far too small to have produced the surface lava. Astronomers wonder how these large craters could have formed.

Could debris launched from the earth have hit the moon? If so, it would have hit primarily the near side. The hits on the far side could have occurred as the moon moved through space, its path intersecting the paths of other bits of debris. The impact of the hits caused tremendous heat, melting rock and resulting in lava flows. (Impacts of one kind or another are frequently used to explain mysterious features of the solar system. They are used to explain the origin of asteroids and comets, the backwards rotation of Venus, the tipped axis of Uranus, the extreme density of Mercury, water on Earth, the formation of our moon, the disappearance of water on Mars, and more.) Earth's gravitational field should have protected the moon from asteroids coming from behind the earth. But if the earth itself produced the impacts, the moon had no protection.

In Galileo's time, the dark spots on the moon were thought to be seas and were named "maria," meaning "seas." We now understand that they are ancient lava flows. They make the moon's "man in the moon."

HYDROTHERMAL VENTS

We briefly looked at hydrothermal vents in chapter 7. These are places where hot mineral-rich water is leaking out of oceanic crust. Just as stalactites form in caves, and travertine formations occur around hot springs, so these vents cause minerals to precipitate out of the water and make formations called "chimneys." Black smokers have iron and sulfur in their water. White smokers have calcium, barium and silicon.

There is a problem with the current theory about how these chimneys are forming. Ocean water can't possibly migrate down into the basaltic crust. We know the densities of water, basalt, and magma, and we can calculate the pressure that all these experience at the bottom of the ocean. Some basic calculations show that the water coming out of the vents can't possibly have come from the ocean above. Geologists ignore these calculations and go right on telling everyone the water is circulating down through the rock. But if the water is not coming from the ocean, then where is it coming from?

The hydroplate theory proposes that there are still remnants of the water that used to be between the crust and the mantle. (Remember what they found at the bottom of the Kola Superdeep Borehole? Hot salty water.) Is it possible that pockets of salt water still exist near the top of the mantle? If so, hot salty water could be leaking upwards through cracks in the ocean floor.

These are "white smokers." The white color comes from calcium and barium.

CELLULOSE DUST IN SPACE

Astronomers have discovered that some of the dust in the solar system appears to have a chemical signature similar to that of cellulose. Cellulose is the type of starch that plants are made of. Why in the world would we see anything resembling plant starch in outer space? That's a question that remains completely unanswered at this point. However, hydroplate theory advocates boldly venture to guess that perhaps some of the ancient earth's plant material got launched into space along with the water that spewed out of the crack. The plant material would have been pulverized in the process, so we can't imagine leaves floating around in the asteroid belt. Will we ever know for sure where this cellulose-like dust came from? No, we will always be stuck with theories.

LINEAMENTS

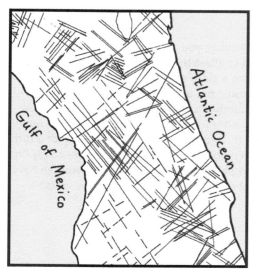

Lineaments of central Florida

The crust of the earth is covered with tiny fracture lines that most people are unaware of. These "lineaments" *(LIN-ee-ah-ments)* are not categorized as true "faults" because they don't seem to shift around and they don't have offset edges. Earthquakes do occur along them sometimes, but it's not because the rocks are slipping like in normal faults. In some places, water or natural gas can seep out of lineaments. Their main significance is that they can give geologists clues about the structure of aquifers down below. People who work with water wells and gas wells are interested in lineaments.

Lineaments can't be seen on the surface. They were only discovered relatively recently because you need radars and computers to see them. The U.S. Geological Survey makes maps of lineaments by using computer-processed data taken by side-looking radar in a plane flying 5 miles above the ground. The radar reflections are digitalized and then processed by a computer that draws an image from the data.

This picture shows lineaments in central Florida. You can't see them when you are walking around on the ground. They are often under plants and soil, but even if they are right under your feet you still can't see them. Geologists are not sure what kind of tectonic motion would produce such amazingly straight, and often parallel, lines.

Interestingly, fracture lines like this would be exactly what you'd expect if the crust had indeed experienced what the hydroplate theory proposes. As water moved along under the crust, rushing to escape out of the giant crack, the crust would have undulated up and down, sort of like a flag flapping in the wind. Imagine waves going through an entire continent. At the top of each wave, the continental rock was being stretched beyond the breaking point and little cracks popped open. Then, as the wave went back down, the cracks closed up. Each wave would have produced lots of thin, parallel cracks.

RADIOACTIVITY AT THE EARTH'S SURFACE

Scientists at the Proton 21 lab in the Ukraine prepare an experiment where they will shoot a lightning bolt into a tiny piece of metal. Sandia lab in the USA does similar experiments, hoping to find a way to harness fusion power.

If the solar system formed the way most books and websites describe it, radioactive elements were formed along with all other elements inside stars billions of years ago. The mix of elements and isotopes would have been random and fairly uniform, with everything being everywhere. Then, as clouds of stuff began to condense and cool, some blobs of hot stuff eventually formed planets. The hot planets have been cooling ever since. Radioactive elements would have been part of this mix of cooling stuff, and therefore we should find it everywhere inside the earth. Or, if gravitational settling occurred, heavy radioactive elements would have migrated down to the core. In fact, we find neither of these to be true.

Radioactivity is a phenomenon of the continental crust. Oceanic basalt has very little radioactivity. Also, not only are radioactive elements found only in continental crust, but they are concentrated very close to the surface in certain places in the crust, such as in and around the Rocky Mountains. Why is radioactivity only in the granite crust? Is there something special about granite?

Yes, granite can do something that basalt cannot. We learned in chapter 3 that quartz has a special property called the piezoelectric effect. When quartz is squeezed or stretched, it generates an electric current. Granite has a lot of quartz in it, but basalt does not. Granite might be capable of producing electrical currents if stretched and/or squeezed. The fluttering of the crust that was described in the LINEAMENTS section would not only have produced those cracks on the surface, but also tremendous amounts of electricity deep down, like lightning bolts traveling through the granite. Even today, electricity can be detected in the crust during earthquakes. Flexing granite does produce electrical currents. The bending and flexing of the crust as described by the hydroplate theory would result in almost unimaginable amounts of electricity going through the granite crust. Lab experiments have shown that powerful surges of electricity can strip electrons away from atoms, allowing the nuclei to break apart. In other words, extreme levels of electricity can cause atoms to become radioactive.

Secondarily, the extreme levels of electricity also cause very fast decay. Atoms become radioactive, but experience decay at an accelerated rate. By our current standards of measurement, they experience millions of years worth of decay in only minutes. Some have proposed that we use electricity to speed up the decay of the radioactive waste from our nuclear reactors. Theoretically, this would work, but the energy it would take to produce this amount of electricity makes this idea too expensive to be practical. The connection between electricity and radioactive is a frontier branch of science.

7) QUIZ GAME: Final review quiz game

Here is a review game that can be used by individuals or groups. For an individual, you can just photocopy the answers onto the back of the cards and lay the cards out question-side up. Allow the student to guess the answers and then turn over cards. If they are correct, they score that number of points. Keep track of the score on a piece of paper.

For groups, you will have to decide what works best in your situation. You could put up cards that have just the categories and the numbers 100 to 500, and have a emcee read the question when the number is called, as in the television show Jeopardy. (Unlike Jeopardy, you don't have to give your answer in question format.) After reading the question, take the number card down so that students can see what questions remain. If you want to have three or more players competing for a turn to answer, you can have some kind of signal, like raising hand or slapping hand down on desk, to see which player responds first (like buzzers on Jeopardy). If you want a slightly less competitive version, you can simply take turns answering, or you can have teams instead of individuals answering questions. If you have teams, have the members take turns being the spokesperson who actually says the answer. Again, though, do what works best in your situation.

MINERAL CHEMISTRY	SOIL	ROCKS	INSIDE THE EARTH	MINES AND QUARRIES
100 What do diamonds, coal, and graphite have in common?	**100** Which horizon contains the most organic material: O, A, B or C?	**100** Rocks that started out as magma are called ___.	**100** This type of seismic wave can't travel through liquids.	**100** Name a type of rock that is quarried. (If you can name a second, double your score.)
200 Halite, NaCl, is often called "table salt." The atoms in this mineral arrange themselves into what shape?	**200** Which horizon is called the "sub-soil"?	**200** The ocean floor is mostly made of this type of rock.	**200** How deep is the Kola Superdeep Borehole?	**200** Anthracite mines are digging out a very high grade of what energy-rich rock?
300 Minerals or rocks that contain a lot of iron and magnesium are called ___.	**300** What percentage of soil is just air?	**300** Name a rock that will float in water.	**300** What is the dividing line called between the crust and the mantle?	**300** Poor people in Europe used to quarry this soft, brown substance because it was cheaper than coal.
400 $CaCO_3$ is the chemical recipe for what mineral? (If you can name a second one, double your score.)	**400** When water flushes minerals out of the soil we call this ___.	**400** Clastic sedimentary rock made out of clay particles is called ___.	**400** What is the area called where seismic waves don't show up on the other side of the earth?	**400** A mine in Detroit, Michigan goes under Lake Erie! What edible rock is being mined here?
500 What mineral is made of SiO_4 molecules surrounded by Mg and Fe ions? Hint: "stone soup"	**500** When soil contains sand, silt and clay, all three of these, it is called ___.	**500** When granite gets squeezed hard enough it can turn into this metamorphic rock.	**500** This geologic layer was named after Wales, the place where it was first discovered by Adam Sedgwick.	**500** Quarries often crush rocks into small pieces called a ___.

MINES AND QUARRIES	INSIDE THE EARTH	ROCKS	SOIL	MINERAL CHEMISTRY
100 — Intended answers: limestone, marble, granite, Also acceptable: salt, gypsum, chalk	100 — S waves	100 — igneous	100 — O horizon	100 — All are made of carbon.
200 — coal	200 — 7 miles / 12 km	200 — basalt	200 — B horizon	200 — cube
300 — peat	300 — Moho	300 — pumice	300 — 25%	300 — mafic
400 — salt	400 — shadow zone	400 — shale	400 — leaching	400 — calcite, aragonite
500 — aggregate	500 — Cambrian	500 — gneiss	500 — loam	500 — olivine

MINERAL PROPERTIES

	SOIL	MAGMA	SILICATES	LIMESTONE
100 What rock or mineral is number 10 on the Mohs hardness scale?	**100** This soft, brown fluffy stuff is the leftovers from bacteria and fungi eating vegetation.	**100** If magma gets lots of air bubbles in it and turns frothy, it could harden into what type of rock?	**100** This silicate mineral is made of stacks of shiny black sheets.	**100** When water drips from the ceiling of a cave, it can make limestone formations called ___.
200 Name an organic mineral. (Double your score if you can name another one.)	**200** What percentage of soil is organic?	**200** Magma that flows into the ocean and cools very quickly can form this black, glassy rock.	**200** When silicon bonds with four oxygen atoms it makes a molecule in this shape.	**200** Landscapes in places where the rock is mostly limestone are called ___.
300 The words "shiny, dull, waxy, metallic, glassy" are all descriptions of ___	**300** On what part of a plant will you find nitrogen-fixing bacteria?	**300** Which type of magma is more runny, mafic or felsic?	**300** What silicate mineral has the property of piezoelectricity?	**300** When water is added to calcium oxide, CaO, there is an exothermic reaction that produces ___.
400 Name a mineral that has a hexagonal crystal shape. (Double your score if you can name another one.)	**400** Name a macronutrient found in soil.	**400** Another word for intrusive igneous rock, named after a Roman god.	**400** What spherical rock formation has spectacular silicate crystals inside?	**400** When magnesium atoms replace more than half of the calcium atoms in limestone, its name is changed to ___.
500 Fool's gold does not streak gold; it streaks this color.	**500** What is it called when something sticks to the outside of something else?	**500** This is what Hawaiians call runny lava.	**500** This green and red silicate was once thought to have magical powers.	**500** Chalk differs from limestone because it is much softer and because it contains these little things.

MINERAL PROPERTIES	SOIL	MAGMA	SILICATES	LIMESTONE
100 diamond	100 humus	100 pumice	100 biotite	100 stalactites
200 coal, amber	200 5%	200 obsidian	200 tetrahedron	200 karst
300 luster	300 roots	300 mafic	300 quartz	300 heat
400 expected answer: quartz others: tourmaline, beryl, ruby, sapphire, corundum, amethyst, hematite	400 nitrogen, N phosphorus, P potassium, K	400 plutonic	400 geode	400 dolomite
500 black	500 adsorption	500 pahoehoe	500 heliotrope	500 coccolithophores